GROUP THEORY AND ELECTRONIC ENERGY BANDS IN SOLIDS

SERIES OF MONOGRAPHS ON SELECTED TOPICS IN SOLID STATE PHYSICS

Editor: E. P. WOHLFARTH

Volume 1: W. F. BROWN, Jr.
Magnetostatic principles in ferromagnetism

Volume 2: C. D. MEE
The physics of magnetic recording

Volume 3: R. R. BIRSS
Symmetry and magnetism

Volume 4: D. J. CRAIK and R. S. TEBBLE
Ferromagnetism and ferromagnetic domains

Volume 5: J. C. BRICE
The growth of crystals from the melt

Volume 6: R. J. WEISS
X-ray determination of electron distributions

Volume 7: E. FATUZZO and W. J. MERZ
Ferroelectricity

Volume 8: M. M. SCHIEBER
Experimental magnetochemistry

Volume 9: H. ZIJLSTRA
Experimental methods in magnetism

Volume 10: J. F. CORNWELL
Group theory and electronic energy bands in solids

Group Theory and Electronic Energy Bands in Solids

BY

J. F. CORNWELL

Lecturer in Theoretical Physics,
University of St. Andrews

1969

NORTH-HOLLAND PUBLISHING COMPANY
AMSTERDAM · LONDON

WILEY INTERSCIENCE DIVISION
JOHN WILEY & SONS, INC. – NEW YORK

Library of Congress Catalog Card Number 72–79723

SBN 7204 1460 1

PUBLISHERS:

NORTH-HOLLAND PUBLISHING CO. – AMSTERDAM, LONDON

SOLE DISTRIBUTORS FOR THE WESTERN HEMISPHERE:

WILEY INTERSCIENCE DIVISION

JOHN WILEY & SONS, INC. – NEW YORK

PRINTED IN THE NETHERLANDS

To my wife Elizabeth

PREFACE

This book has three main aims. Firstly, it is intended to provide a thorough and self-contained introduction to the use of group theory in the calculation and classification of electronic energy bands in solids. It is hoped that this will be useful both for those who intend to calculate energy bands and for those who have to interpret calculations and relate them to the experimental situation. The book has been laid out in such a way as to assist in making the subject more accessible to this latter group. In particular, the theory of groups and its role in quantum mechanics is developed from scratch. Moreover, in the first five chapters only the absolutely essential group theoretical concepts needed for symmorphic space groups are introduced, the more difficult concepts being treated later. To dispel the slightly abstract air which sometimes surrounds this subject, a number of concrete examples are treated in detail. The second aim has been to make a close study of the more advanced aspects of the subject, again treating all the more difficult points in some detail. These aspects include non-symmorphic space groups, time-reversal symmetry, and double groups and spin–orbit coupling. The third and final aim has been to give a summary of the considerable recent work on the subject.

It is a pleasure to acknowledge the fruitful discussions the author has had with Professor E. P. Wohlfarth both concerning this book and topics dealt with in it, and also the careful typing by Mrs C. G. MacArthur of the manuscript.

J. F. CORNWELL

CONTENTS

LIST OF MOST IMPORTANT SYMBOLS

Only the symbols that are very frequently used are listed here. The brief descriptions of them are supplemented by a note of the section or equation in which they are defined. Many other symbols are used from time to time, and are defined as they occur. The notation for matrices is described in appendix 4.

$\boldsymbol{a}_1, \boldsymbol{a}_2, \boldsymbol{a}_3$	Basic lattice vectors of the crystal (ch. 1 § 3.1)
$\boldsymbol{b}_1, \boldsymbol{b}_2, \boldsymbol{b}_3$	Basic lattice vectors of the reciprocal lattice [eq. (4.6)]
C_{ni}	Proper rotation through $2\pi/n$ about the axis Oi (ch. 1 § 1)
$\bar{C}_{ni}$	Generalized proper rotation through $2\pi/n$ about the axis Oi (ch. 8 § 3, appendix 2)
$\mathscr{C}_i$	ith class of a group (ch. 1 § 2.4)
E	Identity transformation (ch. 1 § 1)
$\bar{E}$	Generalized identity transformation (ch. 8 § 3, appendix 2)
$E_n(\boldsymbol{k})$	nth energy level at point $\boldsymbol{k}$ (ch. 4 § 6). (The suffix n is sometimes omitted)
$\mathscr{G}$	In purely group theoretical developments, such as in ch. 1 (except § 3), ch. 2 and ch. 6, merely denotes a group. In ch. 3 $\mathscr{G}$ denotes the group of the Schrödinger equation, and in all other places $\mathscr{G}$ denotes more specifically the space group of the crystal
$\mathscr{G}_o$	The point group of the space group $\mathscr{G}$ (ch. 1 § 3)
$\mathscr{G}(\boldsymbol{k})$	The group of the wave vector $\boldsymbol{k}$ (ch. 7 § 1)
$\mathscr{G}_o(\boldsymbol{k})$	The point group of the wave vector $\boldsymbol{k}$ (ch. 5 § 1)

$H(\boldsymbol{r})$	Hamiltonian operator
I	Inversion operator (ch. 1 § 1)
$\bar{I}$	Generalized inversion operator (ch. 8 § 3, appendix 2)
$\boldsymbol{K}_m$	Reciprocal lattice vector [eq. (4.13)]
$\boldsymbol{k}$	Allowed wave vector [eq. (4.8)]
$\boldsymbol{k}_1, \boldsymbol{k}_2, \ldots$	Vectors of the star of $\boldsymbol{k}$ (ch. 5 § 1, ch. 7 § 1)
$M(\boldsymbol{k})$	Number of vectors in the star of $\boldsymbol{k}$
$O(T)$, $O(\bar{T})$	Spinor transformation operators (ch. 8 § 3)
$P(T)$	Scalar transformation operator (ch. 3 § 2)
$\mathscr{P}^p_{mn}$	Projection operator [eq. (3.22)]
$\boldsymbol{R}$, $\boldsymbol{R}(T)$	Transformation matrix [eq. (1.1)]
$\boldsymbol{R}_1, \boldsymbol{R}_2, \ldots$	Transformation matrices generating star of $\boldsymbol{k}$ (ch. 5 § 1, ch. 7 § 1)
$\mathscr{S}$	Subgroup of $\mathscr{G}$
T	A transformation (ch. 1 § 1)
$\bar{T}$	A generalized transformation (ch. 8 § 3)
$\boldsymbol{t}$	A translation. (This symbol frequently appears with subscripts or superscripts attached)
$\boldsymbol{t}_n$	Lattice vector of the crystal [eq. (1.9)]
$\mathscr{T}$	The subgroup of pure primitive translations of the space group $\mathscr{G}$ (ch. 1 § 3.1, ch. 4 § 2)
$\mathscr{T}(\boldsymbol{k})$	Subgroup of $\mathscr{T}$ corresponding to $\boldsymbol{k}$ (ch. 7 § 2, ch. 9 § 9.1)
$\boldsymbol{u}$, $\boldsymbol{u}(\boldsymbol{R})$	SU_2 matrix corresponding to proper rotation $\boldsymbol{R}$ (ch. 8 § 2)
$\boldsymbol{\Gamma}$	Matrix of a representation (ch. 2 § 1). (This symbol frequently appears with superscripts attached)
χ	Character of representation (ch. 2 § 5). (This symbol frequently appears with superscripts attached)

CHAPTER 1

BASIC CONCEPTS

§ 1. COORDINATE TRANSFORMATIONS

This book is concerned with the study of the symmetry properties of functions having physical significance. These can be described by stating how the functions transform under coordinate transformations.

Consider the following example of such a transformation. Suppose that Ox, Oy, Oz are three mutually perpendicular axes, and Ox', Oy', Oz' are another set of mutually perpendicular axes with the same origin O, which could, for example, be obtained from the first set by a rotation about some axis through O. Suppose that (x, y, z) and (x', y', z') are the coordinates of any point with respect to these two sets of axes. (That is, both sets of coordinates represent the *same* point.) Then the relationship between the two sets of coordinates can be written in the form

$$\boldsymbol{r}' = \boldsymbol{R}(T)\,\boldsymbol{r}, \tag{1.1}$$

where $\boldsymbol{r}=(x, y, z)$ and $\boldsymbol{r}'=(x', y', z')$, all such vectors being treated as 3×1 column matrices in matrix expressions unless otherwise indicated, and $\boldsymbol{R}(T)$ is a 3×3 matrix with real coefficients which depend *only* on the rotation, and not on the particular point under consideration. (The definitions, notations and properties of matrices used in this book are summarized in appendix 4.) $\boldsymbol{R}(T)$ will be called the transformation matrix corresponding to the transformation or symmetry operation T. It will sometimes be written merely as $\boldsymbol{R}$. As an example, suppose that Oz and Oz'

coincide, and Ox', Oy' are obtained from Ox, Oy by a rotation through an angle ϕ in the right-hand screw sense about Oz, as shown in figs. 1.1 and 1.2. Then

$$\begin{aligned} x' &= x \cos\phi + y \sin\phi \\ y' &= -x \sin\phi + y \cos\phi, \\ z' &= z \end{aligned}$$

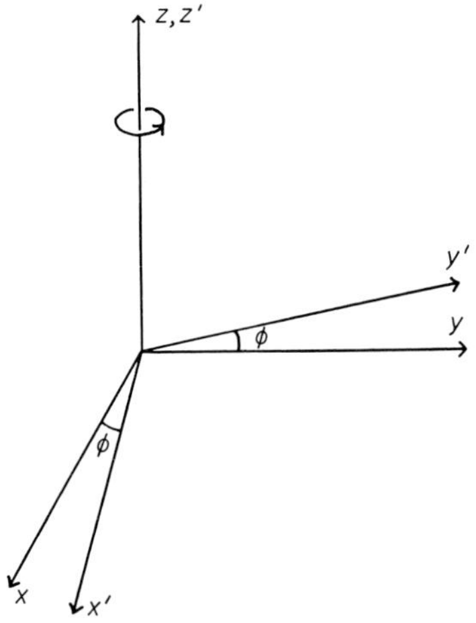

Fig. 1.1

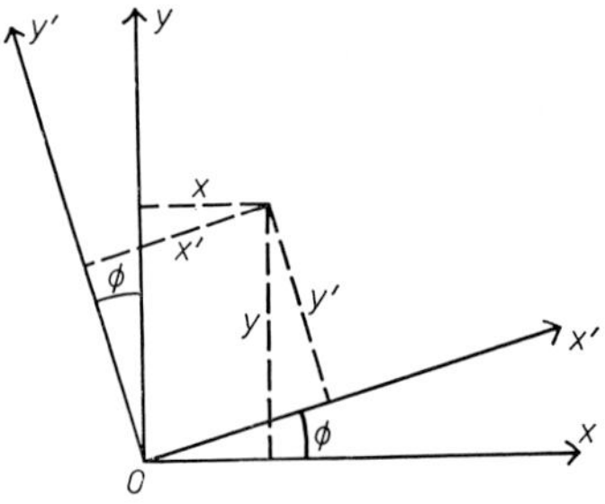

Fig. 1.2

so that in this case

$$\boldsymbol{R}(T)=\begin{pmatrix} \cos\phi & \sin\phi & 0 \\ -\sin\phi & \cos\phi & 0 \\ 0 & 0 & 1 \end{pmatrix}.$$

The relation (1.1) can be inverted so that

$$\boldsymbol{r}=\boldsymbol{R}(T)^{-1}\,\boldsymbol{r}'. \tag{1.2}$$

Requiring that the length of every vector and the angle between any two vectors be unchanged in a transformation implies that $\boldsymbol{R}^{-1}=\tilde{\boldsymbol{R}}$, so that $\boldsymbol{R}$ is described as an 'orthogonal' matrix. For suppose that $\boldsymbol{r}_1$, $\boldsymbol{r}_2$ are any two vectors defined with respect to $Oxyz$, and $\boldsymbol{r}_1'=\boldsymbol{R}\boldsymbol{r}_1$ and $\boldsymbol{r}_2'=\boldsymbol{R}\boldsymbol{r}_2$ are the same vectors defined with respect to $Ox'y'z'$. Then their lengths and the angle between them are unchanged if the scalar product of the vectors is unchanged, that is, if $\boldsymbol{r}_1'\cdot\boldsymbol{r}_2'=\boldsymbol{r}_1\cdot\boldsymbol{r}_2$. But $\boldsymbol{r}_1'\cdot\boldsymbol{r}_2'=\tilde{\boldsymbol{r}}_1'\,\boldsymbol{r}_2'=\tilde{\boldsymbol{r}}_1\tilde{\boldsymbol{R}}\boldsymbol{R}\boldsymbol{r}_2$, which is equal to $\boldsymbol{r}_1\cdot\boldsymbol{r}_2=\tilde{\boldsymbol{r}}_1\boldsymbol{r}_2$ if $\tilde{\boldsymbol{R}}\boldsymbol{R}=\boldsymbol{E}$, the 3×3 unit matrix, from which the above result follows.

As $\det\tilde{\boldsymbol{R}}=\det\boldsymbol{R}$, $\tilde{\boldsymbol{R}}\boldsymbol{R}=\boldsymbol{E}$ implies that $(\det\boldsymbol{R})^2=1$, so that $\det\boldsymbol{R}=\pm 1$. Those transformations of the form (1.1) for which $\det\boldsymbol{R}=+1$ and $\det\boldsymbol{R}=-1$ are called 'proper rotations' and 'improper rotations' respectively. Proper rotations are rotations in the usual sense, of which the transformation of fig. 1.1 and fig. 1.2 is an example. The set of improper rotations consists of the inversion operation I, for which the transformation matrix is

$$\boldsymbol{I}\equiv\boldsymbol{R}(I)=\begin{pmatrix} -1 & 0 & 0 \\ 0 & -1 & 0 \\ 0 & 0 & -1 \end{pmatrix},$$

so that $\boldsymbol{r}'=-\boldsymbol{r}$, together with the set of reflections in planes. For example, the transformation matrix corresponding to reflection in the Oxz plane, for which $x'=x$, $y'=-y$ and $z'=z$ is

$$\begin{pmatrix} 1 & 0 & 0 \\ 0 & -1 & 0 \\ 0 & 0 & 1 \end{pmatrix}.$$

The word 'rotation', when it appears unqualified, will be used to describe both proper and improper rotations.

The 'product' $T_1 T_2$ of two rotations, whose transformation matrices are $\boldsymbol{R}(T_1)$ and $\boldsymbol{R}(T_2)$ respectively, is *defined* to be that rotation for which the transformation matrix is $\boldsymbol{R}(T_1)\,\boldsymbol{R}(T_2)$. That is

$$T = T_1 T_2 \quad \text{if} \quad \boldsymbol{R}(T) = \boldsymbol{R}(T_1)\,\boldsymbol{R}(T_2)\,. \tag{1.3}$$

Thus, if $\boldsymbol{r}' = \boldsymbol{R}(T_2)\,\boldsymbol{r}$ and $\boldsymbol{r}'' = \boldsymbol{R}(T_1)\,\boldsymbol{r}'$, then $\boldsymbol{r}'' = \boldsymbol{R}(T_1)\,\boldsymbol{R}(T_2)\,\boldsymbol{r}$, so the interpretation is that the rotation T_1 takes place *after* the rotation T_2. [Other similar but non-equivalent definitions of the product may be given, but the above definition is the only one that is consistent with the theory developed in ch. 3, particularly eq. (3.5)]. Note that in general $\boldsymbol{R}(T_1)\,\boldsymbol{R}(T_2) \neq \boldsymbol{R}(T_2)\,\boldsymbol{R}(T_1)$, so that $T_1 T_2 \neq T_2 T_1$.

A reflection can then be considered to be the product of the inversion operator and a proper rotation. For example, for the reflection in the Oxz plane considered previously,

$$\begin{pmatrix} 1 & 0 & 0 \\ 0 & -1 & 0 \\ 0 & 0 & 1 \end{pmatrix} = \begin{pmatrix} -1 & 0 & 0 \\ 0 & -1 & 0 \\ 0 & 0 & -1 \end{pmatrix} \begin{pmatrix} -1 & 0 & 0 \\ 0 & 1 & 0 \\ 0 & 0 & -1 \end{pmatrix},$$

where the second transformation matrix corresponds to a rotation through π about Oy.

A proper rotation through an angle $2\pi/n$ in the right-hand screw sense about an axis Oi will be denoted by C_{ni}. The identity transformation will be denoted by E, so that $\boldsymbol{R}(E) = \boldsymbol{E}$, the 3×3 unit matrix. Reflections are sometimes denoted by σ, with subscripts attached to distinguish between different reflections.

In addition to rotations, transformations T of the form

$$\boldsymbol{r}' = \boldsymbol{R}(T)\,\boldsymbol{r} + \boldsymbol{t}(T)\,, \tag{1.4}$$

where $\boldsymbol{t}(T)$ is a real 3×1 column vector, will often be encountered. This corresponds to a rotation of the coordinate axes described by the matrix $\boldsymbol{R}(T)$ followed by a 'translation' of the origin through the vector $-\boldsymbol{t}(T)$. Eq. (1.4) will be rewritten as

$$\boldsymbol{r}' = \{\boldsymbol{R}(T) \mid \boldsymbol{t}(T)\}\, \boldsymbol{r}, \tag{1.5}$$

thereby defining the operator $\{\boldsymbol{R}(T) \mid \boldsymbol{t}(T)\}$, which will often be written merely as $\{\boldsymbol{R} \mid \boldsymbol{t}\}$. Indeed it is sometimes convenient to go a stage further and write $\{\boldsymbol{R}(C_{ni}) \mid \boldsymbol{t}\}$ as $\{C_{ni} \mid \boldsymbol{t}\}$, with a similar notation for improper rotations. It follows from (1.4) and (1.5) that if $\boldsymbol{r}' = \{\boldsymbol{R}(T_2) \mid \boldsymbol{t}(T_2)\}\, \boldsymbol{r}$ and $\boldsymbol{r}'' = \{\boldsymbol{R}(T_1) \mid \boldsymbol{t}(T_1)\}\, \boldsymbol{r}'$ then $\boldsymbol{r}'' = \{\boldsymbol{R}(T_1)\, \boldsymbol{R}(T_2) \mid \boldsymbol{R}(T_1)\, \boldsymbol{t}(T_2) + \boldsymbol{t}(T_1)\}\, \boldsymbol{r}$. The definition (1.3) of the product of two transformations is therefore generalized so that

$$T = T_1 T_2$$

if

$$\{\boldsymbol{R}(T) \mid \boldsymbol{t}(T)\} = \{\boldsymbol{R}(T_1)\, \boldsymbol{R}(T_2) \mid \boldsymbol{R}(T_1)\, \boldsymbol{t}(T_2) + \boldsymbol{t}(T_1)\}. \tag{1.6}$$

It also follows from (1.4) that the inverse relation to (1.5) is

$$\boldsymbol{r} = \{\boldsymbol{R}(T)^{-1} \mid -\boldsymbol{R}(T)^{-1}\, \boldsymbol{t}(T)\}\, \boldsymbol{r}',$$

so that one may define the inverse operator by

$$\{\boldsymbol{R}(T) \mid \boldsymbol{t}(T)\}^{-1} = \{\boldsymbol{R}(T)^{-1} \mid -\boldsymbol{R}(T)^{-1}\, \boldsymbol{t}(T)\}. \tag{1.7}$$

A transformation such that $\boldsymbol{r}' = \{\boldsymbol{E} \mid \boldsymbol{t}\}\, \boldsymbol{r}$ is known as a *pure* translation.

§ 2. GROUP THEORY

§ 2.1. *Definition of a group*

A group is a mathematical concept that has a very precise meaning. It is a set of elements that must obey *all* of four 'group postulates'. From these basic postulates can be built an elaborate and fascinating theory, though only a small part is covered in this book. The abstract development of the theory does not depend on the nature of the elements themselves, which, as will be seen later, can be very different. Herein lies the great power of the theory.

The precise abstract definition of a group and some related terms will now be given, and these will be illustrated in §2.2 by some examples.

Definition of a group. A group $\mathscr{G}$ is a set of distinct elements A, B, C, ... for which an operation of combining is defined, which will be called 'multiplication', and which has the following properties:

(a) The product of any two elements of $\mathscr{G}$ is itself an element of $\mathscr{G}$.

(b) For any three elements, A, B, and C of $\mathscr{G}$, $(AB)\,C = A\,(BC)$, so that this product can be written unambiguously as ABC. (This is called the associative law.)

(c) $\mathscr{G}$ contains an element E, called the identity element, such that for any element A of $\mathscr{G}$, $AE = EA = A$.

(d) For every element A of $\mathscr{G}$ there exists an element denoted by A^{-1}, and called the inverse of A, that is also a member of $\mathscr{G}$ and is such that $AA^{-1} = A^{-1}A = E$.

The number of elements in $\mathscr{G}$ is called the *order* of the group, which may be finite or infinite. It should be noted that it is *not* required that the elements of $\mathscr{G}$ should commute, that is, in general $AB \neq BA$. If all the elements of a group do commute, then the group is said to be *Abelian*.

The properties of a particular group are completely determined by its 'group multiplication table', which shows the product of all pairs of elements, so that the first stage in the analysis of a group is the construction of this table.

A minor point worth noting is that the inverse of a product of elements is the product of inverses taken in the *reverse* order. For example, $(AB)^{-1} = B^{-1}A^{-1}$. (This follows as $B^{-1}A^{-1}AB = B^{-1}EB = B^{-1}B = E$.)

§ 2.2. *Some simple examples of groups*

Two simple examples will be given here of the type of group that occurs in solid state physics. They will be used to explain various concepts as they are developed in later sections.

The first group consists of the following six matrices,

$$\boldsymbol{M}_1 = \begin{pmatrix} 1 & 0 \\ 0 & 1 \end{pmatrix}, \quad \boldsymbol{M}_2 = \begin{pmatrix} -\frac{1}{2} & \frac{1}{2}\sqrt{3} \\ -\frac{1}{2}\sqrt{3} & -\frac{1}{2} \end{pmatrix}, \quad \boldsymbol{M}_3 = \begin{pmatrix} -\frac{1}{2} & -\frac{1}{2}\sqrt{3} \\ \frac{1}{2}\sqrt{3} & -\frac{1}{2} \end{pmatrix},$$

$$\boldsymbol{M}_4 = \begin{pmatrix} -1 & 0 \\ 0 & 1 \end{pmatrix}, \quad \boldsymbol{M}_5 = \begin{pmatrix} \frac{1}{2} & -\frac{1}{2}\sqrt{3} \\ -\frac{1}{2}\sqrt{3} & -\frac{1}{2} \end{pmatrix}, \quad \boldsymbol{M}_6 = \begin{pmatrix} \frac{1}{2} & \frac{1}{2}\sqrt{3} \\ \frac{1}{2}\sqrt{3} & -\frac{1}{2} \end{pmatrix},$$

with matrix multiplication as the operation of group multiplication. Table 1.1 gives the multiplication table. The order of multiplication in the table is that the element specified in the left-hand column is *followed* by the element specified in the top row, so that for example

TABLE 1.1

	$\boldsymbol{M}_1$	$\boldsymbol{M}_2$	$\boldsymbol{M}_3$	$\boldsymbol{M}_4$	$\boldsymbol{M}_5$	$\boldsymbol{M}_6$
$\boldsymbol{M}_1$	$\boldsymbol{M}_1$	$\boldsymbol{M}_2$	$\boldsymbol{M}_3$	$\boldsymbol{M}_4$	$\boldsymbol{M}_5$	$\boldsymbol{M}_6$
$\boldsymbol{M}_2$	$\boldsymbol{M}_2$	$\boldsymbol{M}_3$	$\boldsymbol{M}_1$	$\boldsymbol{M}_6$	$\boldsymbol{M}_4$	$\boldsymbol{M}_5$
$\boldsymbol{M}_3$	$\boldsymbol{M}_3$	$\boldsymbol{M}_1$	$\boldsymbol{M}_2$	$\boldsymbol{M}_5$	$\boldsymbol{M}_6$	$\boldsymbol{M}_4$
$\boldsymbol{M}_4$	$\boldsymbol{M}_4$	$\boldsymbol{M}_5$	$\boldsymbol{M}_6$	$\boldsymbol{M}_1$	$\boldsymbol{M}_2$	$\boldsymbol{M}_3$
$\boldsymbol{M}_5$	$\boldsymbol{M}_5$	$\boldsymbol{M}_6$	$\boldsymbol{M}_4$	$\boldsymbol{M}_3$	$\boldsymbol{M}_1$	$\boldsymbol{M}_2$
$\boldsymbol{M}_6$	$\boldsymbol{M}_6$	$\boldsymbol{M}_4$	$\boldsymbol{M}_5$	$\boldsymbol{M}_2$	$\boldsymbol{M}_3$	$\boldsymbol{M}_1$

$$\boldsymbol{M}_5\boldsymbol{M}_4 = \begin{pmatrix} \frac{1}{2} & -\frac{1}{2}\sqrt{3} \\ -\frac{1}{2}\sqrt{3} & -\frac{1}{2} \end{pmatrix}\begin{pmatrix} -1 & 0 \\ 0 & 1 \end{pmatrix} = \begin{pmatrix} -\frac{1}{2} & -\frac{1}{2}\sqrt{3} \\ \frac{1}{2}\sqrt{3} & -\frac{1}{2} \end{pmatrix} = \boldsymbol{M}_3.$$

The table demonstrates that the product of any two matrices of this set is also a member of the set, so that the first group postulate is satisfied. The second group postulate is automatically satisfied by matrix multiplication. Clearly, the unit matrix $\boldsymbol{M}_1$ is the identity element, and every element has an inverse which can be read off from the table. In fact, $\boldsymbol{M}_1^{-1}=\boldsymbol{M}_1$, $\boldsymbol{M}_2^{-1}=\boldsymbol{M}_3$, $\boldsymbol{M}_3^{-1}=\boldsymbol{M}_2$, $\boldsymbol{M}_4^{-1}=\boldsymbol{M}_4$, $\boldsymbol{M}_5^{-1}=\boldsymbol{M}_5$, $\boldsymbol{M}_6^{-1}=\boldsymbol{M}_6$. As all the group postulates are satisfied, this set of elements with this definition of group multiplication therefore forms a group.

Not all the elements of this group commute. For example, $\boldsymbol{M}_5\boldsymbol{M}_6=\boldsymbol{M}_2$, whereas $\boldsymbol{M}_6\boldsymbol{M}_5=\boldsymbol{M}_3$. The group is therefore non-Abelian.

The second example of a group is known to crystallographers as the point group C_{3v}. Imagine 3 mutually perpendicular axes Ox, Oy, Oz, and four other axes OA, OB, OC, OD lying in the plane Oxy as shown in fig. 1.3. The elements of the group are the following symmetry operations:

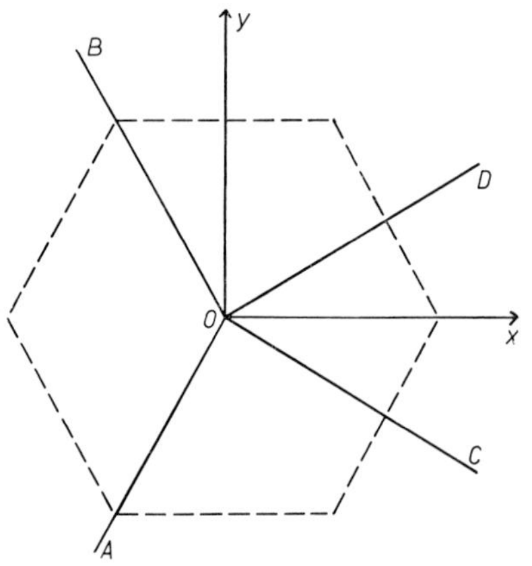

Fig. 1.3

E: the identity operation;

C_{3z}, C_{3z}^{-1}: rotations through $\frac{2}{3}\pi$ about Oz in the right-hand and left-hand screw senses respectively;

IC_{2y}, IC_{2C}, IC_{2D}: reflections in the planes through O whose normals are Oy, OC and OD respectively.

These rotations would transform an equilateral triangle lying in the plane Oxy, with vertices initially along Ox, OA, OB, into an equivalent position, with its vertices again lying along these three axes.

The transformation matrices corresponding to these rotations are

$$R(E)=\begin{pmatrix}1&0&0\\0&1&0\\0&0&1\end{pmatrix},\qquad R(C_{3z})=\begin{pmatrix}-\frac{1}{2}&\frac{1}{2}\sqrt{3}&0\\-\frac{1}{2}\sqrt{3}&-\frac{1}{2}&0\\0&0&1\end{pmatrix},\qquad R(C_{3z}^{-1})=\begin{pmatrix}-\frac{1}{2}&-\frac{1}{2}\sqrt{3}&0\\\frac{1}{2}\sqrt{3}&-\frac{1}{2}&0\\0&0&1\end{pmatrix}.$$

$$R(IC_{2y})=\begin{pmatrix}1&0&0\\0&-1&0\\0&0&1\end{pmatrix},\qquad R(IC_{2C})=\begin{pmatrix}-\frac{1}{2}&\frac{1}{2}\sqrt{3}&0\\\frac{1}{2}\sqrt{3}&\frac{1}{2}&0\\0&0&1\end{pmatrix},\qquad R(IC_{2D})=\begin{pmatrix}-\frac{1}{2}&-\frac{1}{2}\sqrt{3}&0\\-\frac{1}{2}\sqrt{3}&\frac{1}{2}&0\\0&0&1\end{pmatrix}.$$

The definition (1.3) of the product of two rotations then gives the multiplication table, table 1.2, in which, for example $(IC_{2y})\,C_{3z}=IC_{2C}$.

TABLE 1.2

	E	C_{3z}	C_{3z}^{-1}	IC_{2y}	IC_{2C}	IC_{2D}
E	E	C_{3z}	C_{3z}^{-1}	IC_{2y}	IC_{2C}	IC_{2D}
C_{3z}	C_{3z}	C_{3z}^{-1}	E	IC_{2D}	IC_{2y}	IC_{2C}
C_{3z}^{-1}	C_{3z}^{-1}	E	C_{3z}	IC_{2C}	IC_{2D}	IC_{2y}
IC_{2y}	IC_{2y}	IC_{2C}	IC_{2D}	E	C_{3z}	C_{3z}^{-1}
IC_{2C}	IC_{2C}	IC_{2D}	IC_{2y}	C_{3z}^{-1}	E	C_{3z}
IC_{2D}	IC_{2D}	IC_{2y}	IC_{2C}	C_{3z}	C_{3z}^{-1}	E

It is easily verified that the group postulates are all satisfied, with E as the identity element, and $(C_{3z})^{-1}=C_{3z}$, $(C_{3z}^{-1})^{-1}=C_{3z}$, $(IC_{2y})^{-1}=IC_{2y}$, $(IC_{2C})^{-1}=IC_{2C}$ and $(IC_{2D})^{-1}=IC_{2D}$. This group also is non-Abelian.

§ 2.3. *Subgroups*

Definition of a subgroup. Any set of elements of a group which *itself* obeys all the group postulates is called a subgroup of the group.

It is clear that a subgroup must contain the identity element of the group, and that the product of any two elements of a subgroup must also be a member of the subgroup.

The point group C_{3v} described in §2.2 has 5 subgroups, apart from the subgroup consisting of the whole of C_{3v}, namely: E; E, IC_{2y}; E, IC_{2C}; E, IC_{2D}; E, C_{3z}, C_{3z}^{-1}. This may be verified by inspection of the multiplication table 1.2.

A subgroup whose order is less than that of the group will be called a *proper* subgroup.

§ 2.4. *Classes*

A class is another concept that has a precise mathematical meaning. As will be shown in ch. 2, it plays an extremely important role in the theory of matrix representations. It involves the following idea of conjugate elements.

Definition of conjugate elements. A group element B is said to be conjugate to A if there exists a group element X such that

$$B = XAX^{-1}. \tag{1.8}$$

If B is conjugate to A, then A is conjugate to B, for $A=X^{-1}B(X^{-1})^{-1}$. Moreover, if two elements B and C are both conjugate to A, then B is conjugate to C. This follows because there must exist elements of the group X and Y such that $B=XAX^{-1}$ and $C=YAY^{-1}$. Then

$$B = X(Y^{-1}CY)X^{-1} = (XY^{-1})C(XY^{-1})^{-1},$$

which is of the form of (1.8) as XY^{-1} is a group element.

Definition of a class. A class is a collection of mutually conjugate elements of a group.

Obviously a class can be constructed from any element A of the group simply by forming the set of distinct products XAX^{-1} for every X of the group. As $A=EAE^{-1}$, A itself is a member of this class. (In this procedure for constructing a class some elements may appear more than once, but only those elements that are different are retained.) For example, in the group C_{3v}, applying this procedure to C_{3z} gives

$$\begin{aligned} EC_{3z}E^{-1} &= C_{3z}, & C_{3z}C_{3z}(C_{3z})^{-1} &= C_{3z}, \\ C_{3z}^{-1}C_{3z}(C_{3z}^{-1})^{-1} &= C_{3z}, & (IC_{2y})\,C_{3z}(IC_{2y})^{-1} &= C_{3z}^{-1}, \\ (IC_{2C})\,C_{3z}(IC_{2C})^{-1} &= C_{3z}^{-1}, & (IC_{2D})\,C_{3z}(IC_{2D})^{-1} &= C_{3z}^{-1}. \end{aligned}$$

Thus the class formed from C_{3z} consists only of C_{3z} and C_{3z}^{-1}, and will be written as $\{C_{3z}, C_{3z}^{-1}\}$. Starting from C_{3z}^{-1} would give exactly the same class. The group C_{3v} has also two other classes that can be found in the same way, namely $\{E\}$ and $\{IC_{2y}, IC_{2C}, IC_{2D}\}$.

It will be observed that E is in a class of its own, and that no group element lies in two different classes. Both of these results are true for *all* groups. The first follows because $XEX^{-1}=E$ for every group element X. The second follows because, if an element A lies in both a class containing an element B and a class containing an element C, then A is conjugate to both B and C, so that B is conjugate to C, and therefore B lies in the same class as C.

It will also be observed that the elements in a class are all of the same type. For example, $\{C_{3z}, C_{3z}^{-1}\}$ contains only rotations through $\frac{2}{3}\pi$, while $\{IC_{2y}, IC_{2C}, IC_{2D}\}$ contains only reflections. This is a general property of groups of rotations. However, it is not necessarily true that all the elements of the same type lie in the same class.

One further result will be used later, namely, that in an Abelian group each element is in a class of its own. The reason is simply that in an Abelian group, for any group elements A and X, it is obvious that $XAX^{-1}=XX^{-1}A=EA=A$.

The classes of a group will be denoted by $\mathscr{C}_1, \mathscr{C}_2, \ldots$.

§ 2.5. *Isomorphic and homomorphic groups*

These are two closely related concepts upon which the theory of matrix representations is built.

Definition of isomorphic groups. Consider a group $\mathscr{G}$ of order g with elements, E, A, B,... and another group $\mathscr{G}'$ of the *same* order with elements E', A', B',.... These two groups are said to be isomorphic if

(a) to each element of $\mathscr{G}$ there corresponds one and only one element of $\mathscr{G}'$, and to each element of $\mathscr{G}'$ there corresponds one and only one element of $\mathscr{G}$ (that is, there is a *one-to-one* correspondence between the elements of $\mathscr{G}$ and $\mathscr{G}'$),

and

(b) if A' corresponds to A and B' corresponds to B, then $A'B'$ corresponds to AB, and similarly for all other elements of $\mathscr{G}$ and $\mathscr{G}'$.

Isomorphic groups therefore have exactly the same multiplication table, and hence exactly the same properties. The two examples considered in §2.2 are isomorphic, the correspondence being

$$\boldsymbol{M}_1 \leftrightarrow E, \qquad \boldsymbol{M}_2 \leftrightarrow C_{3z}, \qquad \boldsymbol{M}_3 \leftrightarrow C_{3z}^{-1},$$
$$\boldsymbol{M}_4 \leftrightarrow IC_{2y}, \qquad \boldsymbol{M}_5 \leftrightarrow IC_{2C}, \qquad \boldsymbol{M}_6 \leftrightarrow IC_{2D}.$$

Definition of homomorphic groups. Consider a group $\mathscr{G}$ with elements E, A, B,... and another group $\mathscr{G}'$ with elements E', A', B',.... Then $\mathscr{G}'$ is said to be homomorphic to $\mathscr{G}$ if

(a) to each element of $\mathscr{G}$ there corresponds one and only one element of $\mathscr{G}'$,

and

(b) if A' corresponds to A and B' corresponds to B, then $A'B'$ corresponds to AB, and similarly for all other elements of $\mathscr{G}$ and $\mathscr{G}'$.

The difference between the two concepts is that for a *homomorphism* each element of $\mathscr{G}'$ may correspond to *several* elements of $\mathscr{G}$, so that the order of $\mathcal{G}'$ may be less than that of $\mathcal{G}$, whereas for an *isomorphism* each element of $\mathcal{G}'$ must correspond to *only one* element of $\mathcal{G}$. An isomorphism will therefore be regarded merely as a special case of a homomorphism.

To illustrate the concept of homomorphic groups consider the group that consists merely of the two 1×1 matrices $\boldsymbol{M}_1'=(+1)$ and $\boldsymbol{M}_2'=(-1)$, with matrix multiplication as the group multiplication operation. The multiplication table is given in table 1.3. This group is homomorphic to the point group C_{3v} of §2.2, the correspondence being

$$E\to \boldsymbol{M}_1', \qquad C_{3z}\to \boldsymbol{M}_1', \qquad C_{3z}^{-1}\to \boldsymbol{M}_1',$$
$$IC_{2y}\to \boldsymbol{M}_2', \qquad IC_{2C}\to \boldsymbol{M}_2', \qquad IC_{2D}\to \boldsymbol{M}_2'.$$

TABLE 1.3

	$\boldsymbol{M}_1'$	$\boldsymbol{M}_2'$
$\boldsymbol{M}_1'$	$\boldsymbol{M}_1'$	$\boldsymbol{M}_2'$
$\boldsymbol{M}_2'$	$\boldsymbol{M}_2'$	$\boldsymbol{M}_1'$

For example, $C_{3z}(IC_{2C})=IC_{2y}$ (from table 1.2), and the element corresponding to $C_{3z}(IC_{2C})$ is $\boldsymbol{M}_1'\boldsymbol{M}_2'=\boldsymbol{M}_2'$ (from table 1.3), which also corresponds to IC_{2y}.

§ 3. SPACE GROUPS

§ 3.1. *General description of space groups*

An infinite three-dimensional lattice may be defined in terms of three non-coplanar *basic lattice vectors* $\boldsymbol{a}_1, \boldsymbol{a}_2, \boldsymbol{a}_3$. The set of lattice vectors is then given by

$$\boldsymbol{t}_{\boldsymbol{n}} = n_1\boldsymbol{a}_1 + n_2\boldsymbol{a}_2 + n_3\boldsymbol{a}_3\,, \tag{1.9}$$

where $\boldsymbol{n}=(n_1, n_2, n_3)$ and n_1, n_2, n_3 are *integers*. Points having lattice vectors as their position vectors are called lattice points, and a pure translation through a lattice vector, $\{\boldsymbol{E} \mid \boldsymbol{t}_{\boldsymbol{n}}\}$, is called a *primitive* translation.

Definition of a space group. A space group $\mathscr{G}$ is a group of symmetry operations $\{\boldsymbol{R} \mid \boldsymbol{t}\}$, which contains as a subgroup the set of all pure primitive translations of a lattice, $\mathscr{T}$, but which contains no other *pure* translations.

The physical significance of a space group lies in the fact that the nuclei of an ordered solid all lie (at the absolute zero of temperature) on a three-dimensional *array* of points, and the set of all operations which transform the array into itself form a space group. (Any proper subgroup of this space group is also a space group whose operations transform the array into itself, but the maximum information clearly comes from considering the maximal space group that corresponds to the array.) A short description will be given here of the classification of space groups. For further details and proofs the reader is referred to the accounts of SCHÖNFLIESS [1923], KOSTER [1957] and LYUBARSKII [1960], and to the very comprehensive description of the 'International tables for X-ray crystallography' [1965].

There is a restriction on the possible rotations $\boldsymbol{R}$ that can occur in the operations of a space group, namely that if $\{\boldsymbol{R}' \mid \boldsymbol{t}'\}$ is a member of the space group and $\boldsymbol{t}_{\boldsymbol{n}}$ is any lattice vector, then $\boldsymbol{R}'\boldsymbol{t}_{\boldsymbol{n}}$ must also be a lattice vector. (This result follows because if $\{\boldsymbol{R}' \mid \boldsymbol{t}'\}$ and $\{\boldsymbol{E} \mid \boldsymbol{t}_{\boldsymbol{n}}\}$ are both members of the space group, then so must be $\{\boldsymbol{R}' \mid \boldsymbol{t}'\}\,\{\boldsymbol{E} \mid \boldsymbol{t}_{\boldsymbol{n}}\}\,\{\boldsymbol{R} \mid \boldsymbol{t}'\}^{-1} = \{\boldsymbol{E} \mid \boldsymbol{R}'\boldsymbol{t}_{\boldsymbol{n}}\}$, which, being a pure translation, must be a primitive translation.) Detailed investigations show that the proper rotations can only be through multiples of $\frac{1}{3}\pi$ or $\frac{1}{2}\pi$, and the improper rotations can only be products of these proper rotations with the inversion operator.

The set of operations $\{\boldsymbol{R} \mid \boldsymbol{0}\}$, where $\boldsymbol{R}$ ranges over the distinct rotational parts of the operations of a space group $\mathscr{G}$, form a group known as the *point group* of the space group, which will be denoted by $\mathscr{G}_o$. (This definition does not imply that every $\{\boldsymbol{R} \mid \boldsymbol{0}\}$ of $\mathscr{G}_o$ is a member of $\mathscr{G}$. It merely implies that if $\{\boldsymbol{R} \mid \boldsymbol{0}\}$ is a member of $\mathscr{G}_o$ then there exists a $\boldsymbol{t}$, which is not necessarily a lattice vector, such that $\{\boldsymbol{R} \mid \boldsymbol{t}\}$ is a member of $\mathscr{G}$.) There are only 32 crystallographic point groups. They are all finite groups. The group C_{3v} considered in §2.2 is one example. A complete description of all the crystallographic point groups (including in particular the notation of SCHÖNFLIESS [1923]) is given in appendix 1.

Space groups having the same point group $\mathscr{G}_o$ are said to belong to the same *crystal class*, so that there are 32 different crystal classes. In the classification of space groups by SCHÖNFLIESS [1923], a space group is denoted by the Schönfliess symbol for its point group $\mathscr{G}_o$ together with a superscript. Thus, for example, the space group of the face-centred cubic structure with point group O_h is denoted by O_h^5. (The assignment of superscripts by Schönfliess is rather arbitrary). The number of space groups in each crystal class is indicated in appendix 3.

The conditions on rotations mentioned above also impose corresponding restrictions on the possible lattice vectors, and, in fact, only 14 different lattices are allowed. They are known as the *Bravais Lattices*. The Bravais lattices may be classified in terms of the *maximal* point group that is compatible with the lattice. Lattices corresponding to the same maximal point group are said to belong to the same *symmetry system* (or syngony), of which there are only seven. A symmetry system α is said to be subordinate to a symmetry system β if the maximal point group corresponding to α is a subgroup of that corresponding to β, and if every Bravais lattice of β can be transformed into one of the Bravais lattices of α by an arbitrarily small deformation of the basic lattice vectors.

For a given Bravais lattice, any point group that is a proper subgroup of the maximal point group of that lattice is clearly also compatible with that lattice, and will therefore lead to another set of space groups based on that lattice. However, if such a subgroup is also a subgroup of the maximal point group of a subordinate lattice, then these space groups really belong to the subordinate symmetry system. Thus a space group of a crystal class with point group $\mathscr{G}_o$ is considered to belong to the symmetry system with maximal point group $\mathscr{G}_o^{\max}$ if, and only if, $\mathscr{G}_o$ is a subgroup of $\mathscr{G}_o^{\max}$ and $\mathscr{G}_o$ is not a subgroup of the maximal point group of a subordinate symmetry system.

The Bravais lattices and crystal classes belonging to each symmetry system will now be specified. In each case the first-named crystal class corresponds to the maximal point group. The parameters a, b, c and d are arbitrary.

(1) Triclinic symmetry system. This system has only one Bravais lattice, namely the simple triclinic lattice, Γ_t, for which $\boldsymbol{a}_1$, $\boldsymbol{a}_2$ and $\boldsymbol{a}_3$ are arbitrary. The crystal classes are C_i and C_1.

(2) Monoclinic symmetry system. This system has the following two Bravais lattices:

simple monoclinic, Γ_m,
$\boldsymbol{a}_3$ perpendicular to $\boldsymbol{a}_1$ and $\boldsymbol{a}_2$;
base-centered monoclinic, Γ_m^b,
$\boldsymbol{a}_1 = (a, b, 0)$, $\boldsymbol{a}_2 = (a, -b, 0)$, $\boldsymbol{a}_3 = (c, 0, d)$.

The crystal classes are C_{2h}, C_2 and C_s.

(3) Orthorhombic symmetry system. This system has the following four Bravais lattices:

simple orthorhombic, Γ_o,
$\boldsymbol{a}_1 = (a, 0, 0)$, $\boldsymbol{a}_2 = (0, b, 0)$, $\boldsymbol{a}_3 = (0, 0, c)$;
base-centred orthorhombic, Γ_o^b,
$\boldsymbol{a}_1 = (a, b, 0)$, $\boldsymbol{a}_2 = (a, -b, 0)$, $\boldsymbol{a}_3 = (0, 0, c)$;
body-centred orthorhombic, Γ_o^v,
$\boldsymbol{a}_1 = (a, b, c)$, $\boldsymbol{a}_2 = (a, b, -c)$, $\boldsymbol{a}_3 = (a, -b, -c)$;
face-centred orthorhombic, Γ_o^f,
$\boldsymbol{a}_1 = (a, b, 0)$, $\boldsymbol{a}_2 = (0, b, c)$, $\boldsymbol{a}_3 = (a, 0, c)$.

The crystal classes are D_{2h}, C_{2v} and D_2.

(4) Tetragonal symmetry system. This system has the following two Bravais lattices:

$$\begin{aligned}&\text{simple tetragonal, } \Gamma_q,\\ &\quad \boldsymbol{a}_1=(a,0,0),\ \boldsymbol{a}_2=(0,a,0),\ \boldsymbol{a}_3=(0,0,b);\\ &\text{body-centred tetragonal, } \Gamma_q^v,\\ &\quad \boldsymbol{a}_1=(a,a,b),\ \boldsymbol{a}_2=(a,a,-b),\ \boldsymbol{a}_3=(a,-a,b).\end{aligned}$$

The crystal classes are D_{4h}, D_{2d}, S_4, C_4, C_{4h}, D_4 and C_{4v}.

(5) The cubic symmetry system. This system has the following three Bravais lattices:

$$\begin{aligned}&\text{simple cubic, } \Gamma_c,\\ &\quad \boldsymbol{a}_1=(a,0,0),\ \boldsymbol{a}_2=(0,a,0),\ \boldsymbol{a}_3=(0,0,a);\\ &\text{body-centred cubic, } \Gamma_c^v,\\ &\quad \boldsymbol{a}_1=\tfrac{1}{2}a(1,1,1),\ \boldsymbol{a}_2=\tfrac{1}{2}a(1,1-1),\ \boldsymbol{a}_3=\tfrac{1}{2}a(1,-1,-1);\\ &\text{face-centred cubic, } \Gamma_c^f,\\ &\quad \boldsymbol{a}_1=\tfrac{1}{2}a(1,1,0),\ \boldsymbol{a}_2=\tfrac{1}{2}a(0,1,1),\ \boldsymbol{a}_3=\tfrac{1}{2}a(1,0,1).\end{aligned}$$

The crystal classes are O_h, T, T_h, T_d and O.

(6) The rhombohedral (or trigonal) symmetry system. This system has only one Bravais lattice, namely the simple rhombohedral lattice, Γ_{rh}, for which $\boldsymbol{a}_1=(a,0,b)$, $\boldsymbol{a}_2=(\tfrac{1}{2}a\sqrt{3},-\tfrac{1}{2}a,b)$, $\boldsymbol{a}_3=(-\tfrac{1}{2}a\sqrt{3},-\tfrac{1}{2}a,b)$. The crystal classes are D_{3d}, C_3, C_{3v}, D_3 and S_6.

(7) The hexagonal symmetry system. This system has only one Bravais lattice, namely the simple hexagonal, Γ_h, for which $\boldsymbol{a}_1=(0,0,c)$, $\boldsymbol{a}_2=(a,0,0)$, $\boldsymbol{a}_3=(-\tfrac{1}{2}a,-\tfrac{1}{2}a\sqrt{3},0)$. The crystal classes are D_{6h}, C_3, C_{3v}, D_3, S_6, D_{3d}, C_{3h}, D_{3h}, C_6, C_{6h}, C_{6v} and D_6.

The scheme of subordination for symmetry systems is as follows:

$$\begin{aligned}&\text{cubic} > \text{tetragonal} > \text{orthorhombic} > \text{monoclinic} > \text{triclinic};\\ &\text{rhombohedral} > \text{monoclinic};\quad \text{hexagonal} > \text{orthorhombic}.\end{aligned}$$

Here $\beta > \alpha$ indicates that α is subordinate to β. Although the point group D_{3d} is a subgroup of D_{6h}, the second condition for subordination implies that the rhombohedral system is not subordinate to the hexagonal system.

In the theory of the matrix representations of space groups there appears a further vital distinction between two types of space groups, namely the *symmorphic* space groups and the *non-symmorphic* space groups. A symmorphic space group is defined to be such that every symmetry operation consists of a rotation followed by a *primitive* translation, that is, it is of the form $\{\boldsymbol{R} \mid \boldsymbol{t_n}\}$.

The physical significance of a symmorphic space group is as follows. If the solid is such that only one chemical element is present, and the nuclei lie only at the lattice points of a Bravais lattice, then the space group is symmorphic. If the crystal contains more than one chemical element, but the arrays of nuclei of each element each form a Bravais lattice, then the space group is again symmorphic. Examples of both situations are given in §3.2. Non-symmorphic space groups correspond to structures where all the nuclei of one element cannot be put at the lattice points of a Bravais lattice. An example of this structure is given in §3.3.

In a symmorphic space group $\mathscr{G}$, the point group $\mathscr{G}_0$ of the space group is a *subgroup* of $\mathscr{G}$. This follows because if $\{\boldsymbol{R} \mid \boldsymbol{0}\}$ is a member of $\mathscr{G}_0$ and $\mathscr{G}$ is symmorphic then there must exist a transformation of the form $\{\boldsymbol{R} \mid \boldsymbol{t_n}\}$ that is a member of $\mathscr{G}$. Then $\{\boldsymbol{R} \mid \boldsymbol{0}\}$ must also be a member of $\mathscr{G}$, as one can write $\{\boldsymbol{R} \mid \boldsymbol{0}\} = \{\boldsymbol{E} \mid -\boldsymbol{t_n}\}\, \{\boldsymbol{R} \mid \boldsymbol{t_n}\}$, where $\{\boldsymbol{E} \mid -\boldsymbol{t_n}\}$ is a member of $\mathscr{G}$. There are 73 symmorphic space groups. They are listed in table 1.4. In most cases, a symmorphic space group is completely determined by specifying its crystal class and its Bravais lattice. However, there are seven cases of pairs of symmorphic space groups having the same crystal class and Bravais lattice, but differing in the orientation of the rotational axes of the point group to the basic lattice vectors of the Bravais lattice.

It will be shown in ch. 5 that there is no difficulty in carrying through a complete investigation of the representations of any symmorphic space group, because these representations are all

TABLE 1.4

Symmorphic space groups

Symmetry system	Bravais lattice	Symmorphic space groups
Triclinic	Γ_t	C_1^1, C_i^1
Monoclinic	Γ_m	C_s^1, C_2^1, C_{2h}^1
	Γ_m^b	C_s^3, C_2^3, C_{2h}^3
Orthorhombic	Γ_o	$C_{2v}^1, D_2^1, D_{2h}^1$
	Γ_o^b	$C_{2v}^{11}, C_{2v}^{14}, D_2^6, D_{2h}^{17}$
	Γ_o^v	$C_{2v}^{20}, D_2^8, D_{2h}^{25}$
	Γ_o^f	$C_{2v}^{18}, D_2^7, D_{2h}^{23}$
Tetragonal	Γ_q	$D_{2d}^1, D_{2d}^5, S_4^1, C_4^1, C_{4h}^1, D_4^1, C_{4v}^1, D_{4h}^1$
	Γ_q^v	$D_{2d}^9, D_{2d}^{11}, S_4^2, C_4^5, C_{4h}^5, D_4^9, C_{4v}^9, D_{4h}^{17}$
Cubic	Γ_c	$T^1, T_h^1, T_d^1, O^1, O_h^1$
	Γ_c^v	$T^3, T_h^5, T_d^3, O^5, O_h^9$
	Γ_c^f	$T^2, T_h^3, T_d^2, O^3, O_h^5$
Rhombohedral	Γ_{rh}	$C_3^4, C_{3v}^5, D_3^7, S_6^2, D_{3d}^5$
Hexagonal	Γ_h	$C_3^1, C_{3v}^1, C_{3v}^2, D_3^1, D_3^2, S_6^1, D_{3d}^1, D_{3d}^3, C_{3h}^1, D_{3h}^1, D_{3h}^3, C_6^1, C_{6h}^1, C_{6v}^1, D_6^1, D_{6h}^1$

closely related to those of the crystallographic point groups, which are well known and are given in appendix 1. This is not true of the non-symmorphic space groups.

In a non-symmorphic space group $\mathscr{G}$, the point group $\mathscr{G}_o$ is not a subgroup of $\mathscr{G}$, as some operations (namely those relating two different lattices) consist of a rotation followed by a non-primitive translation. The specification of a non-symmorphic space group requires not only a specification of the crystal class and Bravais lattice (and the relative orientation of rotational axes and basic lattice vectors) but also the translations that correspond to each rotation of the point group. A relationship between $\mathscr{G}$, $\mathscr{T}$ and $\mathscr{G}_0$ that holds for all space groups is described in ch. 6 §4.

A complete description of all 230 space groups may be found in the 'International tables for X-ray crystallography', Vol. I [1965], which employs both the Schönfliess notation and the so-called 'international notation'. A simple prescription for determining the symmetry elements of a space group from the 'general position' listed in the International Tables has been given by WONDRATSCHEK and NEUBÜSER [1967]. The book by KOVALEV [1965] and the appendix of the book by LYUBARSKII [1960] also contain a complete specification. Lists of the space groups to which elements and compounds belong have been compiled by WYCKOFF [1963, 1964, 1965] and by DONNAY and NOWACKI [1954].

The more complicated structure of the non-symmorphic space groups requires the development of several rather abstract concepts that are not needed in the presentation of the theory for the symmorphic groups. Accordingly, the theory is first presented for the symmorphic case. Then, in ch. 6, the additional group theoretical concepts are introduced, and applied in ch. 7 to the non-symmorphic case. Chs. 2 to 4 apply equally to both cases, while many of the ideas appearing in ch. 7 are just generalizations of those appearing in ch. 5. The account in ch. 8 of double symmorphic space groups can be read without previous reference to chs. 6 and 7.

§ 3.2. *Some examples of symmorphic space groups.* O_h^9, O_h^5 *and* T_d^2

The body-centred cubic lattice Γ_c^v consists of a repeated cubic array of lattice points, with lattice points also occurring at the centres of the cubes. A section of the lattice is shown in fig. 1.4. The basic lattice vectors may be taken to be

$$\left.\begin{aligned} \boldsymbol{a}_1 &= \tfrac{1}{2}a(1, 1, 1), \\ \boldsymbol{a}_2 &= \tfrac{1}{2}a(1, 1, -1), \\ \boldsymbol{a}_3 &= \tfrac{1}{2}a(1, -1, -1), \end{aligned}\right| \tag{1.10}$$

where a is the length of the cube edge, and the vectors are referred

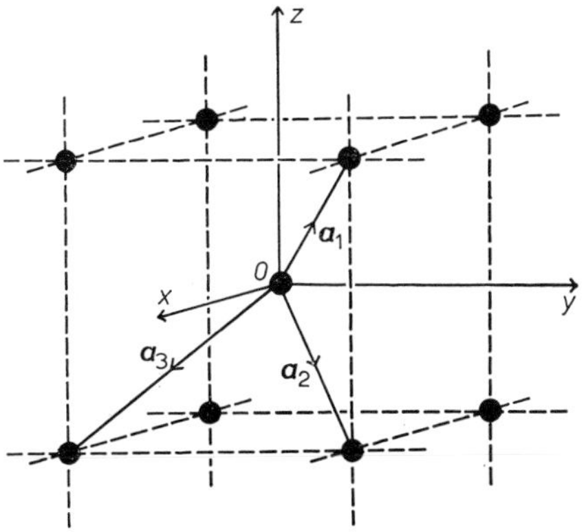

Fig. 1.4. The body-centred cubic lattice Γ_c^v.

to mutually perpendicular axes Ox, Oy, and Oz, which lie parallel to a set of cube edges.

The maximal point group compatible with the body-centred cubic lattice is the full cubic point group O_h, which consists of 48 operations, 24 proper rotations and 24 improper rotations obtained by multiplying the proper rotations by the inversion operator. The proper rotations, defined in terms of axes shown in fig. 1.5 and in the notation established in § 1, are:

E: the identity operation;

$C_{3\alpha}$, $C_{3\beta}$, $C_{3\gamma}$, $C_{3\delta}$: rotations through $\frac{2}{3}\pi$ in the right-hand screw sense about $O\alpha$, $O\beta$, $O\gamma$, $O\delta$ respectively;

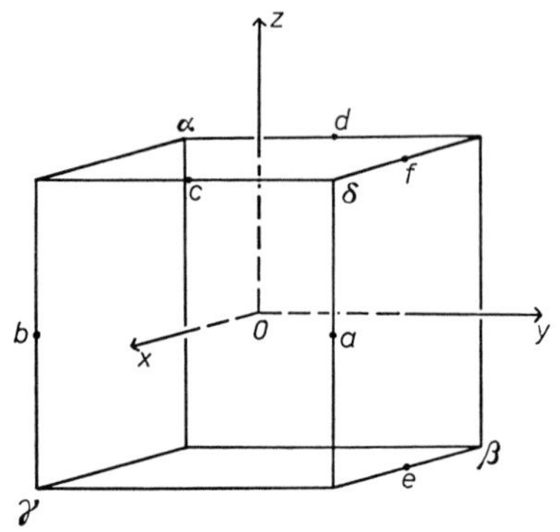

Fig. 1.5

$C_{3\alpha}^{-1}$, $C_{3\beta}^{-1}$, $C_{3\gamma}^{-1}$, $C_{3\delta}^{-1}$: rotations through $\frac{2}{3}\pi$ in the left-hand screw sense about $O\alpha$, $O\beta$, $O\gamma$, $O\delta$ respectively;

C_{2x}, C_{2y}, C_{2z}, C_{2a}, C_{2b}, C_{2c}, C_{2d}, C_{2e}, C_{2f}: rotations through π about Ox, Oy, Oz, Oa, Ob, Oc, Od, Oe, Of respectively;

C_{4x}, C_{4y}, C_{4z}: rotations through $\frac{1}{2}\pi$ in the right-hand screw sense about Ox, Oy, Oz respectively;

C_{4x}^{-1}, C_{4y}^{-1}, C_{4z}^{-1}: rotations through $\frac{1}{2}\pi$ in the left-hand screw sense about Ox, Oy, Oz respectively.

All these operations transform a cube into itself. The transformation matrices for these proper rotations are contained in table 1.5. They will be used frequently in later chapters. The point group O_h contains 10 classes, the contents of which are detailed in appendix 1.

The space group O_h^9 corresponds to a crystal containing only one chemical element whose nuclei all lie at the lattice points of the body-centred cubic lattice. The alkali metals are important examples of this structure.

The face-centred cubic lattice Γ_c^f consists of a repeated cubic array of lattice points, with lattice points also occurring at the mid-point of the faces of the cubes. A section of the lattice is shown

TABLE 1.5

$$\boldsymbol{R}(E) = \begin{pmatrix} 1 & 0 & 0 \\ 0 & 1 & 0 \\ 0 & 0 & 1 \end{pmatrix}, \qquad \boldsymbol{R}(C_{3\alpha}) = \begin{pmatrix} 0 & 1 & 0 \\ 0 & 0 & -1 \\ -1 & 0 & 0 \end{pmatrix},$$

$$\boldsymbol{R}(C_{3\beta}) = \begin{pmatrix} 0 & -1 & 0 \\ 0 & 0 & -1 \\ 1 & 0 & 0 \end{pmatrix}, \qquad \boldsymbol{R}(C_{3\gamma}) = \begin{pmatrix} 0 & -1 & 0 \\ 0 & 0 & 1 \\ -1 & 0 & 0 \end{pmatrix},$$

$$\boldsymbol{R}(C_{3\delta}) = \begin{pmatrix} 0 & 1 & 0 \\ 0 & 0 & 1 \\ 1 & 0 & 0 \end{pmatrix}, \qquad \boldsymbol{R}(C_{3\alpha}^{-1}) = \begin{pmatrix} 0 & 0 & -1 \\ 1 & 0 & 0 \\ 0 & -1 & 0 \end{pmatrix},$$

$$\boldsymbol{R}(C_{3\beta}^{-1}) = \begin{pmatrix} 0 & 0 & 1 \\ -1 & 0 & 0 \\ 0 & -1 & 0 \end{pmatrix}, \qquad \boldsymbol{R}(C_{3\gamma}^{-1}) = \begin{pmatrix} 0 & 0 & -1 \\ -1 & 0 & 0 \\ 0 & 1 & 0 \end{pmatrix},$$

$$\boldsymbol{R}(C_{3\delta}^{-1}) = \begin{pmatrix} 0 & 0 & 1 \\ 1 & 0 & 0 \\ 0 & 1 & 0 \end{pmatrix}, \qquad \boldsymbol{R}(C_{2x}) = \begin{pmatrix} 1 & 0 & 0 \\ 0 & -1 & 0 \\ 0 & 0 & -1 \end{pmatrix},$$

$$\boldsymbol{R}(C_{2y}) = \begin{pmatrix} -1 & 0 & 0 \\ 0 & 1 & 0 \\ 0 & 0 & -1 \end{pmatrix}, \qquad \boldsymbol{R}(C_{2z}) = \begin{pmatrix} -1 & 0 & 0 \\ 0 & -1 & 0 \\ 0 & 0 & 1 \end{pmatrix},$$

$$\boldsymbol{R}(C_{4x}) = \begin{pmatrix} 1 & 0 & 0 \\ 0 & 0 & 1 \\ 0 & -1 & 0 \end{pmatrix}, \qquad \boldsymbol{R}(C_{4y}) = \begin{pmatrix} 0 & 0 & -1 \\ 0 & 1 & 0 \\ 1 & 0 & 0 \end{pmatrix},$$

$$\boldsymbol{R}(C_{4z}) = \begin{pmatrix} 0 & 1 & 0 \\ -1 & 0 & 0 \\ 0 & 0 & 1 \end{pmatrix}, \qquad \boldsymbol{R}(C_{4x}^{-1}) = \begin{pmatrix} 1 & 0 & 0 \\ 0 & 0 & -1 \\ 0 & 1 & 0 \end{pmatrix},$$

$$\boldsymbol{R}(C_{4y}^{-1}) = \begin{pmatrix} 0 & 0 & 1 \\ 0 & 1 & 0 \\ -1 & 0 & 0 \end{pmatrix}, \qquad \boldsymbol{R}(C_{4z}^{-1}) = \begin{pmatrix} 0 & -1 & 0 \\ 1 & 0 & 0 \\ 0 & 0 & 1 \end{pmatrix},$$

$$\boldsymbol{R}(C_{2a}) = \begin{pmatrix} 0 & 1 & 0 \\ 1 & 0 & 0 \\ 0 & 0 & -1 \end{pmatrix}, \qquad \boldsymbol{R}(C_{2b}) = \begin{pmatrix} 0 & -1 & 0 \\ -1 & 0 & 0 \\ 0 & 0 & -1 \end{pmatrix},$$

$$\boldsymbol{R}(C_{2c}) = \begin{pmatrix} 0 & 0 & 1 \\ 0 & -1 & 0 \\ 1 & 0 & 0 \end{pmatrix}, \qquad \boldsymbol{R}(C_{2d}) = \begin{pmatrix} 0 & 0 & -1 \\ 0 & -1 & 0 \\ -1 & 0 & 0 \end{pmatrix},$$

$$\boldsymbol{R}(C_{2e}) = \begin{pmatrix} -1 & 0 & 0 \\ 0 & 0 & 1 \\ 0 & 1 & 0 \end{pmatrix}, \qquad \boldsymbol{R}(C_{2f}) = \begin{pmatrix} -1 & 0 & 0 \\ 0 & 0 & -1 \\ 0 & -1 & 0 \end{pmatrix},$$

$$\boldsymbol{R}(C_{3z}) = \begin{pmatrix} -\frac{1}{2} & \frac{1}{2}\sqrt{3} & 0 \\ -\frac{1}{2}\sqrt{3} & -\frac{1}{2} & 0 \\ 0 & 0 & 1 \end{pmatrix}, \qquad \boldsymbol{R}(C_{3z}^{-1}) = \begin{pmatrix} -\frac{1}{2} & -\frac{1}{2}\sqrt{3} & 0 \\ \frac{1}{2}\sqrt{3} & -\frac{1}{2} & 0 \\ 0 & 0 & 1 \end{pmatrix},$$

$$\boldsymbol{R}(C_{6z}) = \begin{pmatrix} \frac{1}{2} & \frac{1}{2}\sqrt{3} & 0 \\ -\frac{1}{2}\sqrt{3} & \frac{1}{2} & 0 \\ 0 & 0 & 1 \end{pmatrix}, \qquad \boldsymbol{R}(C_{6z}^{-1}) = \begin{pmatrix} \frac{1}{2} & -\frac{1}{2}\sqrt{3} & 0 \\ \frac{1}{2}\sqrt{3} & \frac{1}{2} & 0 \\ 0 & 0 & 1 \end{pmatrix},$$

$$\boldsymbol{R}(C_{2A}) = \begin{pmatrix} -\frac{1}{2} & \frac{1}{2}\sqrt{3} & 0 \\ \frac{1}{2}\sqrt{3} & \frac{1}{2} & 0 \\ 0 & 0 & -1 \end{pmatrix}, \qquad \boldsymbol{R}(C_{2B}) = \begin{pmatrix} -\frac{1}{2} & -\frac{1}{2}\sqrt{3} & 0 \\ -\frac{1}{2}\sqrt{3} & \frac{1}{2} & 0 \\ 0 & 0 & -1 \end{pmatrix},$$

$$\boldsymbol{R}(C_{2C}) = \begin{pmatrix} \frac{1}{2} & -\frac{1}{2}\sqrt{3} & 0 \\ -\frac{1}{2}\sqrt{3} & -\frac{1}{2} & 0 \\ 0 & 0 & -1 \end{pmatrix}, \qquad \boldsymbol{R}(C_{2D}) = \begin{pmatrix} \frac{1}{2} & \frac{1}{2}\sqrt{3} & 0 \\ \frac{1}{2}\sqrt{3} & -\frac{1}{2} & 0 \\ 0 & 0 & -1 \end{pmatrix},$$

in fig. 1.6. The basic lattice vectors may then be taken to be

$$\left.\begin{aligned} \boldsymbol{a}_1 &= \tfrac{1}{2}a\,(1,1,0), \\ \boldsymbol{a}_2 &= \tfrac{1}{2}a\,(0,1,1), \\ \boldsymbol{a}_3 &= \tfrac{1}{2}a\,(1,0,1), \end{aligned}\right| \tag{1.11}$$

with the same conventions as in (1.10). The maximal point group compatible with the face-centred cubic lattice is again the full cubic

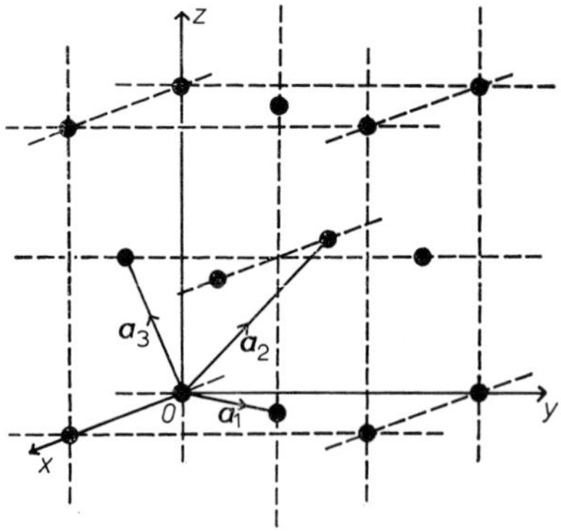

Fig. 1.6. The face-centred cubic lattice Γ_c^f.

point group O_h that is described above. The space group O_h^5 can correspond to a crystal containing only one chemical element, whose nuclei all lie at the lattice points of the face-centred cubic lattice, the noble metals providing examples of this structure.

The space groups O_h^5 and O_h^9 will be frequently used to illustrate in detail various concepts as they are developed in later chapters, particularly in chs. 5 and 8. The original investigation of the representations of these space groups was by BOUCKAERT *et al.* [1936], their 'double groups' being considered later by ELLIOTT [1954].

An example of a symmorphic space group whose point group is a proper subgroup of the maximal point group compatible with its lattice is provided in nature by the zinc blende structure. In crystals of zinc blende, the zinc nuclei occupy the lattice points of a face-centred cubic lattice Γ_c^f, and the sulphur nuclei occupy the lattice points of another face-centred cubic lattice that is simply displaced relative to the 'zinc' lattice by one quarter of a cube diagonal. Fig. 1.7 is a projection of lattice points onto a cube face

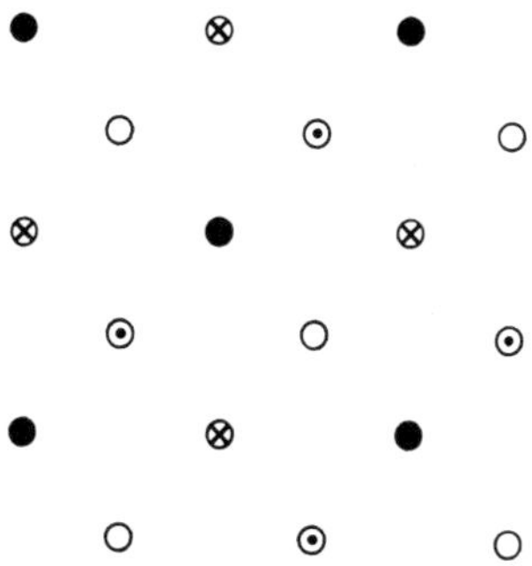

Fig. 1.7. The zinc blende structure T_d^2. The full and crossed circles represent zinc nuclei in the plane $z=0$ and $z=\frac{1}{2}a$ respectively, and the empty and dotted circles represent sulphur nuclei in the planes $z=\frac{1}{4}a$ and $z=\frac{3}{4}a$ respectively.

($z=0$) showing the relative configuration of the two lattices. The translational symmetry of the structure is clearly that of the face-centred cubic lattice. However, the point group $\mathscr{G}_o$ is not O_h, for some members of O_h do not transform the array into itself. An example of such a member is provided by the proper rotation C_{4z} through $\frac{1}{2}\pi$ about an axis perpendicular to the planes of fig. 1.7 and passing through a nucleus. Close investigation shows that the point group G_o is actually T_d, so by table 1.4 the space group is T_d^2. A study of the representations of the 'single' groups corresponding to this structure has been made by PARMENTER [1955]. The 'double' groups have been investigated both by PARMENTER [1955] and DRESSELHAUS [1955].

§ 3.3. *A common example of a non-symmorphic space group.* D_{6h}^4

In the hexagonal close-packed structure D_{6h}^4 the nuclei lie at the lattice points of *two* identical simple hexagonal lattices Γ_h. A section of a simple hexagonal lattice is shown in fig. 1.8. Its lattice

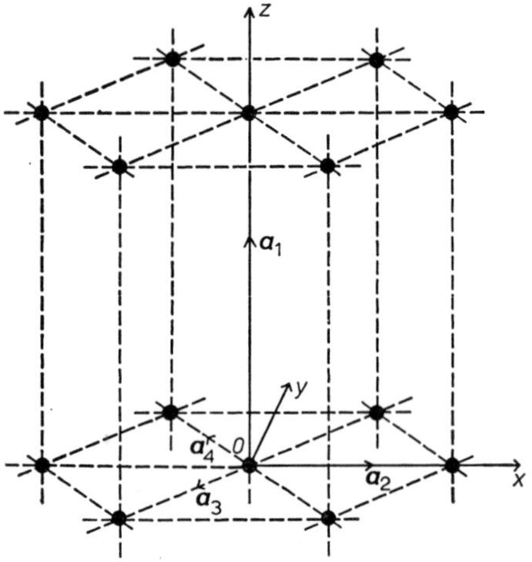

Fig. 1.8. The simple hexagonal lattice Γ_h.

points all lie in a set of equidistant parallel planes, which will be described as being horizontal. In each horizontal plane the points form a triangular 'net', and lattice points in different horizontal planes lie vertically above each other. In the hexagonal close-packed structure the nuclei lie on equidistant horizontal planes that belong alternately to the two simple lattices. The configuration of the second lattice relative to the first is shown in fig. 1.9, in which full circles denote nuclei lying in a plane of the first lattice, and empty circles denote nuclei lying in an adjacent plane of the second lattice.

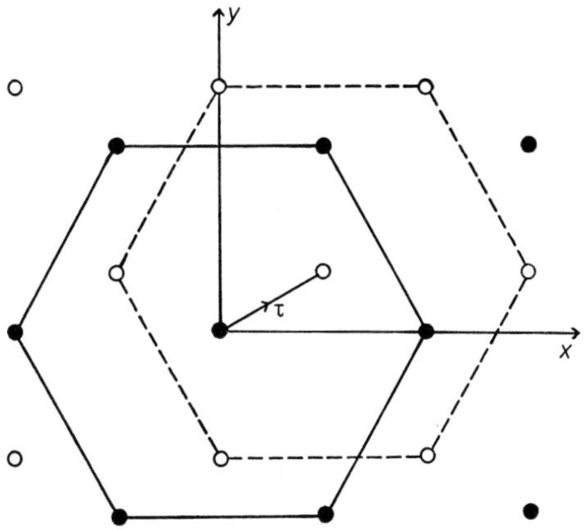

Fig. 1.9. The hexagonal closed-packed structure D_{6h}^4. The full and empty circles represent nuclei lying in the planes $z=0$ and $z=\frac{1}{2}c$ respectively.

The basic lattice vectors of the hexagonal lattice may be taken to be

$$\left.\begin{aligned} \boldsymbol{a}_1 &= (0, 0, c), \\ \boldsymbol{a}_2 &= (a, 0, 0), \\ \boldsymbol{a}_3 &= (-\tfrac{1}{2}a, -\tfrac{1}{2}a\sqrt{3}, 0). \end{aligned}\right| \tag{1.12}$$

It is useful also to define the vector $\boldsymbol{a}_4 = -\boldsymbol{a}_2 - \boldsymbol{a}_3 = (-\frac{1}{2}a, \frac{1}{2}a\sqrt{3}, 0)$. These four vectors are shown in fig. 1.8.

The point group $\mathscr{G}_o$ is the group D_{6h}, which consists of 12 proper rotations and 12 improper rotations obtained by multiplying the proper rotations by the inversion operator. The proper rotations, defined in terms of axes shown in figs. 1.3 and 1.5, are:

E: the identity operation;

C_{6z}, C_{3z}: rotations through $\frac{1}{3}\pi$ and $\frac{2}{3}\pi$ respectively in the right-hand screw sense about Oz;

C_{6z}^{-1}, C_{3z}^{-1}: rotations through $\frac{1}{3}\pi$ and $\frac{2}{3}\pi$ respectively in the left-hand screw sense about Oz;

$C_{2x}, C_{2y}, C_{2z}, C_{2A}, C_{2B}, C_{2C}, C_{2D}$: rotations through π about Ox, Oy, Oz, OA, OB, OC, OD respectively.

The transformation matrices for these proper rotations are contained in table 1.5. The point group D_{6h} contains 12 classes, as listed in appendix 1.

Unfortunately, in the original analysis of the hexagonal close-packed structure by HERRING [1942], a different notation is employed. The connection between Herring's notation for rotations and that used here is given in table 1.6. The relationship between

TABLE 1.6

Herring's notation	Present notation	Herring's notation	Present notation
ε	E	i	I
δ_6	C_{6z}	σ_3	IC_{6z}
δ_6^{-1}	C_{6z}^{-1}	σ_3^{-1}	IC_{6z}^{-1}
δ_3	C_{3z}	σ_6	IC_{3z}
δ_3^{-1}	C_{3z}^{-1}	σ_6^{-1}	IC_{3z}^{-1}
δ_2	C_{2z}	ϱ	IC_{2z}
δ'_{22}	C_{2x}	ϱ'_2	IC_{2x}
δ'_{23}	C_{2A}	ϱ'_3	IC_{2A}
δ'_{24}	C_{2B}	ϱ'_4	IC_{2B}
δ''_{22}	C_{2y}	ϱ''_2	IC_{2y}
δ''_{23}	C_{2C}	ϱ''_3	IC_{2C}
δ''_{24}	C_{2D}	ϱ''_4	IC_{2D}

the vectors $\boldsymbol{t}_i$ introduced by Herring and the $\boldsymbol{a}_i$ of (1.12) is simply that $\boldsymbol{t}_i = \boldsymbol{a}_i$, $i = 1, 2, 3, 4$.

In the space group, the rotations $E, C_{3z}, C_{3z}^{-1}, C_{2y}, C_{2C}, C_{2D}, IC_{6z}, IC_{6z}^{-1}, IC_{2z}, IC_{2x}, IC_{2A}, IC_{2B}$ always appear in the form $\{\boldsymbol{R} \mid \boldsymbol{t}_n\}$, that is, they are always associated with *primitive* translations, whereas all the other rotations of the point group always appear in the form $\{\boldsymbol{R} \mid \boldsymbol{t}_n + \boldsymbol{\tau}\}$, where

$$\boldsymbol{\tau} = (\tfrac{1}{2}a, \tfrac{1}{6}a\sqrt{3}, \tfrac{1}{2}c) \tag{1.13}$$

is a vector from one nucleus to one of the nearest nuclei in the plane above.

One possibly puzzling point is worth mentioning. As may be seen from fig. 1.9, the pure translation $\{E \mid \boldsymbol{\tau}\}$ also transforms the first hexagonal lattice into the second. However, $\{E \mid \boldsymbol{\tau}\}$ is *not* a member of the space group. The reason is that if it were, then so would be $\{E \mid \boldsymbol{\tau}\}\{E \mid \boldsymbol{\tau}\} = \{E \mid 2\boldsymbol{\tau}\}$, by the first of the group postulates. But this corresponds to a vertical translation of a horizontal plane of the first lattice into the next horizontal plane of the *same* lattice, together with a horizontal translation which does *not* transform lattice points into lattice points, so that $\{E \mid 2\boldsymbol{\tau}\}$ is not a member of the space group. This difficulty does not arise with $\{\boldsymbol{R}(C_{6z}) \mid \boldsymbol{\tau}\}$ for example, for $\{\boldsymbol{R}(C_{6z}) \mid \boldsymbol{\tau}\}$ also transforms the first lattice into the second, but

$$\begin{aligned}\{\boldsymbol{R}(C_{6z}) \mid \boldsymbol{\tau}\}\{\boldsymbol{R}(C_{6z}) \mid \boldsymbol{\tau}\} &= \{\boldsymbol{R}(C_{3z}) \mid \boldsymbol{R}(C_{6z})\boldsymbol{\tau} + \boldsymbol{\tau}\} \\ &= \{\boldsymbol{R}(C_{3z}) \mid \boldsymbol{a}_1 + \boldsymbol{a}_2\},\end{aligned}$$

which *is* clearly a member of the space group.

The hexagonal close-packed structure will be used in ch. 7 to illustrate the theory of non-symmorphic space groups.

CHAPTER 2

THE THEORY OF MATRIX REPRESENTATIONS OF A GROUP

§ 1. INTRODUCTION

As will be shown in the next chapter, the matrix representations of groups play an all important role when concepts of symmetry are introduced into the quantum theory, the precise connection being established in ch. 3 §4.2. In applications to solid state physics, it is the space groups and point groups that are most relevant, but the theory that will be developed in this chapter applies to any *finite* group (and to some other types of group as well). Actually, a space group is not a finite group, for the lattice or lattices involved are infinite in extent, so that a space group contains an infinite number of pure translations. However, as will be demonstrated in ch. 4 §2, by imposing the Born cyclic boundary conditions, it is possible to work with a closely related finite group.

A number of the theorems quoted in this chapter have quite long proofs, often consisting largely of matrix manipulation, which do not themselves provide much additional insight into the theory. In order to concentrate on the essential results, these proofs have been omitted, but the interested reader may find them in most of the works quoted in the bibliography.

Definition of a matrix representation of a group. A group of square matrices, with matrix multiplication as the group multiplication operation, which is *homomorphic* to a group $\mathscr{G}$, is said to form a matrix representation of $\mathscr{G}$. That is, to each element A of $\mathscr{G}$ there corresponds a square matrix $\boldsymbol{\Gamma}(A)$ (from a group of square matrices), and

$$\boldsymbol{\Gamma}(A)\,\boldsymbol{\Gamma}(B)=\boldsymbol{\Gamma}(AB)$$

for every A and B of $\mathscr{G}$.

As the correspondence is merely a homomorphism, several elements of $\mathscr{G}$ may be represented by the same matrix. In the special case of an isomorphism, the representation is said to be 'faithful' or 'true'.

The number of rows (or columns) of the matrices is called the *dimension*, l, of the representation. It is clear that $\boldsymbol{\Gamma}(E)=\boldsymbol{E}$, the $l\times l$ unit matrix, and so $\boldsymbol{\Gamma}(A^{-1})=[\boldsymbol{\Gamma}(A)]^{-1}$ for each A of $\mathscr{G}$. Also, a matrix representation of a group is automatically a matrix representation of a subgroup.

These ideas will now be illustrated for the point group C_{3v} of ch. 1 §2.2. Several matrix representations of C_{3v} have in fact already been discussed. The transformation matrices $\boldsymbol{R}(T)$ of ch. 1 §2.2 themselves form a three-dimensional representation by virtue of the definition (1.3) of the product of two rotations. As noted in ch. 1 §2.5, the matrices $\boldsymbol{M}_1, \boldsymbol{M}_2, \boldsymbol{M}_3, \boldsymbol{M}_4, \boldsymbol{M}_5$ and $\boldsymbol{M}_6$ form a faithful two-dimensional representation, and the matrices $\boldsymbol{M}_1', \boldsymbol{M}_2'$ form a one-dimensional representation.

C_{3v} has another one-dimensional representation, in which *every* element of C_{3v} is represented by the 1×1 matrix whose only component is 1, that is,

$$\boldsymbol{\Gamma}(E)=\boldsymbol{\Gamma}(C_{3z})=\cdots=(1).$$

This representation is called the *identity* representation. Clearly, *every* group has an identity representation.

The theory that will be developed in subsequent sections shows that every group has actually an infinite number of representations, but all these representations can be constructed from a number of basic representations, namely the 'irreducible representations' that will be defined in ch. 2 §4.

§ 2. SIMILARITY TRANSFORMATIONS AND EQUIVALENT REPRESENTATIONS

Suppose that for every element A of a group $\mathscr{G}$ there is defined the l-dimensional matrix representation $\boldsymbol{\Gamma}(A)$. Then if $\boldsymbol{S}$ is any non-singular $l \times l$ matrix, the set of matrices $\boldsymbol{\Gamma}'(A)$ defined for each A of $\mathscr{G}$ by

$$\boldsymbol{\Gamma}'(A) = \boldsymbol{S}^{-1}\boldsymbol{\Gamma}(A)\,\boldsymbol{S} \tag{2.1}$$

also form an l-dimensional matrix representation of $\mathscr{G}$. This follows because for any A and B of $\mathscr{G}$,

$$\begin{aligned}\boldsymbol{\Gamma}'(A)\,\boldsymbol{\Gamma}'(B) &= \boldsymbol{S}^{-1}\boldsymbol{\Gamma}(A)\,\boldsymbol{S}\boldsymbol{S}^{-1}\boldsymbol{\Gamma}(B)\,\boldsymbol{S} = \boldsymbol{S}^{-1}\boldsymbol{\Gamma}(A)\,\boldsymbol{\Gamma}(B)\,\boldsymbol{S}\\ &= \boldsymbol{S}^{-1}\boldsymbol{\Gamma}(AB)\,\boldsymbol{S} = \boldsymbol{\Gamma}'(AB).\end{aligned}$$

The two representations $\boldsymbol{\Gamma}$ and $\boldsymbol{\Gamma}'$ are said to be *equivalent*, and the transformation (2.1) is called a *similarity transformation*. For one-dimensional representations $\boldsymbol{\Gamma}(A)=\boldsymbol{\Gamma}'(A)$ for all A of G and any 1×1 matrix $\boldsymbol{S}$, but for representations of more than one dimension it is clear that, by varying $\boldsymbol{S}$, similarity transformations can generate an infinite number of equivalent representations. However, equivalent representations can be regarded to a large extent as being essentially the same, in a sense that will become clearer as the theory is developed.

In practice, in order to establish whether the representations are equivalent, it is *not* necessary to actually find a matrix $\boldsymbol{S}$ that induces the similarity transformation, for in §6 a very simple direct test will be given.

§ 3. UNITARY REPRESENTATIONS

Definition of a unitary matrix. If the inverse $\boldsymbol{U}^{-1}$ of a matrix $\boldsymbol{U}$ is equal to the complex conjugate transpose $\tilde{\boldsymbol{U}}^*$ of $\boldsymbol{U}$, then $\boldsymbol{U}$ is said to be unitary.

It should be noted that any real orthogonal matrix $\boldsymbol{R}$ is unitary, for if it is orthogonal $\boldsymbol{R}^{-1}=\tilde{\boldsymbol{R}}$, and if it is real $\tilde{\boldsymbol{R}}^*=\tilde{\boldsymbol{R}}$.

Definition of a unitary representation. A matrix representation of a group $\mathscr{G}$ in which all the matrices are unitary is said to be a unitary representation of $\mathscr{G}$.

The following theorem may then be proved.

Theorem. Any representation of a finite group $\mathscr{G}$ is equivalent to a unitary representation of $\mathscr{G}$. (That is, it can be transformed into a unitary representation by a suitable similarity transformation.)

It will be seen shortly that there are considerable advantages in working with unitary representations, for many formulae take a particularly simple form for such representations, and the above theorem shows that one may always do so for a finite group. For representations of more than one dimension the restriction to unitary matrices still does not uniquely specify the matrices of the representation, for it may easily be verified that inducing a similarity transformation by a *unitary* matrix $\boldsymbol{S}$ on a unitary representation $\boldsymbol{\Gamma}$ produces another (equivalent) unitary representation $\boldsymbol{\Gamma}'$.

§ 4. REDUCIBLE AND IRREDUCIBLE REPRESENTATIONS

Suppose that $\boldsymbol{\Gamma}^1$ and $\boldsymbol{\Gamma}^2$ are two matrix representations of a group $\mathscr{G}$ of dimensions l_1 and l_2 respectively, and for each element A of $\mathscr{G}$ an (l_1+l_2)-dimensional matrix $\boldsymbol{\Gamma}(A)$ is defined by

$$\boldsymbol{\Gamma}(A)=\left(\begin{array}{c:c}\boldsymbol{\Gamma}^1(A) & \mathbf{0}\\ \hdashline \mathbf{0} & \boldsymbol{\Gamma}^2(A)\end{array}\right). \tag{2.2}$$

That is, $\boldsymbol{\Gamma}(A)$ is a partitioned matrix in which the elements of the $l_1\times l_2$ top right-hand block and the $l_2\times l_1$ bottom left-hand block are all zero. (For a brief description of partitioned matrices see appendix 4.) Using the multiplication property of partitioned matrices, it follows that for any A and B of $\mathscr{G}$,

$$\begin{aligned}\boldsymbol{\Gamma}(A)\,\boldsymbol{\Gamma}(B)&=\left(\begin{array}{c:c}\boldsymbol{\Gamma}^1(A)\,\boldsymbol{\Gamma}^1(B) & \mathbf{0}\\ \hdashline \mathbf{0} & \boldsymbol{\Gamma}^2(A)\,\boldsymbol{\Gamma}^2(B)\end{array}\right)\\ &=\left(\begin{array}{c:c}\boldsymbol{\Gamma}^1(AB) & \mathbf{0}\\ \hdashline \mathbf{0} & \boldsymbol{\Gamma}^2(AB)\end{array}\right)=\boldsymbol{\Gamma}(AB),\end{aligned}$$

so that these matrices $\boldsymbol{\Gamma}$ also form a representation of $\mathscr{G}$. This process could be repeated indefinitely, giving rise to an infinite number of representations.

The representations formed in this way are said to be *reducible*, and the form (2.2) may be described as 'block form'. The representation $\boldsymbol{\Gamma}$ of (2.2) is described as the *direct sum* of the representations $\boldsymbol{\Gamma}^1$ and $\boldsymbol{\Gamma}^2$, and one may accordingly write

$$\boldsymbol{\Gamma} = \boldsymbol{\Gamma}^1 \oplus \boldsymbol{\Gamma}^2 .$$

$\boldsymbol{\Gamma}^1$ and $\boldsymbol{\Gamma}^2$ are said to appear 'in the reduction' of $\boldsymbol{\Gamma}$. A similarity transformation applied to (2.2) could obscure the block form by making the zero matrix elements disappear, but the representation is still considered to be reducible, that is, a representation is reducible if it can be put into block form by an appropriate similarity transformation.

If a representation is not reducible, then it is said to be an *irreducible representation*. These are clearly the 'basic' representations from which all others can be constructed.

In §6 a very simple direct test will be given for determining whether a representation is reducible or not. It does not involve actually finding a similarity transformation which puts the representation in block form. Another less simple test is provided by the following theorem.

Theorem (Schur's Lemma). A matrix which commutes with all the matrices of an irreducible representation must be multiple of the unit matrix. Thus, if some *other* matrix commutes with all the matrices of a representation, the representation is reducible.

Schur's Lemma appears at an intermediate stage of the proof of the following orthogonality theorem for matrices. This extremely important theorem is the basis for the rest of the theory.

The orthogonality theorem for matrices. Suppose that $\boldsymbol{\Gamma}^i$ and $\boldsymbol{\Gamma}^j$ are two unitary irreducible representations of a group, and that they are not equivalent if $i \neq j$ and are identical if $i = j$. Then

$$\sum_T \Gamma^i(T)^*_{\mu\nu}\, \Gamma^j(T)_{\alpha\beta} = (g/l_i)\, \delta_{ij}\delta_{\mu\alpha}\delta_{\nu\beta},$$

where the summation is over all the elements T of the group, g is the order of the group, l_i is the dimension of $\boldsymbol{\Gamma}^i$, and $\delta_{ij} = 1$ if $i = j$ and 0 if $i \neq j$.

§ 5. THE CHARACTERS OF A REPRESENTATION

Because representations related by similarity transformations are equivalent, there is a considerable degree of arbitrariness in the actual forms of the matrices. However, there does exist a set of quantities which do not change under similarity transformations, and which therefore provides a unique way of characterizing a representation. These quantities are appropriately called the characters of the representation.

Definition of the characters of a representation. Suppose that $\boldsymbol{\Gamma}$ is an l-dimensional representation of a group $\mathscr{G}$. Then

$$\chi(A) = \sum_{j=1}^{l} \Gamma(A)_{jj}$$

is defined to be the *character* of the group element A in this representation (that is, it is the 'trace' of $\boldsymbol{\Gamma}(A)$, the sum of the diagonal matrix elements). The set of characters corresponding to a representation is called the *character system* of the representation.

The major properties of the characters depend on a simple result of matrix algebra, namely that the trace of a matrix is unaltered by a similarity transformation, that is, if $\boldsymbol{B}' = \boldsymbol{S}^{-1}\boldsymbol{B}\boldsymbol{S}$, then $\sum_{j=1}^{l} B'_{jj} = \sum_{j=1}^{l} B_{jj}$, for any non-singular $\boldsymbol{S}$.

It follows immediately that the character $\chi(A)$ of a group element A is unaltered by a similarity transformation on the matrices of the representation, so equivalent representations have the same character system.

Another important consequence of this result is that the characters of group elements in the same class are all equal. This follows from the fact that if A and A' are group elements in the same class, then by definition there exists a group element X such that $A'=X^{-1}AX$. Then $\boldsymbol{\Gamma}(A')=[\boldsymbol{\Gamma}(X)]^{-1}\,\boldsymbol{\Gamma}(A)\,\boldsymbol{\Gamma}(X)$, and hence $\chi(A')=\chi(A)$.

A further important result comes straight from the definition. This is that the dimension of the representation is equal to the character of the identity element of the group, simply because the identity element is represented by a unit matrix.

§ 6. THEOREMS INVOLVING CHARACTERS

In this section a number of very important theorems will be formulated, which can be used to give very useful results from a knowledge of the character systems *alone*, without requiring explicit expressions for the actual matrices. (These theorems are not quoted in the order in which they can be proved.)

Theorem. A necessary and sufficient condition for the equivalence of two representations is the equality of their character systems.

This necessary condition was actually demonstrated in the previous section. The sufficient condition constitutes the simple direct text for equivalence of two representations that was alluded to in §2.

Theorem. A necessary and sufficient condition for a representation to be *irreducible* is

$$\sum_T |\chi(T)|^2 = g\,,$$

where the summation is over all the elements of the group, which is of order g.

This is the simple direct test mentioned in §4 for establishing whether a representation is irreducible or not.

Theorem. The number of times n_i that an irreducible representation

$\boldsymbol{\Gamma}^i$, or a representation equivalent to $\boldsymbol{\Gamma}^i$, appears in the reduction of a reducible representation $\boldsymbol{\Gamma}$ is given by

$$n_i = g^{-1} \sum_T \chi(T)\, \chi^i(T)^*,$$

where the summation is over all the elements T of the group, g is the order of the group, and $\chi(T)$ and $\chi^i(T)$ denote the characters of the group element T in $\boldsymbol{\Gamma}$ and $\boldsymbol{\Gamma}^i$ respectively.

This theorem is particularly useful in discussing the effects of perturbation on a system, as will be demonstrated in ch. 3 §§6.4, 7.3.

Theorem. The number of inequivalent irreducible representations of a group $\mathscr{G}$ is equal to the number of classes of $\mathscr{G}$.

It follows that a finite group has only a *finite* number of inequivalent irreducible representations. The importance of the concept of a class is now obvious.

Theorem. The sum of the squares of the dimensions of the inequivalent irreducible representations is equal to the order of the group.

When combined with the previous theorem, this is usually sufficient to determine all the possible dimensions of the irreducible representations.

There are also two orthogonality theorems for characters which are used in actually constructing character systems.

Orthogonality theorems for characters.

$$\sum_k \chi^i(\mathscr{C}_k)^*\, \chi^j(\mathscr{C}_k)\, N_k = g\delta_{ij}$$

and

$$\sum_i \chi^i(\mathscr{C}_k)^*\, \chi^i(\mathscr{C}_l)\, N_k = g\delta_{kl},$$

where $\chi^i(\mathscr{C}_k)$ and $\chi^j(\mathscr{C}_k)$ denote the characters of the elements of the class $\mathscr{C}_k$ in the irreducible representations $\boldsymbol{\Gamma}^i$ and $\boldsymbol{\Gamma}^j$ respectively (it being assumed that they are inequivalent if $i \neq j$), N_k is the number of elements in $\mathscr{C}_k$, g is the order of the group, the first summation is over all the classes of the group, and the second summation is over all the inequivalent irreducible representations.

§ 7. CHARACTER TABLES

The character systems of the irreducible representations are conveniently displayed in the form of a character table. The classes of the group are usually listed along the top of the table, and the irreducible representations down the left-hand side. As a consequence of the fourth theorem of the previous section, this table is always square.

For example, the character table for the point group C_{3v} is given in table 2.1. It will be observed that $\boldsymbol{\Gamma}^1$ and $\boldsymbol{\Gamma}^2$ are one-dimensional,

TABLE 2.1

	E	C_{3z}, C_{3z}^{-1}	$IC_{2y}, IC_{2C}, IC_{2D}$
$\boldsymbol{\Gamma}^1$	1	1	1
$\boldsymbol{\Gamma}^2$	1	1	-1
$\boldsymbol{\Gamma}^3$	2	-1	0

that $\boldsymbol{\Gamma}^3$ is two-dimensional, and all the relevant theorems of the previous section are satisfied.

It is interesting to relate $\boldsymbol{\Gamma}^1$, $\boldsymbol{\Gamma}^2$ and $\boldsymbol{\Gamma}^3$ to the representations of C_{3v} discussed previously in §1. $\boldsymbol{\Gamma}^1$ is clearly the 'identity' representation, $\boldsymbol{\Gamma}^2$ is the representation consisting of $\boldsymbol{M}_1'$ and $\boldsymbol{M}_2'$, and $\boldsymbol{\Gamma}^3$ is the representation consisting of $\boldsymbol{M}_1$, $\boldsymbol{M}_2$, $\boldsymbol{M}_3$, $\boldsymbol{M}_4$, $\boldsymbol{M}_5$ and $\boldsymbol{M}_6$. Thus

$$\boldsymbol{\Gamma}^1(E) = \boldsymbol{\Gamma}^1(C_{3z}) = \boldsymbol{\Gamma}^1(C_{3z}^{-1}) = \boldsymbol{\Gamma}^1(IC_{2y}) = \boldsymbol{\Gamma}^1(IC_{2C})$$
$$= \boldsymbol{\Gamma}^1(IC_{2D}) = (1),$$

$$\boldsymbol{\Gamma}^2(E) = \boldsymbol{\Gamma}^2(C_{3z}) = \boldsymbol{\Gamma}^2(C_{3z}^{-1}) = (1),$$
$$\boldsymbol{\Gamma}^2(IC_{2y}) = \boldsymbol{\Gamma}^2(IC_{2C}) = \boldsymbol{\Gamma}^2(IC_{2D}) = (-1),$$

and

$$\boldsymbol{\Gamma}^3(E) = \begin{pmatrix} 1 & 0 \\ 0 & 1 \end{pmatrix}, \quad \boldsymbol{\Gamma}^3(C_{3z}) = \begin{pmatrix} -\frac{1}{2} & \frac{1}{2}\sqrt{3} \\ -\frac{1}{2}\sqrt{3} & -\frac{1}{2} \end{pmatrix}, \quad \boldsymbol{\Gamma}^3(C_{3z}^{-1}) = \begin{pmatrix} -\frac{1}{2} & -\frac{1}{2}\sqrt{3} \\ \frac{1}{2}\sqrt{3} & -\frac{1}{2} \end{pmatrix},$$
$$\boldsymbol{\Gamma}^3(IC_{2y}) = \begin{pmatrix} -1 & 0 \\ 0 & 1 \end{pmatrix}, \quad \boldsymbol{\Gamma}^3(IC_{2C}) = \begin{pmatrix} \frac{1}{2} & -\frac{1}{2}\sqrt{3} \\ -\frac{1}{2}\sqrt{3} & -\frac{1}{2} \end{pmatrix}, \quad \boldsymbol{\Gamma}^3(IC_{2D}) = \begin{pmatrix} \frac{1}{2} & \frac{1}{2}\sqrt{3} \\ \frac{1}{2}\sqrt{3} & -\frac{1}{2} \end{pmatrix}.$$

As C_{3v} has no three-dimensional irreducible representations, the transformation matrices $\boldsymbol{R}(T)$ constitute a reducible representation. In fact the third theorem of §6, together with table 2.1 and the explicit expressions given for $\boldsymbol{R}(T)$ in ch. 1 §2.2, shows that this representation is the direct sum of $\boldsymbol{\Gamma}^1$ and $\boldsymbol{\Gamma}^3$. In this particular case, it is obvious that $\boldsymbol{R}(T)$ is reducible, for the matrices of ch. 1 §2.2 are already in block form. The 2×2 top left-hand blocks of $\boldsymbol{R}(T)$ form a representation that is equivalent, but not identical, to that formed by the matrices $\boldsymbol{M}_1$, $\boldsymbol{M}_2$, $\boldsymbol{M}_3$, $\boldsymbol{M}_4$, $\boldsymbol{M}_5$ and $\boldsymbol{M}_6$.

It is possible to calculate character tables directly, without first obtaining explicit forms for the matrices. The character tables of all the 32 point groups were first given by WIGNER [1930]. They are listed in appendix 1. As will be shown in ch. 5, for the *symmorphic* space groups *all* the necessary information is provided by these point groups. For the *non-symmorphic* space groups, the character tables of another type of group have to be calculated, namely the factor groups $\mathscr{G}(\boldsymbol{k})/\mathscr{T}(\boldsymbol{k})$ of ch. 7 §2. The methods of calculating these tables will not be described here, but they are referred to in ch. 7 §2. A list of the published character tables is given in appendix 3.

Although a number of results of physical significance follow immediately from a knowledge of the characters, it is often necessary to obtain explicit expressions for the matrices of the representations. A method for constructing such explicit expressions from the characters is described in ch. 3 §5.3. Of course, for one-dimensional representations, the characters themselves *are* the matrix elements.

CHAPTER 3

GROUP THEORY IN QUANTUM MECHANICS

§ 1. SCALAR AND SPINOR WAVE FUNCTIONS

In the absence of spin–orbit coupling, ferromagnetism, anti-ferromagnetism and external magnetic fields, the spin of an electron does not play any role apart from simply doubling the degeneracy of every 'orbital' energy level, for in such a situation the wave function of an electron can be taken to be the product of an 'orbital' wave function with one of two possible spin functions. In effect, the spin of an electron can then be conveniently neglected, and all the symmetry arguments can be concentrated on the orbital part of the wave function, which is a *scalar* quantity. Accordingly, this and succeeding chapters up to and including ch. 7 will be devoted to a study of the orbital part of the wave function. Every degeneracy mentioned will be an 'orbital' degeneracy, which is doubled by spin.

In ch. 8 spin–orbit coupling will be considered, and it will be necessary then to consider the properties of *spinor* functions. However, the theory for these in many respects runs parallel to that for scalar functions, although it is more complicated.

This present chapter contains the central part of the theory in which the group theoretical ideas of the previous two chapters are related to quantum mechanical concepts and then applied to several general physical problems. The solid state theory of the succeeding chapters is then based on the ideas developed here.

§ 2. THE TRANSFORMATION PROPERTIES OF SCALAR WAVE FUNCTIONS

A scalar quantity is by definition a quantity that has a definite *value* at every point in space, this value being completely independent of any coordinate system set up in the space. The *same* scalar quantity is therefore represented by *different* functions in different coordinate systems. Suppose that the quantity is represented by the function $f(\boldsymbol{r})$ in a Cartesian coordinate system with axes Ox, Oy and Oz. Then in another Cartesian coordinate system with axes $O'x'$, $O'y'$ and $O'z'$, obtained from Ox, Oy, and Oz by a transformation T, the same quantity will be represented by a different function, which may be denoted by $P(T)\,f(\boldsymbol{r}')$. The relationship between these two function is

$$P(T)\,f(\boldsymbol{r}') = f(\boldsymbol{r}), \tag{3.1}$$

where $\boldsymbol{r}'$ and $\boldsymbol{r}$ are the position vectors of the *same* point referred to the two different coordinate systems, for *every* point in space, as the *value* of the quantity is independent of the coordinate system.

The most general form for the transformation T is a rotation followed by a translation, so that, as in eq. (1.5),

$$\boldsymbol{r}' = \{\boldsymbol{R}(T)\,|\,\boldsymbol{t}(T)\}\,\boldsymbol{r},$$

which will sometimes be written more briefly as

$$\boldsymbol{r}' = T\boldsymbol{r},$$

so that (3.1) becomes

$$P(T)\,f(T\boldsymbol{r}) = f(\boldsymbol{r}). \tag{3.2}$$

Eq. (3.1) can be rewritten as

$$P(T)\,f(\boldsymbol{r}') = f(\{\boldsymbol{R}(T)\,|\,\boldsymbol{t}(T)\}^{-1}\,\boldsymbol{r}'), \tag{3.3}$$

or equivalently and more briefly as

$$P(T)\,f(\boldsymbol{r}') = f(T^{-1}\boldsymbol{r}'). \tag{3.4}$$

Eq. (3.3) can be used to actually find the function $P(T)\,f$ for a given function f, for it merely implies that the components of $\boldsymbol{r}$ are to be replaced everywhere they appear in f by the components of $\{\boldsymbol{R}(T)\,|\,\boldsymbol{t}(T)\}^{-1}\,\boldsymbol{r}'$. For example, if $f(\boldsymbol{r})=x^3$ and T is the pure rotation of figs. 1.1 and 1.2, as $\boldsymbol{r}=\boldsymbol{R}(T)^{-1}\,\boldsymbol{r}'=\tilde{\boldsymbol{R}}(T)\,\boldsymbol{r}'$,

$$x = x' \cos\phi - y' \sin\phi,$$

and so $P(T)\,f(\boldsymbol{r}')=(x' \cos\phi - y' \sin\phi)^3$.

The function $P(T)f$ can be very profitably considered to be obtained from f by acting on f with the 'function' operator $P(T)$. That is, to each transformation T there corresponds an operator $P(T)$, *which acts on scalar functions*, $P(T)$ being defined to be such that eq. (3.1) and its equivalent forms hold for *every* function f.

It follows from (3.3) and the definition (1.6) of the product T of two transformations T_1 and T_2 that if

$$T = T_1 T_2$$

then

$$P(T) = P(T_1)\,P(T_2). \tag{3.5}$$

This latter operator equation is to be interpreted to mean that

$$P(T)\,f = P(T_1)\,P(T_2)\,f$$

for *any* function f, where on the right-hand side $P(T_2)$ acts first on f and $P(T_1)$ then acts on the result. The proof of (3.5) is as follows. For *any* function $f(\boldsymbol{r}'')$, by (1.6) and (3.3)

$$P(T)\,f\,(\boldsymbol{r}'') = P(T_1 T_2)\,f\,(\boldsymbol{r}'') = f\,((T_1 T_2)^{-1}\,\boldsymbol{r}'') = f\,(T_2^{-1} T_1^{-1} \boldsymbol{r}'').$$

However

$$P(T_1)\,P(T_2)\,f\,(\boldsymbol{r}'') = P(T_1)\,f\,(T_2^{-1}\boldsymbol{r}''),$$

and as the effect of $P(T_1)$ on a function is to replace $\boldsymbol{r}''$ everywhere by $T_1^{-1}\boldsymbol{r}''$, then this is equal to $f(T_2^{-1}T_1^{-1}\boldsymbol{r}'')$. Thus for any $f(\boldsymbol{r}'')$,

$$P(T)\,f\,(\boldsymbol{r}'') = P(T_1)\,P(T_2)\,f\,(\boldsymbol{r}'').$$

This very important result implies that if the set of transformations T form a group, then the corresponding set of function operators $P(T)$ also form a group that is *isomorphic* to the group of transformations. That is, the two groups have the same multiplication table, and hence the same subgroups, classes, and, most important of all, the same representations. In fact, the two groups can be thought of as corresponding to the same abstract group. It is easily established that the identity of the function operator group is $P(E)$, where E is the identity transformation which leaves the coordinate system unchanged, and that $P(T)^{-1}=P(T^{-1})$ for all transformations T, where the inverse operator $P(T)^{-1}$ has the property that

$$P(T)^{-1} f(\boldsymbol{r}) = f(T\boldsymbol{r}), \tag{3.6}$$

so that $P(T)^{-1} P(T)=P(E)$.

The operators $P(T)$ have two other important properties. Firstly, they are *linear* operators, which means that

$$P(T)\{af(\boldsymbol{r}') + bg(\boldsymbol{r}')\} = aP(T) f(\boldsymbol{r}') + bP(T) g(\boldsymbol{r}')$$

for any constants a and b and any functions $f(\boldsymbol{r}')$ and $g(\boldsymbol{r}')$, as follows immediately from (3.1). Secondly, they are *unitary* operators, which means that if the '*inner product*' (f, g) of two functions $f(\boldsymbol{r})$ and $g(\boldsymbol{r})$ is defined by

$$(f, g) = \iiint f(\boldsymbol{r})^* g(\boldsymbol{r})\, \mathrm{d}x\, \mathrm{d}y\, \mathrm{d}z, \tag{3.7}$$

where the integration is over all space, then for *any* functions $f(\boldsymbol{r})$ and $g(\boldsymbol{r})$,

$$(P(T) f, P(T) g) = (f, g). \tag{3.8}$$

In eq. (3.7) $f(\boldsymbol{r})^*$ is the complex conjugate of $f(\boldsymbol{r})$. This latter result simply follows from the fact that $P(T) f$ and f are defined to have the same values at the same points of space, and so are $P(T) g$ and g.

At this point it is worth mentioning the terminology that is used in connection with inner products. If $(f, g)=0$, the functions $f(\boldsymbol{r})$ and $g(\boldsymbol{r})$ are said to be *orthogonal.* If (f,f) is finite, $f(\boldsymbol{r})$ is said to be *normalizable*, and if $(f,f)=1$, $f(\boldsymbol{r})$ is said to be *normalized.* Clearly a normalizable function $f(\boldsymbol{r})$ can be normalized simply by multiplying it by the constant factor $(f,f)^{-\frac{1}{2}}$.

When Born cyclic boundary conditions are introduced, as in ch. 4 §2, the integration over all space in the definition (3.7) may be replaced by an integration over a very large but finite region of the crystal. Any continuous function $f(\boldsymbol{r})$ is then normalizable.

The development given above is directly applicable to a one-particle theory, such as the 'one-electron' theory of solids that is the concern of this book. However, it is possible to consider a system of N particles by a generalization of the definition (3.4) of the operator $P(T)$. Thus, if $\boldsymbol{r}_1, \boldsymbol{r}_2, \ldots, \boldsymbol{r}_N$ are the position vectors of the N particles, the generalized $P(T)$ is defined by

$$P(T)\, f(\boldsymbol{r}_1, \boldsymbol{r}_2, \ldots, \boldsymbol{r}_N) = f(T^{-1}\boldsymbol{r}_1, T^{-1}\boldsymbol{r}_2, \ldots, T^{-1}\boldsymbol{r}_N)$$

for any function $f(\boldsymbol{r}_1, \boldsymbol{r}_2, \ldots, \boldsymbol{r}_N)$, and has very similar properties to that of the one-particle theory. The lattice vibrations of a system of N nuclei are an example of a phenomenon that may be dealt with in this generalization. However, the usual procedure, as followed for example by WIGNER [1930] and CHEN [1967], and as reviewed by MARADUDIN and VOSKO [1968] and WARREN [1968], is to apply group theory to the *classical* formulation, and make the transition to quantum mechanics *after* the normal modes of vibration have been found.

§ 3. THE GROUP OF THE SCHRÖDINGER EQUATION

§ 3.1. *The transformation properties of the Hamiltonian operator*

In this section the vital link between quantum mechanics and group theory will be established. This link comes about through the transformation properties of the Hamiltonian operator, whose eigenfunctions and eigenvalues play an extremely important role in quantum mechanics.

Suppose that the Hamiltonian operator defined with respect to a set of Cartesian axes Ox, Oy and Oz is $H(\boldsymbol{r})$. In general the Hamiltonian operator is a function of the derivatives with respect to the coordinates x, y and z as well as of the coordinates x, y and z themselves, so that writing the Hamiltonian as $H(\boldsymbol{r})$ merely expresses the dependence on the coordinate system Ox, Oy and Oz, and does *not* imply that $H(\boldsymbol{r})$ is only a function of the co-ordinates.

To establish how the Hamiltonian transforms under a transformation T of the coordinate system, let $f(\boldsymbol{r})$ be *any* function of $\boldsymbol{r}$, not necessarily an eigenfunction of $H(\boldsymbol{r})$, and define $g(\boldsymbol{r})$ by

$$g(\boldsymbol{r}) = H(\boldsymbol{r})\, f(\boldsymbol{r}). \tag{3.9}$$

By eq. (3.2),

$$g(\boldsymbol{r}) = P(T)\, g(T\boldsymbol{r}),$$

and as $g(T\boldsymbol{r})$ is merely obtained from $g(\boldsymbol{r})$ by replacing $\boldsymbol{r}$ everywhere by $T\boldsymbol{r}$, from eq. (3.9),

$$g(\boldsymbol{r}) = P(T)\, H(T\boldsymbol{r})\, f(T\boldsymbol{r}).$$

But from eq. (3.6), $f(T\boldsymbol{r}) = P(T)^{-1} f(\boldsymbol{r})$, so using (3.9) again gives

$$H(\boldsymbol{r})\, f(\boldsymbol{r}) = P(T)\, H(T\boldsymbol{r})\, P(T)^{-1}\, f(\boldsymbol{r}).$$

As this is true for *any* function $f(\boldsymbol{r})$ the operator equation

$$H(\boldsymbol{r}) = P(T)\, H(T\boldsymbol{r})\, P(T)^{-1} \tag{3.10}$$

can be abstracted. Eq. (3.10) gives the required transformation properties of the Hamiltonian operator.

Particular interest attaches to those transformations T which leave the Hamiltonian invariant, that is, to transformations such that

$$H(T\boldsymbol{r}) = H(\boldsymbol{r}). \tag{3.11}$$

The meaning of eq. (3.11) is as follows. By definition, $H(T\boldsymbol{r})$ is the operator obtained from $H(\boldsymbol{r})$ by replacing $\boldsymbol{r}$ everywhere by $T\boldsymbol{r}$, that is, it is the same function of $T\boldsymbol{r}$ as $H(\boldsymbol{r})$ is of $\boldsymbol{r}$. Using eq. (1.5), $H(T\boldsymbol{r})$ can then be rewritten as a function of $\boldsymbol{r}$ and its derivatives. If this *latter* function is exactly the same function as $H(\boldsymbol{r})$, then eq. (3.11) holds true.

An example may make this clearer. Suppose that the system consists of an electron moving in the field of a proton fixed at O, so that the Hamiltonian for the electron is

$$H(\boldsymbol{r}) = -\frac{\hbar^2}{2m}\left\{\frac{\partial^2}{\partial x^2} + \frac{\partial^2}{\partial y^2} + \frac{\partial^2}{\partial z^2}\right\} - e^2/\{x^2 + y^2 + z^2\}^{\frac{1}{2}}.$$

Then for *any* transformation $\boldsymbol{r}' = T\boldsymbol{r}$, by definition

$$H(T\boldsymbol{r}) = -\frac{\hbar^2}{2m}\left\{\frac{\partial^2}{\partial x'^2} + \frac{\partial^2}{\partial y'^2} + \frac{\partial^2}{\partial z'^2}\right\} - e^2/\{x'^2 + y'^2 + z'^2\}^{\frac{1}{2}}.$$

If the transformation T is the pure translation $x' = x + a$, $y' = y$, $z' = z$, then rewriting $H(T\boldsymbol{r})$ as a function of $\boldsymbol{r}$ gives

$$H(T\boldsymbol{r}) = -\frac{\hbar^2}{2m}\left\{\frac{\partial^2}{\partial x^2} + \frac{\partial^2}{\partial y^2} + \frac{\partial^2}{\partial z^2}\right\} - e^2/\{(x + a)^2 + y^2 + z^2\}^{\frac{1}{2}},$$

which is *not* identical to $H(\boldsymbol{r})$, so that this transformation does *not* leave the Hamiltonian invariant. On the other hand, if the transformation T is a pure rotation about O, such as that of figs. 1.1 and 1.2, then after a little algebra one finds that

$$H(T\boldsymbol{r}) = -\frac{\hbar^2}{2m}\left\{\frac{\partial^2}{\partial x^2} + \frac{\partial^2}{\partial y^2} + \frac{\partial^2}{\partial z^2}\right\} - e^2/\{x^2 + y^2 + z^2\}^{\frac{1}{2}},$$

which *is* identical to $H(\boldsymbol{r})$, so that this transformation does leave the Hamiltonian invariant.

Returning to the general theory, it follows from (3.10) and (3.11) that if T leaves the Hamiltonian invariant then

$$P(T)\,H(\boldsymbol{r}) = H(\boldsymbol{r})\,P(T), \tag{3.12}$$

that is, $P(T)$ and $H(\boldsymbol{r})$ commute – a very important result.

The set of all transformations leaving the Hamiltonian invariant form a group. It is easily verified that the four group postulates are satisfied. Firstly it is clear that the product of two transformations which each leave the Hamiltonian invariant also leaves the Hamiltonian invariant. Secondly the identity transformation leaves the Hamiltonian invariant and so is a member of the group. Finally the inverse of a transformation which leaves the Hamiltonian invariant itself leaves the Hamiltonian invariant.

As noted in the previous section, this implies that the set of function operators $P(T)$ corresponding to these transformations also form a group which is isomorphic to the group of transformations. This group of function operators, known as *the group of the Schrödinger equation*, is the basic group of the quantum mechanical system. Eq. (3.12) shows that every operator $P(T)$ of this group commutes with the Hamiltonian operator.

§ 3.2. *The group of the Schrödinger equation for electrons in a crystalline solid*

Within the 'one-electron' approximation every electron of mass m has associated with it a Hamiltonian of the simple form

$$H(\boldsymbol{r}) = -\frac{\hbar^2}{2m}\left\{\frac{\partial^2}{\partial x^2} + \frac{\partial^2}{\partial y^2} + \frac{\partial^2}{\partial z^2}\right\} + V(\boldsymbol{r}),$$

where the potential energy term $V(\boldsymbol{r})$ contains the field caused by the nuclei and the average field of all the other electrons acting on the electron under consideration. $V(\boldsymbol{r})$ can be calculated in principle using the Hartree–Fock approximation, as described for example by REITZ [1955], although it is usually obtained by superimposing and slightly modifying the corresponding atomic potentials, and it is usually assumed to be the same for all the valence electrons.

The fact that allows one to draw very important conclusions without knowing the precise form of $V(\boldsymbol{r})$ is that

$$V(\{\boldsymbol{R}(T) \mid \boldsymbol{t}(T)\}\, \boldsymbol{r}) = V(\boldsymbol{r})$$

for every transformation T of the space group of the crystal, $\mathcal{G}$. It follows then that for every T of $\mathcal{G}$,

$$H(T\boldsymbol{r}) = H(\boldsymbol{r}),$$

and hence, for an electron in a crystal, the group of the Schrödinger equation is the space group of the crystal.

§ 4. BASIS FUNCTIONS AND ENERGY EIGENFUNCTIONS

In the previous section the relationship between group theory and the Schrödinger equation was found. This section will be devoted to establishing the vital role of the matrix representations.

§ 4.1. *Properties of basis functions*

Definition of a set of basis functions. Suppose that the set of l-dimensional matrices $\boldsymbol{\Gamma}(T)$ form a representation of the group of the Schrödinger equation, and that $\phi_1(\boldsymbol{r}), \phi_2(\boldsymbol{r}), \ldots, \phi_l(\boldsymbol{r})$ are a set of linearly independent functions such that

$$P(T)\,\phi_n(\boldsymbol{r}) = \sum_{m=1}^{l} \Gamma(T)_{mn}\,\phi_m(\boldsymbol{r}) \qquad n = 1, 2, \ldots, l. \tag{3.13}$$

Then the functions $\phi_n(\boldsymbol{r})$ are said to be partners in a set of basis functions for the representation $\boldsymbol{\Gamma}$, the function $\phi_n(\boldsymbol{r})$ being said to transform as the nth row of this representation.

There is a very close connection between basis functions and the concepts of similarity transformations and unitary representations that were introduced in ch. 2. In fact, any set of linearly independent functions $\psi_m(\boldsymbol{r})$ defined as linear combinations of the basis functions $\phi_n(\boldsymbol{r})$ by

$$\psi_m(\boldsymbol{r}) = \sum_{n=1}^{l} S_{nm}\phi_n(\boldsymbol{r}), \qquad m = 1, 2, \ldots, l, \tag{3.14}$$

forms a basis of a representation $\boldsymbol{\Gamma}'$ that is equivalent to $\boldsymbol{\Gamma}$, in the sense of ch. 2 §2, the relationship being

$$\boldsymbol{\Gamma}'(T) = \boldsymbol{S}^{-1}\boldsymbol{\Gamma}(T)\,\boldsymbol{S}, \tag{3.15}$$

where $\boldsymbol{S}$ in the $l \times l$ matrix whose n, m element is the $\boldsymbol{S}_{nm}$ of eq. (3.14). The proof is as follows:

$$\begin{aligned} P(T)\,\psi_m(\boldsymbol{r}) &= \sum_{n=1}^{l} S_{nm} P(T)\,\phi_n(\boldsymbol{r}) \\ &= \sum_{p,\,n=1}^{l} S_{nm} \Gamma(T)_{pn}\,\phi_p(\boldsymbol{r}) \\ &= \sum_{p,\,q,\,n=1}^{l} S_{nm} \Gamma(T)_{pn}\, S_{qp}^{-1} \psi_q(\boldsymbol{r}) \\ &= \sum_{q=1}^{l} \Gamma'(T)_{qm}\,\psi_q(\boldsymbol{r}), \end{aligned}$$

where $\boldsymbol{\Gamma}'(T)$ is defined by eq. (3.15). Thus it will be seen that similarity transformations are induced by rearranging basis functions.

In solid state physics, where the group of the Schrödinger equation is of finite order, there are an *infinite* number of sets of basis functions that are *not* related by transformations of the form (3.14) for each representation of the group. That is, there is an infinite number of *essentially different* sets of basis functions corresponding to each representation of the group of the Schrödinger equation.

The representation is unitary if the basis functions $\phi_n(\boldsymbol{r})$ are such that

$$(\phi_n, \phi_m) = C\delta_{nm} \tag{3.16}$$

where the inner product is defined in (3.7) and C is a positive constant independent of n and m. (It is sometimes convenient to multiply each of the basis functions by $C^{-\frac{1}{2}}$, in which case

$$(\phi_n, \phi_m) = \delta_{nm},$$

so that the basis functions form an ortho-normal set, but this is not essential.) The proof involves the unitary property of $P(T)$, eq. (3.8), for by (3.16),

$$\begin{aligned} C\delta_{nm} = (\phi_n, \phi_m) &= (P(T)\,\phi_n, P(T)\,\phi_m) \\ &= \sum_{p,q=1}^{l} \Gamma(T)^*_{qn}\,\Gamma(T)_{pm}\,(\phi_q, \phi_p) \\ &= C \sum_{p=1}^{l} \Gamma(T)^*_{pn}\,\Gamma(T)_{pm}\,, \end{aligned}$$

using (3.16) again, so that $\tilde{\boldsymbol{\Gamma}}(T)^* \boldsymbol{\Gamma}(T)$ is equal to the $l \times l$ unit matrix, and hence $\boldsymbol{\Gamma}(T)$ is unitary.

Any set of l normalizable basis functions $\phi_n(\boldsymbol{r})$, $n=1, 2, \ldots, l$, can be rearranged to form an orthogonal set $\theta_n(\boldsymbol{r})$ by the following algorithm. Let

$$\begin{aligned} \theta_1(\boldsymbol{r}) &= \phi_1(\boldsymbol{r}), \\ \theta_2(\boldsymbol{r}) &= \phi_2(\boldsymbol{r}) - \{(\theta_1, \phi_2)/(\theta_1, \theta_1)\}\,\theta_1(\boldsymbol{r}), \\ \theta_3(\boldsymbol{r}) &= \phi_3(\boldsymbol{r}) - \{(\theta_1, \phi_3)/(\theta_1, \theta_1)\}\,\theta_1(\boldsymbol{r}) \\ &\quad - \{(\theta_2, \phi_3)/(\theta_2, \theta_2)\}\,\theta_2(\boldsymbol{r}), \end{aligned}$$

and so on. The functions $\theta_1(\boldsymbol{r}), \theta_2(\boldsymbol{r}), \ldots, \theta_l(\boldsymbol{r})$ then form an orthogonal set, as may be verified by inspection. A set of functions $\psi_1(\boldsymbol{r}), \psi_2(\boldsymbol{r}), \ldots, \psi_l(\boldsymbol{r})$ that are each normalized as well as being orthogonal can be obtained by defining

$$\psi_n(\boldsymbol{r}) = (\theta_n, \theta_n)^{-\frac{1}{2}}\,\theta_n(\boldsymbol{r})$$

for $n=1, 2, \ldots, l$. This orthogonalization process is known as the Schmidt process, and it will be used in §5.3. It is clear that the relationship between the ortho-normal set $\psi_n(\boldsymbol{r})$ and the original set of basis functions $\phi_n(\boldsymbol{r})$ is of the form given in eq. (3.14), so that the corresponding similarity transformation (3.15) is one that makes the representation unitary.

§ 4.2. *Energy eigenfunctions as basis functions*

Suppose that the energy E is an l-fold degenerate eigenvalue of the

Hamiltonian operator, and that $\phi_n(\boldsymbol{r})$, $n=1, 2, \ldots, l$, are a set of linearly independent eigenfunctions corresponding to this eigenvalue, so that

$$H(\boldsymbol{r})\,\phi_n(\boldsymbol{r}) = E\phi_n(\boldsymbol{r}), \qquad n = 1, 2, \ldots, l.$$

Now consider the function $P(T)\,\phi_n(\boldsymbol{r})$, where $P(T)$ is a member of the group of the Schrödinger equation. As $P(T)$ commutes with $H(\boldsymbol{r})$,

$$\begin{aligned} H(\boldsymbol{r})\,P(T)\,\phi_n(\boldsymbol{r}) = P(T)\,H(\boldsymbol{r})\,\phi_n(\boldsymbol{r}) &= P(T)\,E\phi_n(\boldsymbol{r}) \\ &= EP(T)\,\phi_n(\boldsymbol{r}), \end{aligned}$$

so that

$$H(\boldsymbol{r})\,\{P(T)\,\phi_n(\boldsymbol{r})\} = E\,\{P(T)\,\phi_n(\boldsymbol{r})\}.$$

This demonstrates that $P(T)\,\phi_n(\boldsymbol{r})$ is *also* an eigenfunction of $H(\boldsymbol{r})$ with the same eigenvalue E. As E was assumed to have only l linearly independent eigenfunctions, $P(T)\,\phi_n(\boldsymbol{r})$ must be some linear combination of the eigenfunctions $\phi_n(\boldsymbol{r})$. (It is very easily shown that any linear combination of eigenfunctions corresponding to the same eigenvalue is also an eigenfunction corresponding to that eigenvalue.) Thus $P(T)\,\phi_n(\boldsymbol{r})$ may be written as

$$P(T)\,\phi_n(\boldsymbol{r}) = \sum_{m=1}^{l} \Gamma(T)_{mn}\,\phi_m(\boldsymbol{r}), \tag{3.17}$$

and this is so for each $n=1, 2, \ldots, l$. Eq. (3.17) defines an array of l^2 numbers $\Gamma(T)_{mn}$, there being one such array for each transformation T of the group of the Schrödinger equation.

As is suggested by the notation, *this set of arrays form an l-dimensional matrix representation* of the group of the Schrödinger equation, for it will be shown that

$$\Gamma(T_1 T_2)_{mn} = \sum_{p=1}^{l} \Gamma(T_1)_{mp}\,\Gamma(T_2)_{pn} \tag{3.18}$$

for any two transformations T_1 and T_2 of the group. The proof of this vital statement is as follows. By the definition (3.17) of the arrays corresponding to T_1, T_2 and T_1T_2,

$$P(T_1)\,\phi_p(\boldsymbol{r}) = \sum_{m=1}^{l} \Gamma(T_1)_{mp}\,\phi_m(\boldsymbol{r}),$$

$$P(T_2)\,\phi_n(\boldsymbol{r}) = \sum_{p=1}^{l} \Gamma(T_2)_{pn}\,\phi_p(\boldsymbol{r}),$$

and

$$P(T_1T_2)\,\phi_n(\boldsymbol{r}) = \sum_{m=1}^{l} \Gamma(T_1T_2)_{mn}\,\phi_m(\boldsymbol{r}), \tag{3.19}$$

and hence

$$P(T_1)\,P(T_2)\,\phi_n(\boldsymbol{r}) = \sum_{m=1}^{l}\sum_{p=1}^{l} \Gamma(T_1)_{mp}\,\Gamma(T_2)_{pn}\,\phi_m(\boldsymbol{r}). \tag{3.20}$$

As $P(T_1T_2)\,\phi_n(\boldsymbol{r}) = P(T_1)\,P(T_2)\,\phi_n(\boldsymbol{r})$ by eq. (3.5), the right-hand sides of eqs. (3.19) and (3.20) must be equal. The functions $\phi_m(\boldsymbol{r})$ are linearly independent, so that eq. (3.18) follows on equating coefficients of each $\phi_m(\boldsymbol{r})$ in this equality.

It has therefore been demonstrated that *a set of energy eigenfunctions corresponding to an l-fold degenerate energy eigenvalue form a set of basis functions for an l-dimensional representation of the group of the Schrödinger equation.* Every energy level therefore corresponds to some representation of the group of the Schrödinger equation. This establishes the connection between the theory of matrix representations and quantum theory.

It is worth pointing out that the converse is not necessarily true, for an arbitrarily chosen set of basis functions do not constitute a set of energy eigenfunctions. However, in calculations of energy eigenvalues and eigenfunctions, there are very great advantages in working with sets of basis functions, as will be demonstrated in §6.2.

In the analysis of this section, the representations could be reducible or irreducible, but it will be shown in §6.2 that eigenfunctions usually only form basis functions of *irreducible* representations. An energy eigenvalue corresponding to a reducible representation is then considered to exhibit an '*accidental*' degeneracy between energy levels corresponding to its constituent irreducible representations.

There are an infinite number of energy levels corresponding to each irreducible representation of the group of the Schrödinger equation, but of course only a few of each are occupied at normal temperatures. Interest is concentrated in practice on these low lying levels, and those immediately above them.

§ 5. CONSTRUCTION OF BASIS FUNCTIONS

§ 5.1. *Decomposition of an arbitrary function into basis functions*

In applications of group theory to physical problems, repeated use is made of the following theorem.

Theorem. Any normalizable function $\phi(\boldsymbol{r})$ can be decomposed into a linear combination of basis functions of the irreducible representations of the group of the Schrödinger equation, $\mathscr{G}$. That is, one can write

$$\phi(\boldsymbol{r}) = \sum_{p} \sum_{n=1}^{l_p} \phi_n^p(\boldsymbol{r}), \tag{3.21}$$

where $\phi_n^p(\boldsymbol{r})$ is a function transforming as the nth row of the unitary irreducible representation $\boldsymbol{\Gamma}^p$. The summation over p is over all the non-equivalent unitary irreducible representations of $\mathscr{G}$, so that if $\mathscr{G}$ is of finite order the right-hand side of (3.21) involves only a *finite* number of functions. Some of the functions $\phi_n^p(\boldsymbol{r})$ may be identically zero.

The basis functions $\phi_n^p(\boldsymbol{r})$ of (3.21) depend on $\phi(\boldsymbol{r})$, and in §5.2 it will be shown how they can be constructed from $\phi(\boldsymbol{r})$ by the use of projection operators. In §6 it will then be demonstrated that the use of the expansion (3.21) leads to important simplification of several types of quantum mechanical calculation.

The rest of this subsection will be devoted to a proof of the theorem stated above.

Proof: Construct the functions $P(T)\,\phi(\boldsymbol{r})$ for all the transformations T of $\mathscr{G}$, and abstract the linearly independent functions involved. Using the Schmidt orthogonalization process described in §4.1, an ortho-normal set $\theta_m(\boldsymbol{r})$ can be obtained. If this set contains l linearly independent functions, it forms a basis for an l-dimensional unitary representation $\boldsymbol{\Gamma}$ of $\mathscr{G}$. This follows because each $P(T)\,\phi(\boldsymbol{r})$ is a linear combination of the functions $\theta_m(\boldsymbol{r})$, and, conversely, each $\theta_m(\boldsymbol{r})$ is a linear combination of the functions $P(T)\,\phi(\boldsymbol{r})$, and as $P(T')\,\{P(T)\,\phi(\boldsymbol{r})\} = P(T'T)\,\phi(\boldsymbol{r})$ for every T and T' of $\mathscr{G}$, then $P(T')\,\theta_m(\boldsymbol{r})$ is a linear combination of $\theta_1(\boldsymbol{r}), \ldots, \theta_l(\boldsymbol{r})$, for each $m = 1, 2, \ldots, l$ and each T' of $\mathscr{G}$, as in eq. (3.17).

If this representation $\boldsymbol{\Gamma}$ is irreducible the theorem is proved, and the sum in (3.21) involves only this one irreducible representation.

If this representation $\boldsymbol{\Gamma}$ is reducible it may be put into the block form of ch. 2 §4 by a similarity transformation with a $l \times l$ matrix $\boldsymbol{S}$, the new basis functions $\psi_m(\boldsymbol{r})$ being given by

$$\psi_m(\boldsymbol{r}) = \sum_{n=1}^{l} S_{nm}\theta_n(\boldsymbol{r}),$$

as in eq. (3.14). However, these functions $\psi_m(\boldsymbol{r})$ are basis functions of the irreducible representations appearing in the reduction of $\boldsymbol{\Gamma}$. The original function $\phi(\boldsymbol{r})$ is a linear combination of the $\psi_m(\boldsymbol{r})$, so the theorem is then proved for this case.

§ 5.2. *Projection operators*

The decomposition (3.21) of an arbitrary normalizable function into basis functions can be carried out very simply using certain projection operators that will be described in this subsection. The application of these projection operators is usually an essential part of any group theoretical analysis of a physical problem, so an example will be worked through in detail.

The projection operator $\mathscr{P}^p_{mn}$ is defined by

$$\mathscr{P}^p_{mn} = (l_p/g) \sum_T \Gamma^p(T)^*_{mn}\, P(T), \tag{3.22}$$

where l_p is the dimension of the unitary irreducible representation $\boldsymbol{\Gamma}^p$ of the group of the Schrödinger equation $\mathcal{G}$, g is the order of $\mathcal{G}$, and the summation is over all the transformations T of $\mathcal{G}$.

If $\phi_i^q(\boldsymbol{r})$ transforms as the ith row of $\boldsymbol{\Gamma}^q$ then

$$\mathscr{P}_{mn}^p \, \phi_i^q(\boldsymbol{r}) = \delta_{pq}\delta_{ni} \, \phi_m^p(\boldsymbol{r}). \tag{3.23}$$

This is true because by eqs. (3.13) and (3.22)

$$\begin{aligned}\mathscr{P}_{mn}^p \, \phi_i^q(\boldsymbol{r}) &= (l_p/g) \sum_T \Gamma^p(T)_{mn}^* \sum_{j=1}^{l_q} \Gamma^q(T)_{ji} \, \phi_j^q(\boldsymbol{r}), \\ &= \sum_{j=1}^{l_q} \delta_{pq}\delta_{mj}\delta_{ni} \, \phi_j^q(\boldsymbol{r}),\end{aligned}$$

by the orthogonality theorem for matrices of ch. 2 §4,

$$= \delta_{pq}\delta_{ni} \, \phi_m^p(\boldsymbol{r}),$$

the required result.

It then follows from (3.21) and (3.23) that

$$\mathscr{P}_{nn}^p \, \phi(\boldsymbol{r}) = \phi_n^p(\boldsymbol{r}),$$

that is, $\mathscr{P}_{nn}^p$ *projects out of a normalizable function* $\phi(\boldsymbol{r})$ *the part* $\phi_n^p(\boldsymbol{r})$ *transforming as the* nth *row of* $\boldsymbol{\Gamma}^p$, a *very* useful property. The function $\phi_n^p(\boldsymbol{r})$ will be identically zero if $\phi(\boldsymbol{r})$ contains no part transforming as the nth row of $\boldsymbol{\Gamma}^p$. Should the partners of $\phi_n^p(\boldsymbol{r})$ be required, they could be found by operating on $\phi_n^p(\boldsymbol{r})$ with $\mathscr{P}_{mn}^p$ for $m=1, 2, \ldots, l_p$.

An example will show how easy the projection operators are to apply. Let $\mathcal{G}$ be the point group C_{3v} considered in chs. 1 and 2, whose irreducible representations are specified in ch. 2 §7, and suppose that $\phi(\boldsymbol{r})$ is chosen to be such that $\phi(\boldsymbol{r}) = x^2 g(r)$, where $\boldsymbol{r} = (x, y, z)$, $r^2 = x^2 + y^2 + z^2$ and $g(r)$ is a given function of r such that $\phi(\boldsymbol{r})$ is normalized. (It will be seen that $g(r)$ is invariant under all the transformations of the group C_{3v}, and so appears as a factor in every function without playing any role in the symmetry arguments, apart from ensuring that every function in which it appears is normalizable.) For a transformation T of a point group, eq. (3.3) shows that

$$P(T)\,\phi(\boldsymbol{r}) = \phi(\boldsymbol{R}(T)^{-1}\boldsymbol{r}) = \phi(\tilde{\boldsymbol{R}}(T)\,\boldsymbol{r}),$$

the orthogonal transformation matrices $\boldsymbol{R}(T)$ for C_{3v} being given in ch. 1 §2.2. For example, for $T = C_{3z}$,

$$\tilde{\boldsymbol{R}}(C_{3z})\,\boldsymbol{r} = \begin{pmatrix} -\frac{1}{2} & -\frac{1}{2}\sqrt{3} & 0 \\ \frac{1}{2}\sqrt{3} & -\frac{1}{2} & 0 \\ 0 & 0 & 1 \end{pmatrix} \begin{pmatrix} x \\ y \\ z \end{pmatrix} = \begin{pmatrix} -\frac{1}{2}x - \frac{1}{2}y\sqrt{3} \\ \frac{1}{2}x\sqrt{3} - \frac{1}{2}y \\ z \end{pmatrix},$$

and as $\phi(\tilde{\boldsymbol{R}}(C_{3z})\,\boldsymbol{r})$ is the function in which the x, y and z in $\phi(\boldsymbol{r})$ are replaced by the 11, 21 and 31 components of $\tilde{\boldsymbol{R}}(C_{3z})\,\boldsymbol{r}$ respectively, and here $\phi(\boldsymbol{r}) = x^2 g(r)$, then

$$\begin{aligned} P(C_{3z})\,\phi(\boldsymbol{r}) &= (-\tfrac{1}{2}x - \tfrac{1}{2}y\sqrt{3})^2\, g(r) \\ &= (\tfrac{1}{4}x^2 + \tfrac{1}{2}xy\sqrt{3} + \tfrac{3}{4}y^2)\, g(r). \end{aligned}$$

The following is a complete list of the functions $P(T)\,\phi(\boldsymbol{r})$ obtained this way:

$$\left.\begin{aligned} P(E)\,\phi(\boldsymbol{r}) &= x^2 g(r), \\ P(C_{3z})\,\phi(\boldsymbol{r}) &= (\tfrac{1}{4}x^2 + \tfrac{1}{2}xy\sqrt{3} + \tfrac{3}{4}y^2)\, g(r), \\ P(C_{3z}^{-1})\,\phi(\boldsymbol{r}) &= (\tfrac{1}{4}x^2 - \tfrac{1}{2}xy\sqrt{3} + \tfrac{3}{4}y^2)\, g(r), \\ P(IC_{2y})\,\phi(\boldsymbol{r}) &= x^2 g(r), \\ P(IC_{2C})\,\phi(\boldsymbol{r}) &= (\tfrac{1}{4}x^2 - \tfrac{1}{2}xy\sqrt{3} + \tfrac{3}{4}y^2)\, g(r), \\ P(IC_{2D})\,\phi(\boldsymbol{r}) &= (\tfrac{1}{4}x^2 + \tfrac{1}{2}xy\sqrt{3} + \tfrac{3}{4}y^2)\, g(r). \end{aligned}\right\} \quad (3.24)$$

Then from the definition (3.22) and the matrices given in ch. 2 §7,

$$\begin{aligned} \mathscr{P}^1_{11}\phi(\boldsymbol{r}) &= \tfrac{1}{2}(x^2 + y^2)\, g(r), \\ \mathscr{P}^2_{11}\phi(\boldsymbol{r}) &= 0, \\ \mathscr{P}^3_{11}\phi(\boldsymbol{r}) &= 0, \\ \mathscr{P}^3_{22}\phi(\boldsymbol{r}) &= \tfrac{1}{2}(x^2 - y^2)\, g(r). \end{aligned}$$

This shows that $x^2 g(r)$ is the sum of functions transforming as the first row of $\boldsymbol{\Gamma}^1$ and the second row of $\boldsymbol{\Gamma}^3$.

The partner of $\frac{1}{2}(x^2 - y^2)\, g(r)$ transforming as the first row of

Γ^3 is obtained by operating on $\frac{1}{2}(x^2-y^2)\,g(r)$ with $\mathscr{P}^3_{12}$, which gives $xyg(r)$.

§ 5.3. *A method for the explicit calculation of matrix representations from the character system*

The procedure for constructing basis functions described in §5.2 required an explicit knowledge of the matrix elements of the representations, and not merely a knowledge of the character system alone, which is usually the only information which is given in the published literature. Of course, for one-dimensional representations the characters give the matrix elements immediately, but for other representations some further analysis is needed. This subsection is devoted to a method which can be used in such cases.

Define the 'character projection operator' $\mathscr{P}^p$ by

$$\mathscr{P}^p = (l_p/g) \sum_T \chi^p(T)^* \, P(T), \tag{3.25}$$

where $\chi^p(T)$ is the character of T in the irreducible representation Γ^p, which can be found in the character table. Then

$$\mathscr{P}^p = \sum_{n=1}^{l_p} \mathscr{P}^p_{nn},$$

so that $\mathscr{P}^p$ has the property of projecting out of a function $\phi(\boldsymbol{r})$ the sum of *all* the parts transforming according to the rows of Γ^p.

This implies that if $\mathscr{P}^p\phi(\boldsymbol{r})$ is not identically zero it is a linear combination of basis functions of Γ^p (which are as yet undetermined). However, as noted previously in §4.1, linear combinations of basis functions are themselves basis functions in an equivalent representation, and so the linear combination $\mathscr{P}^p\phi(\boldsymbol{r})$ may be taken to transform as the first row of some form of the pth irreducible representation. This particular form will henceforth be denoted by Γ^p. (Γ^p was previously completely unspecified up to a similarity transformation.) The procedure to be described then generates the explicit matrix elements for this form of $\boldsymbol{\Gamma}^p$, which is, of course, as good as any other equivalent representation.

Having chosen a normalizable $\phi(\boldsymbol{r})$ such that $\mathscr{P}^p\phi(\boldsymbol{r})$ is not identically zero, write $\phi_1^p(\boldsymbol{r})=\mathscr{P}^p\phi(\boldsymbol{r})$, and construct $P(T)\,\phi_1^p(\boldsymbol{r})$ for each T of $\mathscr{G}$. Each of these must be linear combinations of the l_p basis functions of $\boldsymbol{\Gamma}^p$. From these linear combinations it is then possible to abstract l_p linearly independent functions, of which one is $\phi_1^p(\boldsymbol{r})$. The Schmidt orthogonalization process, described in §4.1, can be used on these functions to produce l_p orthogonal functions $\phi_n^p(\boldsymbol{r})$ such that (ϕ_n^p, ϕ_n^p) is a positive constant independent of n. These functions can be taken to be the basis functions of a unitary representation of $\boldsymbol{\Gamma}^p$. The matrix elements can then be found from the definition (3.13), namely

$$P(T)\,\phi_n^p(\boldsymbol{r}) = \sum_{m=1}^{l_p} \Gamma^p(T)_{mn}\,\phi_m^p(\boldsymbol{r}),$$

as $P(T)\,\phi_n^p(\boldsymbol{r})$ can be found for each T using (3.4), as described in detail in §5.2.

This method will be illustrated by using it to obtain a set of matrices for the irreducible representation $\boldsymbol{\Gamma}^3$ of the point group C_{3v}, which is a purely academic exercise here, as a set is already known. Take $\phi(\boldsymbol{r})=x^2 g(r)$, where $g(r)$ is any function of r such that $\phi(\boldsymbol{r})$ is normalized. Then from the character table 2.1, the definition (3.25), and the expressions (3.24), it follows that

$$\phi_1^3(\boldsymbol{r}) = \mathscr{P}^3\phi(\boldsymbol{r}) = \tfrac{1}{2}(x^2 - y^2)\,g(r).$$

Now

$$P(C_{3z})\,\phi_1^3(\boldsymbol{r}) = \phi_1^3(\tilde{\boldsymbol{R}}(C_{3z})\,\boldsymbol{r}) = -\tfrac{1}{4}(x^2 - y^2)\,g(r) + \tfrac{1}{2}xyg(r)\sqrt{3},$$

and as $l_3=2$ there are only two linearly independent basis functions of $\boldsymbol{\Gamma}^3$, of which one is $\frac{1}{2}(x^2-y^2)\,g(r)$, so that the second one can be taken to be $xyg(r)$. It happens here that $\frac{1}{2}(x^2-y^2)\,g(r)$ and $xyg(r)$ are orthogonal, so the Schmidt process is not needed, and one can write $\phi_2^3(\boldsymbol{r})=xyg(r)$. The constants involved in $\phi_1^3(\boldsymbol{r})$ and $\phi_2^3(\boldsymbol{r})$ are such that $(\phi_1^3, \phi_1^3)=(\phi_2^3, \phi_2^3)$. [In calculating (ϕ_m^p, ϕ_n^p) for a point group, it is easiest to use spherical polar coordinates r, θ and ϕ in which

$$x = r \sin\theta \cos\phi\,, \qquad y = r \sin\theta \sin\phi\,, \qquad z = r\cos\theta\,,$$

so that

$$(\phi^p_m, \phi^p_n) = \int_0^{2\pi} \mathrm{d}\phi \int_0^{\pi} \mathrm{d}\theta \int_0^{\infty} \mathrm{d}r\, \phi^p_m(\boldsymbol{r})^* \,\phi^p_n(\boldsymbol{r})\, r^2 \sin\theta\,.$$

It will then be apparent that the integration over r is superfluous in this situation in that it is the same for every m and n.] Then, for example,

$$\begin{aligned} P(IC_{2C})\,\phi^3_1(\boldsymbol{r}) &= \tfrac{1}{2}\{(-\tfrac{1}{2}x + \tfrac{1}{2}y\sqrt{3})^2 - (\tfrac{1}{2}x\sqrt{3} - \tfrac{1}{2}y)^2\} \\ &= -\tfrac{1}{2}\phi^3_1(\boldsymbol{r}) - \tfrac{1}{2}\sqrt{3}\phi^3_2(\boldsymbol{r}), \\ P(IC_{2C})\,\phi^3_2(\boldsymbol{r}) &= (-\tfrac{1}{2}x + \tfrac{1}{2}y\sqrt{3})(\tfrac{1}{2}x\sqrt{3} - \tfrac{1}{2}y) \\ &= -\tfrac{1}{2}\sqrt{3}\phi^3_1(\boldsymbol{r}) + \tfrac{1}{2}\phi^3_2(\boldsymbol{r}), \end{aligned}$$

which on comparison with (3.13) gives as the matrix representing IC_{2C}

$$\begin{pmatrix} -\tfrac{1}{2} & -\tfrac{1}{2}\sqrt{3} \\ -\tfrac{1}{2}\sqrt{3} & \tfrac{1}{2} \end{pmatrix}.$$

The matrices representing the other elements of the group C_{3v} may be found in the same way. They are not identical to those given for $\boldsymbol{\Gamma}^3$ in ch. 2 §7, but could be obtained from them by a similarity transformation, as in eq. (2.1) with

$$\boldsymbol{S} = \begin{pmatrix} 0 & 1 \\ 1 & 0 \end{pmatrix}.$$

§ 6. THE APPLICATION OF BASIS FUNCTIONS TO QUANTUM MECHANICAL PROBLEMS

It will be shown in this section that by using basis functions of representations of the group of the Schrödinger equation it is possible to significantly simplify several important types of quantum mechanical calculation. The procedures depend on the matrix element theorems that are proved in §6.1. The general methods developed in this section will be applied in the following chapters to solid state problems.

§ 6.1. *The matrix element theorems*

Theorem. Suppose that $\phi_1^p(\boldsymbol{r}), \phi_2^p(\boldsymbol{r}), \ldots$ and $\psi_1^q(\boldsymbol{r}), \psi_2^q(\boldsymbol{r}), \ldots$ are respectively sets of basis functions for the unitary irreducible representations $\boldsymbol{\Gamma}^p$ and $\boldsymbol{\Gamma}^q$ of the group of the Schrödinger equation, it being assumed that $\boldsymbol{\Gamma}^p$ and $\boldsymbol{\Gamma}^q$ are not equivalent if $p \neq q$ but are identical if $p=q$. Then

(a) $$(\phi_m^p, \psi_n^q) = 0,$$

unless $p=q$ *and* $m=n$ (that is, functions belonging to different irreducible representations or to different rows of the same irreducible representation are orthogonal), and

(b) (ϕ_m^p, ψ_m^p) is a constant independent of m.

The proof of this theorem is as follows. Using the unitary property (3.8) and the definition (3.13), for any $P(T)$ of the group,

$$\begin{aligned}(\phi_m^p, \psi_n^q) &= (P(T)\,\phi_m^p, P(T)\,\psi_n^q) \\ &= \sum_{i=1}^{l_p} \sum_{j=1}^{l_q} \Gamma^p(T)_{im}^* \,\Gamma^q(T)_{jn} (\phi_i^p, \psi_j^q),\end{aligned}$$

l_p and l_q being the dimensions $\boldsymbol{\Gamma}^p$ and $\boldsymbol{\Gamma}^q$ respectively. Summing this equality over all the transformations T of the group and invoking the orthogonality theorem for matrices of ch. 2 §4 gives

$$g\,(\phi_m^p, \psi_n^q) = (g/l_p)\,\delta_{pq}\delta_{mn} \sum_{i=1}^{l_p} (\phi_i^p, \psi_i^p).$$

Putting $p \neq q$, or $p=q$ and $m \neq n$, immediately gives the result (a) of the theorem. When $p=q$ and $m=n$, the right-hand side of this last equation becomes independent of m, so that (ϕ_m^p, ψ_m^p) is independent of m.

Theorem. With the same notations as in the previous theorem, if H is the Hamiltonian operator then
(a) $(\phi^p_m, H\psi^q_n)=0$ unless $p=q$ *and* $m=n$, and
(b) $(\phi^p_m, H\psi^p_m)$ is independent of m.
The similarity to the previous theorem is obvious. The proof follows from the fact that as $P(T)$ is unitary and commutes with H,

$$(\phi^p_m, H\psi^q_n) = (P(T)\,\phi^p_m, P(T)\,H\psi^q_n) = (P(T)\,\phi^p_m, HP(T)\,\psi^q_n).$$

The rest of the proof is exactly as given for the previous theorem, apart from the presence of the additional factor H in all inner products.

§ 6.2. *Approximate calculation of energy eigenvalues and eigenfunctions*

One of the most valuable applications of group theory is to the approximate solution of the Schrödinger equation. Only for very simple systems, such as the hydrogen atom, is it possible to obtain an exact analytic solution. For other systems it is necessary to resort to numerical calculations, but the work involved can often be very much shortened by the application of group theory. This is particularly true in electronic energy band calculations in solid state physics, where accurate calculations are only feasible when group theoretical arguments are used to exploit the symmetry of the system to the full.

The basic ideas involved are quite simple. Suppose that the Schrödinger equation is

$$H(\boldsymbol{r})\,\psi(\boldsymbol{r}) = E\psi(\boldsymbol{r}), \tag{3.26}$$

and it is required to find the energy eigenvalue E and the corresponding eigenfunction for all low lying eigenvalues. The function $\psi(\boldsymbol{r})$ can be expanded in terms of a complete set of known functions $\phi_1(\boldsymbol{r}), \phi_2(\boldsymbol{r}), \ldots$, so that one can write

$$\psi(\boldsymbol{r}) = \sum_{p=1}^{\infty} C_p\phi_p(\boldsymbol{r}), \tag{3.27}$$

the constants C_p being unknown at this stage. Naturally the functions $\phi_p(\boldsymbol{r})$ are chosen in such a way as to make the series (3.27) converge rapidly. The different types of energy band calculation, described for instance in the article by REITZ [1955], essentially merely differ in this choice of the functions $\phi_p(\boldsymbol{r})$. For example, in solid state problems when the valence electrons are expected to be tightly bound to the ions, it is natural to take the $\phi_p(\boldsymbol{r})$ to be atomic orbitals, thereby giving the so-called 'method of linear combinations of atomic orbitals', often described more briefly as the L.C.A.O. method. At the other extreme, when the valence electrons are expected to be nearly free, it is natural to form the $\phi_p(\boldsymbol{r})$ from plane waves (orthogonalized to the ionic electronic eigenfunctions to stop the expansion giving ionic electron eigenfunctions), thereby giving the so-called 'orthogonalized plane wave method', or O.P.W. method for short.

Substituting (3.27) into (3.26) gives

$$\sum_{p=1}^{\infty} C_p \{H\phi_p(\boldsymbol{r}) - E\phi_p(\boldsymbol{r})\} = 0,$$

and forming the inner product with $\phi_q(\boldsymbol{r})$ gives

$$\sum_{p=1}^{\infty} C_p \{(\phi_q, H\phi_p) - E(\phi_q, \phi_p)\} = 0, \qquad q = 1, 2 \ldots \tag{3.28}$$

This is an infinite set of linear simultaneous algebraic equations for the unknown constants C_p, and the condition for the existence of a solution is that

$$|(\phi_q, H\phi_p) - E(\phi_q, \phi_p)| = 0, \tag{3.29}$$

where the determinant of the left-hand side has an infinite number of rows and columns.

Thus far no approximations have been made. The approximation consists of truncating the series (3.27) so that it only contains N terms, where N is finite. The determinant of (3.29) is then of dimensions $N \times N$, and eq. (3.29) becomes a *polynomial* in E of order N, although the calculation of the coefficient of each power of E is a formidable task if N is large. An equation of the form (3.29) is often called a *secular equation.* The Nth order polynomial has N roots, which constitute approximations to the N lowest energy eigenvalues. Again, the calculation of the roots is lengthy if N is large. With the roots obtained, it is then possible to go back to (3.28), regarded now as a system of N equations, and for each root obtain the corresponding set of constants C_p, thereby giving by (3.27) an approximation to the corresponding eigenfunction $\psi(\boldsymbol{r})$.

The number N is here quite arbitrary, but clearly, as N is increased, two effects follow. Firstly, the accuracy of the approximations to the lower energy eigenvalues is improved, which is very desirable. Secondly, more eigenvalues at higher energies appear, although these are usually less important. However, as noted above, the numerical work involved increases rapidly as N increases, the work being roughly proportional to $N!$.

The matrix element theorems of the previous section allow this numerical work to be cut tremendously without losing any accuracy. All that has to be done is to arrange that the members of the complete set of known functions $\phi_p(\boldsymbol{r})$ of (3.27) are each basis functions of the group of the Schrödinger equation, $\mathscr{G}$. In practice, this is done by applying the projection operators of §5.2 to the atomic orbitals, orthogonalized plane waves, or other given functions that are appropriate to the particular system under consideration. An extra pair of indices m and i have now to be included in the designation of the functions such that $\phi^p_{im}(\boldsymbol{r})$ transforms as the mth row of the irreducible representation $\boldsymbol{\Gamma}^p$ of $\mathscr{G}$, the index i distinguishing the different sets of basis functions having this particular symmetry. Eq. (3.27) is then modified to read

$$\psi(\boldsymbol{r}) = \sum_{i=1}^{\infty} \sum_{p} \sum_{m} C^p_{im} \phi^p_{im}(\boldsymbol{r}),$$

and (3.29) becomes

$$|(\phi^q_{jn}, H\phi^p_{im}) - E(\phi^q_{jn}, \phi^p_{im})| = 0, \qquad (3.30)$$

where, again, the approximation consists of treating this determinant as having finite dimensions. ($\boldsymbol{\Gamma}^p$ can be taken to be unitary.)

The rows and columns of this determinant may be rearranged so that all the terms corresponding to a particular row of a particular irreducible representation are grouped together. As the matrix element theorems of §6.1 show that $(\phi^q_{jn}, \phi^p_{im})$ and $(\phi^q_{jn}, H\phi^p_{im})$ are zero unless $p=q$ and $m=n$, the determinant of (3.30) takes a 'block form' giving

$$\begin{vmatrix} \boldsymbol{D}(1,1) & \boldsymbol{0} & \boldsymbol{0} & \cdots & \boldsymbol{0} & \boldsymbol{0} & \boldsymbol{0} & \cdots \\ \boldsymbol{0} & \boldsymbol{D}(1,2) & \boldsymbol{0} & \cdots & \boldsymbol{0} & \boldsymbol{0} & \boldsymbol{0} & \cdots \\ \boldsymbol{0} & \boldsymbol{0} & \boldsymbol{D}(1,3) & \cdots & \boldsymbol{0} & \boldsymbol{0} & \boldsymbol{0} & \cdots \\ \vdots & \vdots & \vdots & & \vdots & \vdots & \vdots & \\ \boldsymbol{0} & \boldsymbol{0} & \boldsymbol{0} & \cdots & \boldsymbol{D}(1,l_1) & \boldsymbol{0} & \boldsymbol{0} & \cdots \\ \boldsymbol{0} & \boldsymbol{0} & \boldsymbol{0} & \cdots & \boldsymbol{0} & \boldsymbol{D}(2,1) & \boldsymbol{0} & \cdots \\ \boldsymbol{0} & \boldsymbol{0} & \boldsymbol{0} & \cdots & \boldsymbol{0} & \boldsymbol{0} & \boldsymbol{D}(2,2) & \cdots \\ \vdots & \vdots & \vdots & & \vdots & \vdots & \vdots & \end{vmatrix} = 0, \tag{3.31}$$

where

$$\boldsymbol{D}(p,m)_{ji} = (\phi^p_{jm}, H\phi^p_{im}) - E(\phi^p_{jm}, \phi^p_{im})$$

is a matrix involving *only* basis functions corresponding to the mth row of $\boldsymbol{\Gamma}^p$. The matrices $\boldsymbol{0}$ consist entirely of zero elements. Eq. (3.31) implies that

$$|\boldsymbol{D}(1,1)|\,|\boldsymbol{D}(1,2)| \ldots |\boldsymbol{D}(1,l_1)|\,|\boldsymbol{D}(2,1)|\,|\boldsymbol{D}(2,2)| \ldots = 0 .$$

The full set of eigenvalues of (3.31) are then obtained by taking

$$|\boldsymbol{D}(p,m)| = 0 \tag{3.32}$$

for every p and every m. *The energy eigenvalues corresponding to the* mth *row of* $\boldsymbol{\Gamma}^p$ *are thus given by the secular equation* (3.32), *which involves only basis functions corresponding to the* mth *row of* $\boldsymbol{\Gamma}^p$. As the dimensions of $\boldsymbol{D}(p,m)$ are usually very much smaller than those of the determinant of (3.30), very much less numerical work is now needed to find the energy eigenvalues and eigenfunctions for the same degree of accuracy.

A further valuable saving of effort is provided by noting that the matrix element theorems of §6.1 also imply that

$$\boldsymbol{D}(p, 1) = \boldsymbol{D}(p, 2) = \cdots = \boldsymbol{D}(p, l_p),$$

where l_p is the dimension of $\boldsymbol{\Gamma}^p$. *Thus only one secular equation* (3.32) *has to be solved for each irreducible representation* $\boldsymbol{\Gamma}^p$, and each of the resulting energy eigenvalues can therefore be taken to be l_p-fold degenerate.

This analysis shows clearly the difference that was mentioned in §4.2 between a degeneracy caused by symmetry, such as the above l_p-fold degeneracy, and an accidental degeneracy. The above argument shows that an accidental degeneracy occurs *only* if two or more of the secular equations (3.32) corresponding to different irreducible representations *happen* to possess a common eigenvalue, which in general they do not.

§ 6.3. *Selection rules*

Consider the standard problem of time-dependent perturbation theory in which a system is in an initial state with wave function $\psi_i(\boldsymbol{r})$, an eigenfunction of the unperturbed Hamiltonian operator $H(\boldsymbol{r})$, and a perturbation associated with a potential function $V'(\boldsymbol{r})$ is applied to the system. Then the probability for a transition in unit time of the system into a final state with wave function $\psi_f(\boldsymbol{r})$, another eigenfunction of $H(\boldsymbol{r})$, is proportional to $|(\psi_f, V'\psi_i)|^2$, according to the first-order theory. In particular, if $(\psi_f, V'\psi_i)=0$ the transition cannot occur, at least to first-order in perturbation theory. (In this theory $V'(\boldsymbol{r})$ may also vary slowly with time.)

It is therefore of great interest to see which matrix elements $(\psi_f, V'\psi_i)$ must be zero for symmetry reasons alone. The so-called selection rules then consist of a list of those transitions which are not forbidden, to first-order, by symmetry.

The selection rules can easily be found as follows. The wave function $\psi_f(\boldsymbol{r})$, being an energy eigenfunction, transforms according to some irreducible representation $\boldsymbol{\Gamma}^r$ of the group of the unperturbed Schrödinger equation, $\mathscr{G}$. Suppose that the matrices of $\boldsymbol{\Gamma}^r$ are chosen so that $\psi_f(\boldsymbol{r})$ transforms as the nth row of $\boldsymbol{\Gamma}^r$. As shown in §5.1, the function $V'(\boldsymbol{r})\,\psi_i(\boldsymbol{r})$ can be written as a linear combination of parts transforming according to the rows of the irreducible representations of $\mathscr{G}$. If *no* part of this linear combination belongs to the nth row of $\boldsymbol{\Gamma}^r$, then the first matrix element theorem of §6.1 shows that $(\psi_f, V'\psi_i)$ is zero. However, the projection operator $\mathscr{P}^r_{nn}$ defined in (3.22) projects out of any function the part transforming as the nth row of $\boldsymbol{\Gamma}^r$, so that if $\mathscr{P}^r_{nn}\{V'(\boldsymbol{r})\,\psi_i(\boldsymbol{r}\}$ is zero, then $(\psi_f, V'\psi_i)$ is zero.

Exactly the same technique is applicable if $V'(\boldsymbol{r})$ is an *operator* function of $\boldsymbol{r}$, an example of which is given in ch. 5 §4.2. An alternative method involving direct product representations is described in §7.3.

§ 6.4. *Reduction of symmetry due to static perturbations*

Consider a system with a Hamiltonian operator $H(\boldsymbol{r})$, and suppose that a static perturbation $V(\boldsymbol{r})$ is applied giving a new Hamiltonian $H'(\boldsymbol{r})$, where

$$H'(\boldsymbol{r}) = H(\boldsymbol{r}) + V(\boldsymbol{r}).$$

Let $\mathscr{G}$ and $\mathscr{G}'$ be the respective groups of the Schrödinger equation. In general, not every transformation of $\mathscr{G}$ will leave $V(\boldsymbol{r})$ invariant, although every transformation of $\mathscr{G}'$ will leave $H(\boldsymbol{r})$ invariant, so that in general $\mathscr{G}'$ will be a *subgroup* of $\mathscr{G}$.

The Zeeman effect for the hydrogen atom provides a well-known example. In this effect $H(\boldsymbol{r})$ is spherically symmetric, but the application of the static magnetic field in some direction reduces the symmetry of $H'(\boldsymbol{r})$ to that of axial symmetry about that direction, and the result is that some of the degeneracies in the electron energy levels are split, causing a splitting of the spectral lines.

This splitting of energy level degeneracies is a general characteristic of the reduction of symmetry, and can be predicted solely from a knowledge of the character tables of $\mathscr{G}'$ and $\mathscr{G}$. The analysis proceeds as follows. Consider an energy level of the unperturbed system and suppose that it corresponds to an irreducible representation $\boldsymbol{\Gamma}$ of $\mathscr{G}$. As $\mathscr{G}'$ is a subgroup, $\boldsymbol{\Gamma}$ provides a representation of $\mathscr{G}'$, but it is not necessarily an *irreducible* representation of $\mathscr{G}'$. Suppose that for $\mathscr{G}'$ the reduction of $\boldsymbol{\Gamma}$ is given by

$$\boldsymbol{\Gamma} = \boldsymbol{\Gamma}'^1 \oplus \boldsymbol{\Gamma}'^2 \oplus \cdots \oplus \boldsymbol{\Gamma}'^p, \tag{3.33}$$

where $\boldsymbol{\Gamma}'^1, \boldsymbol{\Gamma}'^2, \ldots, \boldsymbol{\Gamma}'^p$ are all irreducible representations of $\mathscr{G}'$ and $\oplus$ indicates the direct sum defined in ch. 2 §4. The eigenfunctions of $H'(\boldsymbol{r})$ corresponding to these p irreducible representations correspond in general to different energies, and hence the effect of reducing the symmetry has been to split the energy level corresponding to $\boldsymbol{\Gamma}$ into p levels. The degeneracies of each of these p levels can be read off from the character table for $\mathscr{G}'$. Moreover, the third theorem of ch. 2 §6 applied to the character table for $\mathscr{G}'$ and the character system of $\boldsymbol{\Gamma}$ shows immediately which irreducible representations of $\mathscr{G}'$ appear in the reduction (3.33). It will be noted that if $\boldsymbol{\Gamma}$ is an *irreducible* representation of $\mathscr{G}'$, then the energy level is not split, although its value may be changed.

§ 7. DIRECT PRODUCT REPRESENTATIONS

Direct product representations can be used to give an alternative derivation of selection rules, as will be shown in §7.3. They are also needed in the theory of double groups, as in ch. 8 §8.4. As a preliminary, the concept of the direct product of two matrices is described in §7.1.

§ 7.1. *Direct products of matrices*

Definition of the direct product of two matrices. Let $\boldsymbol{\alpha}$ be an $m \times m$

matrix and $\boldsymbol{\beta}$ an $n\times n$ matrix. Then their direct product, denoted by $\boldsymbol{\alpha}\otimes\boldsymbol{\beta}$, is defined to be an $mn\times mn$ matrix, whose rows and columns are labelled by a double set of subscripts such that

$$(\boldsymbol{\alpha}\otimes\boldsymbol{\beta})_{is,\,jt}=\alpha_{ij}\beta_{st}\,, \tag{3.34}$$

for $i,j=1,2,\ldots,m$ and $s,t=1,2,\ldots,n$.

For example, the subscripts on the elements of the first row of $\boldsymbol{\alpha}\otimes\boldsymbol{\beta}$ are

$$11{,}11;\ 11{,}12;\ \ldots;\ 11{,}1n;$$
$$11{,}21;\ \ldots;\ 11{,}2n;\quad \ldots;\quad 11{,}m1;\ \ldots;\ 11{,}mn\,.$$

The direct product of two diagonal matrices is also diagonal, as if $\alpha_{ij}=\alpha_i\delta_{ij}$ and $\beta_{st}=\beta_s\delta_{st}$ then $(\boldsymbol{\alpha}\otimes\boldsymbol{\beta})_{is,\,jt}=\alpha_i\beta_s\delta_{ij}\delta_{st}$. This is a diagonal matrix, because the subscripts of the diagonal matrix elements of the direct product are:

$$11{,}11;\ 12{,}12;\ \ldots;\ 1n{,}1n;\quad 21{,}21;\ 22{,}22;\ \ldots;\ 2n{,}2n;\quad \ldots.$$

The direct product also has the very important property that if $\boldsymbol{\alpha}$ and $\boldsymbol{\alpha}'$ are $m\times m$ matrices, and $\boldsymbol{\beta}$ and $\boldsymbol{\beta}'$ are $n\times n$ matrices, then

$$(\boldsymbol{\alpha}\otimes\boldsymbol{\beta})(\boldsymbol{\alpha}'\otimes\boldsymbol{\beta}')=(\boldsymbol{\alpha}\boldsymbol{\alpha}')\otimes(\boldsymbol{\beta}\boldsymbol{\beta}'),$$

all products other than those indicated by the symbol $\otimes$ being ordinary matrix products. This result follows because the is,jt element of the left-hand side is

$$\sum_{k=1}^{m}\sum_{u=1}^{n}(\boldsymbol{\alpha}\otimes\boldsymbol{\beta})_{is,\,ku}(\boldsymbol{\alpha}'\otimes\boldsymbol{\beta}')_{ku,\,jt}$$
$$=\sum_{k=1}^{m}\sum_{u=1}^{n}\alpha_{ik}\beta_{su}\alpha'_{kj}\beta'_{ut}\,,$$

while the is,jt element of the right-hand side is

$$(\boldsymbol{\alpha}\boldsymbol{\alpha}')_{ij}(\boldsymbol{\beta}\boldsymbol{\beta}')_{st}=\sum_{k=1}^{m}\sum_{u=1}^{n}\alpha_{ik}\alpha'_{kj}\beta_{su}\beta'_{ut}\,.$$

§ 7.2. *Direct product representations of a group*

Let $\boldsymbol{\Gamma}^p(T)$ and $\boldsymbol{\Gamma}^q(T)$ be two irreducible representations of a group $\mathscr{G}$, and for each T of $\mathscr{G}$ define a matrix $\boldsymbol{\Gamma}(T)$ by

$$\boldsymbol{\Gamma}(T) = \boldsymbol{\Gamma}^p(T) \otimes \boldsymbol{\Gamma}^q(T).$$

This set of matrices form a representation of $\mathscr{G}$, because there is by construction one such matrix for every element of $\mathscr{G}$, and if T_1 and T_2 are any two elements of $\mathscr{G}$ then

$$\boldsymbol{\Gamma}(T_1 T_2) = \boldsymbol{\Gamma}(T_1)\,\boldsymbol{\Gamma}(T_2).$$

This latter property may be proved as follows. By construction

$$\begin{aligned}\boldsymbol{\Gamma}(T_1)\,\boldsymbol{\Gamma}(T_2) &= \{\boldsymbol{\Gamma}^p(T_1) \otimes \boldsymbol{\Gamma}^q(T_1)\}\,\{\boldsymbol{\Gamma}^p(T_2) \otimes \boldsymbol{\Gamma}^q(T_2)\} \\ &= \{\boldsymbol{\Gamma}^p(T_1)\,\boldsymbol{\Gamma}^p(T_2)\} \otimes \{\boldsymbol{\Gamma}^q(T_1)\,\boldsymbol{\Gamma}^q(T_2)\}\end{aligned}$$

(by the last result of §7.1),

$$= \boldsymbol{\Gamma}^p(T_1 T_2) \otimes \boldsymbol{\Gamma}^q(T_1 T_2)$$

(as $\boldsymbol{\Gamma}^p$ and $\boldsymbol{\Gamma}^q$ are themselves representations),

$$= \boldsymbol{\Gamma}(T_1 T_2)$$

(by construction).

The usefulness of the direct product representation lies in the fact that if $\boldsymbol{\Gamma}^p$ is an m-dimensional irreducible representation of the group $\mathscr{G}$ of the Schrödinger equation with the m basis functions $\phi_1^p(\boldsymbol{r}), \phi_2^p(\boldsymbol{r}), \ldots, \phi_m^p(\boldsymbol{r})$, and $\boldsymbol{\Gamma}^q$ is an n-dimensional irreducible representation of $\mathscr{G}$ with the n basis functions $\phi_1^q(\boldsymbol{r}), \ldots, \phi_n^q(\boldsymbol{r})$, then the set of mn functions $\phi_i^p(\boldsymbol{r})\,\phi_s^q(\boldsymbol{r})$ $(i=1, \ldots, m;\ s=1, 2, \ldots, n)$ form a basis for $\boldsymbol{\Gamma} = \boldsymbol{\Gamma}^p \otimes \boldsymbol{\Gamma}^q$. The proof of this statement is elementary, for

$$\begin{aligned}P(T)\,\{\phi_i^p(\boldsymbol{r})\,\phi_s^q(\boldsymbol{r})\} &= \phi_i^p(T^{-1}\boldsymbol{r})\,\phi_s^q(T^{-1}\boldsymbol{r}) \\ &= \{P(T)\,\phi_i^p(\boldsymbol{r})\}\,\{P(T)\,\phi_s^q(\boldsymbol{r})\} \\ &= \sum_{j=1}^{m} \Gamma^p(T)_{ji}\,\phi_j^p(\boldsymbol{r}) \sum_{t=1}^{n} \Gamma^q(T)_{ts}\,\phi_t^q(\boldsymbol{r}) \\ &= \sum_{j=1}^{m} \sum_{t=1}^{n} \Gamma(T)_{jt,\,is}\,\{\phi_j^p(\boldsymbol{r})\,\phi_t^q(\boldsymbol{r})\}.\end{aligned}$$

The characters $\chi(T)$ of $\boldsymbol{\Gamma}=\boldsymbol{\Gamma}^p\otimes\boldsymbol{\Gamma}^q$ are simply given in terms of the characters $\chi^p(T)$ and $\chi^q(T)$ by

$$\chi(T)=\chi^p(T)\,\chi^q(T)\,. \tag{3.35}$$

This is because

$$\begin{aligned}\chi(T)&=\sum_{i=1}^{m}\sum_{s=1}^{n}\Gamma(T)_{is,\,is}\\&=\sum_{i=1}^{m}\sum_{s=1}^{n}\Gamma^p(T)_{ii}\,\Gamma^q(T)_{ss}=\chi^p(T)\,\chi^q(T)\,.\end{aligned}$$

The character systems of the representations $\boldsymbol{\Gamma}^p\otimes\boldsymbol{\Gamma}^q$ and $\boldsymbol{\Gamma}^q\otimes\boldsymbol{\Gamma}^p$ are therefore equal, and so the first theorem of ch. 2 §6 shows that these two representations are equivalent.

The representation $\boldsymbol{\Gamma}=\boldsymbol{\Gamma}^p\otimes\boldsymbol{\Gamma}^q$ is, in general, a reducible representation of $\mathscr{G}$. The number of times n^r_{pq} that an irreducible representation $\boldsymbol{\Gamma}^r$ of $\mathscr{G}$ appears in its reduction is given by

$$n^r_{pq}=g^{-1}\sum_T\chi(T)\,\chi^r(T)^*\,,$$

according to the third theorem of ch. 2 §6, where g is the order of $\mathscr{G}$ and the summation is over all the elements T of $\mathscr{G}$. Using the result just obtained, this can be written as

$$n^r_{pq}=g^{-1}\sum_T\chi^p(T)\,\chi^q(T)\,\chi^r(T)^*\,. \tag{3.36}$$

§ 7.3. *Application of direct product representations to selection rules*

The above theory of direct product representations provides a slightly more convenient way of obtaining selection rules than the method described in §6.3. Assuming the same notation as in §6.3, suppose that $\psi_i(\boldsymbol{r})$ and $\psi_f(\boldsymbol{r})$ transform as some rows of the irreducible representations $\boldsymbol{\Gamma}^p$ and $\boldsymbol{\Gamma}^r$ of the group of the Schrödinger equation $\mathscr{G}$ respectively. The perturbing potential $V'(\boldsymbol{r})$ in general contains parts transforming according to several irreducible representations of $\mathscr{G}$. However, suppose for simplicity that $V'(\boldsymbol{r})$

transforms only as some row of the irreducible representation $\boldsymbol{\Gamma}^q$ of $\mathscr{G}$. Then $V'(\boldsymbol{r})\,\psi_i(\boldsymbol{r})$ transforms as some row of $\boldsymbol{\Gamma}=\boldsymbol{\Gamma}^p\otimes\boldsymbol{\Gamma}^q$. Thus, by the first matrix element theorem of §6.1, $(\psi_f, V'\psi_i)$ will only be non-zero if the reduction of $\boldsymbol{\Gamma}$ contains $\boldsymbol{\Gamma}^r$, that is, if $n^r_{pq}\neq 0$. Of course, n^r_{pq} is very easily calculated from the character table of $\mathscr{G}$ by use of eq. (3.36), and, moreover, only the *character* projection operators of (3.25) are needed to determine to which irreducible representations $\psi_i(\boldsymbol{r})$, $\psi_f(\boldsymbol{r})$ and $V'(\boldsymbol{r})$ belong.

The advantage of this technique compared with that of §6.3 is that everything can be obtained directly from the character table. The disadvantage is that it may not predict every forbidden transition. The reason for this is that it is possible, even when $\boldsymbol{\Gamma}=\boldsymbol{\Gamma}^p\otimes\boldsymbol{\Gamma}^q$ contains $\boldsymbol{\Gamma}^r$ in its reduction, that the function $V'(\boldsymbol{r})\,\psi_i(\boldsymbol{r})$ does not contain a part transforming as any row of $\boldsymbol{\Gamma}^r$. Moreover, even when $V'(\boldsymbol{r})\,\psi_i(\boldsymbol{r})$ does contain a part transforming as some rows of $\boldsymbol{\Gamma}^r$, $\psi_f(\boldsymbol{r})$ may not transform as one of these rows. In these situations $n^r_{pq}\neq 0$ even though $(\psi_f, V'\psi_i)=0$.

When $V'(\boldsymbol{r})$ is not merely a function of $\boldsymbol{r}$, but is an *operator* (such as a differential operator) which depends on $\boldsymbol{r}$, the analysis has to be slightly modified. It is first of all necessary to define what is meant by an *operator* transforming as a certain row of a certain representation.

Definition. The set of l operators $Q^q_1(\boldsymbol{r}),\ldots, Q^q_l(\boldsymbol{r})$ transform according to the rows of the l-dimensional irreducible representation $\boldsymbol{\Gamma}^q$ of the group of the Schrödinger equation $\mathscr{G}$ if for every T of $\mathscr{G}$

$$P(T)\,Q^q_i(\boldsymbol{r})\,P(T^{-1})=\sum_{j=1}^{l}\Gamma^q(T)_{ji}\,Q^q_j(\boldsymbol{r}),\qquad i=1,2,\ldots,l.\quad(3.37)$$

Eq. (3.37) is to be understood as an operator equation. That is, both sides must produce the same result when acting on *any* scalar function $f(\boldsymbol{r})$. It is also to be understood that in the left-hand side every operator acts on *everything* to its right. This definition is consistent with the usual definition (3.13) if $Q^q_i(\boldsymbol{r})$ is merely the product of a function transforming as the ith row of $\boldsymbol{\Gamma}^q$ with the identity operator, which by definition leaves everything unchanged.

An argument will now be given leading to the conclusion that $(\psi_k^r, Q_j^q \psi_i^p)$ is non-zero only if the reduction of $\boldsymbol{\Gamma}=\boldsymbol{\Gamma}^p \otimes \boldsymbol{\Gamma}^q$ contains $\boldsymbol{\Gamma}^r$, where $\psi_i^p(\boldsymbol{r})$ and $\psi_k^r(\boldsymbol{r})$ are basis functions for $\boldsymbol{\Gamma}^p$ and $\boldsymbol{\Gamma}^r$ respectively. This is the *same* condition as was encountered above, and of course it can be examined in terms of the characters alone using eq. (3.36).

It follows from (3.37) and (3.8) that

$$(\psi_k^r, Q_j^q \psi_i^p) = \sum_{b=1} \Gamma^q(T)_{bj} \, (P(T)\,\psi_k^r, Q_b^q P(T)\,\psi_i^p)$$

$$= \sum_a \sum_b \sum_c \Gamma^q(T)_{bj} \, \Gamma^p(T)_{ai} \, \Gamma^r(T)^*_{ck} \, (\psi_c^r, Q_b^q \psi_a^p)$$

$$= \sum_a \sum_b \sum_c (\boldsymbol{\Gamma}^p(T) \otimes \boldsymbol{\Gamma}^q(T))_{ab,\,ij} \, \Gamma^r(T)^*_{ck} \, (\psi_c^r, Q_b^q \psi_a^p)\,.$$

On summing both sides over all the transformations T of $\mathscr{G}$, and using the orthogonality theorem for matrices of ch. 2 § 4, the right-hand side is found to be non-zero only if $\boldsymbol{\Gamma}^r$ appears in the reduction of $\boldsymbol{\Gamma}=\boldsymbol{\Gamma}^p \otimes \boldsymbol{\Gamma}^q$.

§ 7.4. *'Two-body' matrix elements and direct product representations*

In problems involving two-body interactions there occur matrix elements of the form

$$\iiint \iiint \psi_j^q(\boldsymbol{r})^* \, \psi_l^s(\boldsymbol{r}')^* \, V(\boldsymbol{r}-\boldsymbol{r}') \, \psi_i^p(\boldsymbol{r}) \, \psi_k^r(\boldsymbol{r}') \times \mathrm{d}x \, \mathrm{d}y \, \mathrm{d}z \, \mathrm{d}x' \, \mathrm{d}y' \, \mathrm{d}z' , \tag{3.38}$$

where $V(\boldsymbol{r}-\boldsymbol{r}')$ depends on $|\boldsymbol{r}-\boldsymbol{r}'|$ alone, and where $\psi_i^p(\boldsymbol{r})$, $\psi_j^q(\boldsymbol{r})$, $\psi_k^r(\boldsymbol{r})$ and $\psi_l^s(\boldsymbol{r})$ form bases for the irreducible representations $\boldsymbol{\Gamma}^p$, $\boldsymbol{\Gamma}^q$, $\boldsymbol{\Gamma}^r$ and $\boldsymbol{\Gamma}^s$ of the group of the Schrödinger equation $\mathscr{G}$ respectively. These matrix elements are a generalization of those defined by eq. (3.7), and can be dealt with along the same lines as above.

Suppose that $P(T)$ is the operator corresponding to the transformation T that acts on functions of $\boldsymbol{r}$ as hitherto, and $P'(T)$ denotes the corresponding operator that acts on functions of $\boldsymbol{r}'$. Then $P(T)\,P'(T)$ leaves $V(\boldsymbol{r}-\boldsymbol{r}')$ invariant, and the set of operators $P(T)\,P'(T)$ form a group that is isomorphic to the group of the Schrödinger equation, and hence has the same irreducible representations. The set of functions $\psi_i^p(\boldsymbol{r})\,\psi_k^r(\boldsymbol{r}')$, $i=1,2,\ldots,l_p$, $k=1,2,\ldots l_r$, then form a basis for the representation $\boldsymbol{\Gamma}^p\otimes\boldsymbol{\Gamma}^r$, for

$$\begin{aligned}
P(T)\,P'(T)\,\psi_i^p(\boldsymbol{r})\,\psi_k^r(\boldsymbol{r}') \\
&= \Big\{\sum_{m=1}^{l_p} \Gamma^p(T)_{mi}\,\psi_m^p(\boldsymbol{r})\Big\}\Big\{\sum_{n=1}^{l_r} \Gamma^r(T)_{nk}\,\psi_n^r(\boldsymbol{r}')\Big\} \\
&= \sum_{m=1}^{l_p}\sum_{n=1}^{l_r} (\boldsymbol{\Gamma}^p(T)\otimes\boldsymbol{\Gamma}^r(T))_{mn,\,ik}\,\psi_m^p(\boldsymbol{r})\,\psi_n^r(\boldsymbol{r}').
\end{aligned}$$

It then follows by a generalization of the first matrix element theorem of §6.1 that the matrix element (3.38) is only non-zero if $\boldsymbol{\Gamma}^p\otimes\boldsymbol{\Gamma}^r$ and $\boldsymbol{\Gamma}^q\otimes\boldsymbol{\Gamma}^s$ contain at least one common irreducible representation in their reductions.

CHAPTER 4

TRANSLATIONAL SYMMETRY AND ELEMENTARY ELECTRONIC ENERGY BAND THEORY

§ 1. INTRODUCTION

It was shown in ch. 3 § 3.2 that the group of the Schrödinger equation for electrons in a perfect crystalline solid is the space group of the crystal, $\mathscr{G}$, which of course contains rotations as well as pure translations. However, as a first step in the study of the space group, it is very useful to limit attention to the subgroup of $\mathscr{G}$ that consists only of all pure translations of $\mathscr{G}$. Only the translational symmetry is then being taken into account. The energy eigenfunctions must transform according to the irreducible representations of this subgroup, which is equivalent to saying that they satisfy Bloch's theorem, as will be demonstrated in § 3.

Bloch's theorem has now become so much a part of the theory of solids that it is sometimes forgotten that it is essentially a group theoretical result. The elementary energy band theory based upon Bloch's theorem itself requires no group theoretical knowledge and is presented in most textbooks on solid state theory. For many purposes this elementary theory is sufficient. However, the neglect of the rotational symmetry of the crystal in this elementary theory does mean that in it some phenomena are overlooked, and, in particular, it cannot predict the extra degeneracies which can occur in the electronic energy levels. Moreover, it is only by taking into account the rotational symmetry that it is possible to reduce the numerical work in energy band calculations to a manageable amount and still produce accurate results. Thus, only by using the group theory of the preceding chapters is it possible to extract the full benefit from the rotational symmetries in the space group.

A proof of Bloch's theorem is given in §3 that involves only a very straightforward application of the ideas of the previous chapters. The rest of this chapter is then devoted to a brief account of the elementary electronic energy band theory that is based on this theorem. Chs. 5 and 7 then describe how the theory is modified when the *full* space group $\mathcal{G}$ is introduced in place of its translational subgroup. It will be seen there that the concepts introduced in this chapter still play a fundamental role.

§ 2. THE CYCLIC BOUNDARY CONDITIONS

At this stage, it is very convenient to assume that for each energy eigenfunction $\psi(\boldsymbol{r})$,

$$\psi(\boldsymbol{r}) = \psi(\{\boldsymbol{E} \mid N_1\boldsymbol{a}_1\}\,\boldsymbol{r}) = \psi(\{\boldsymbol{E} \mid N_2\boldsymbol{a}_2\}\,\boldsymbol{r}) = \psi(\{\boldsymbol{E} \mid N_3\boldsymbol{a}_3\}\,\boldsymbol{r}), \quad (4.1)$$

where N_1, N_2, N_3 are very large positive integers and $\boldsymbol{a}_1$, $\boldsymbol{a}_2$, $\boldsymbol{a}_3$ are the basic lattice vectors introduced in ch. 1 §3.1. Define the positive integer N by $N = N_1N_2N_3$. The assumption (4.1) is equivalent to the assumption that the infinite crystal consists of basic blocks in the form of parallelepipeds having sides $N_1\boldsymbol{a}_1$, $N_2\boldsymbol{a}_2$ and $N_3\boldsymbol{a}_3$, and that the physical situation is identical in corresponding points of different blocks. This is a physically reasonable approximation, as the situation in the interior of a crystal must be almost independent of what happens at the boundaries and hence of the boundary conditions. The integers N_1, N_2 and N_3 may be taken to be as large as is desired. The integration involved in the inner product defined in (3.7) is now to be taken as being over just one basic block of the crystal, B. The operators $P(T)$ retain the unitary property (3.8) provided all functions involved satisfy eqs. (4.1). This follows because $(P(T)\,f, P(T)\,g)$ is equal to

$$\iiint_B f(T^{-1}\boldsymbol{r})^*\, g(T^{-1}\boldsymbol{r})\,\mathrm{d}x\,\mathrm{d}y\,\mathrm{d}z = \iiint_{B'} f(\boldsymbol{r})^*\, g(\boldsymbol{r})\,\mathrm{d}x\,\mathrm{d}y\,\mathrm{d}z\,,$$

where B' is obtained from B by the transformation T^{-1}. As every part of B' can be mapped into a part of B by an appropriate combination of translations through $N_1\boldsymbol{a}_1$, $N_2\boldsymbol{a}_2$ and $N_3\boldsymbol{a}_3$, by (4.1) the last integral becomes (f, g). The cyclic boundary conditions (4.1) are often referred to as the Born cyclic boundary conditions. They were proposed originally by BORN and VON KÁRMÁN [1912].

The assumption (4.1) is equivalent to the assumption that

$$P(\{\boldsymbol{E} \mid N_j\boldsymbol{a}_j\}) = P(\{\boldsymbol{E} \mid \boldsymbol{0}\})$$

for every function of interest and for $j=1, 2, 3$. (The function operator $P(\{\boldsymbol{E} \mid \boldsymbol{0}\})$ is of course merely the identity operator.) However,

$$P(\{\boldsymbol{E} \mid N_j\boldsymbol{a}_j\}) = P(\{\boldsymbol{E} \mid \boldsymbol{a}_j\})^{N_j}$$

so that invoking (4.1) is equivalent to working with the *finite* group of function operators $P(\{\boldsymbol{E} \mid n_1\boldsymbol{a}_1+n_2\boldsymbol{a}_2+n_3\boldsymbol{a}_3\})$ such that

$$0 \leqslant n_j \leqslant N_j - 1 \tag{4.2}$$

and

$$P(\{\boldsymbol{E} \mid \boldsymbol{a}_j\})^{N_j} = P(\{\boldsymbol{E} \mid \boldsymbol{0}\}) \tag{4.3}$$

for $j=1, 2, 3$. This group has $N=N_1N_2N_3$ distinct elements. Henceforth, this group will be used in place of the infinite group of pure primitive translations. No confusion will be caused if this group is also denoted by $\mathscr{T}$. As $\mathscr{T}$ and all space groups having $\mathscr{T}$ as their subgroup of pure primitive translations are then of finite order, the theory of chs. 2 and 3 can be applied immediately.

§ 3. BLOCH'S THEOREM

As all transformations consisting of pure translations commute, the group $\mathscr{T}$ of §2 is Abelian, and, as such, it has the property that *all* its irreducible representations are one-dimensional. The proof of this latter assertion is as follows. As shown in ch. 1 §2.4, in an Abelian group each element is in a class of its own, so that the number of classes is equal to the order of the group. However, the number of irreducible representations is equal to the number of classes (the fourth theorem of ch. 2 § 6), and the sum of the squares of the dimensions of the irreducible representations is equal to the order of the group (the fifth theorem of ch. 2 § 6), from which the result follows immediately.

Consider a particular one-dimensional irreducible representation $\boldsymbol{\Gamma}$ of $\mathscr{T}$ and suppose that $\boldsymbol{\Gamma}(\{E \mid \boldsymbol{a}_j\}) = (C_j)$, for $j = 1, 2, 3$. Then from eq. (4.3) it follows that

$$C_j^{N_j} = 1\,, \tag{4.4}$$

so that

$$C_j = \exp(-\,2\pi \mathrm{i} p_j / N_j)\,, \qquad j = 1, 2, 3\,,$$

where p_j is an integer. As $\exp\{-2\pi\mathrm{i}(p_j + N_j)/N_j\} = \exp(-2\pi\mathrm{i} p_j / N_j)$, there are only N_j *different* values of C_j allowed by (4.4), and each of these by convention may be taken to correspond to a p_j having one of the values $0, 1, \ldots N_j - 1$. Then

$$\boldsymbol{\Gamma}(\{E \mid n_j \boldsymbol{a}_j\}) = (\exp(-\,2\pi\mathrm{i} p_j n_j / N_j))$$

and hence

$$\boldsymbol{\Gamma}(\{E \mid \boldsymbol{t}_n\}) = (\exp[-\,2\pi\mathrm{i}\,\{(p_1 n_1/N_1) + (p_2 n_2/N_2) + (p_3 n_3/N_3)\}])\,, \tag{4.5}$$

where $\boldsymbol{t}_\mathbf{n} = n_1\boldsymbol{a}_1 + n_2\boldsymbol{a}_2 + n_3\boldsymbol{a}_3$ as in eq. (1.9). There are $N = N_1 N_2 N_3$ sets of integers p_1, p_2 and p_3 allowed by the above convention which can be used to label the N different irreducible representations of $\mathscr{T}$.

The formula (4.5) can be simplified and given a simple geometric interpretation by introducing the following notation. Define the *basic lattice vectors of the reciprocal lattice* $\boldsymbol{b}_1$, $\boldsymbol{b}_2$ and $\boldsymbol{b}_3$ by

$$\boldsymbol{a}_i \cdot \boldsymbol{b}_j = 2\pi\delta_{ij}\,, \qquad i, j = 1, 2, 3\,, \tag{4.6}$$

so that explicitly

$$\boldsymbol{b}_1 = 2\pi \boldsymbol{a}_2 \wedge \boldsymbol{a}_3 / \{\boldsymbol{a}_1 \cdot (\boldsymbol{a}_2 \wedge \boldsymbol{a}_3)\}\,, \tag{4.7}$$

with similar expressions for $\boldsymbol{b}_2$ and $\boldsymbol{b}_3$. Then define the so-called '*allowed* **k**-*vectors*' by

$$\boldsymbol{k} = k_1\boldsymbol{b}_1 + k_2\boldsymbol{b}_2 + k_3\boldsymbol{b}_3\,, \tag{4.8}$$

where $k_j = p_j/N_j$. Thus $\boldsymbol{k}\cdot\boldsymbol{t}_n = 2\pi\{(n_1 p_1/N_1)+(n_2 p_2/N_2)+(n_3 p_3/N_3)\}$, so that (4.5) becomes

$$\boldsymbol{\Gamma}^{\boldsymbol{k}}(\{E \mid \boldsymbol{t}_n\}) = (\exp(-\,\mathrm{i}\boldsymbol{k}\cdot\boldsymbol{t}_n))\,, \tag{4.9}$$

where the N irreducible representations are now labelled by the N allowed **k**-vectors. These representations are clearly unitary.

Suppose that $\phi_{\boldsymbol{k}}(\boldsymbol{r})$ is a basis function for the irreducible representation $\boldsymbol{\Gamma}^{\boldsymbol{k}}$ of (4.9). Then by eq. (3.13),

$$P(\{E \mid \boldsymbol{t}_n\})\,\phi_{\boldsymbol{k}}(\boldsymbol{r}) = \boldsymbol{\Gamma}^{\boldsymbol{k}}(\{E \mid \boldsymbol{t}_n\})_{11}\,\phi_{\boldsymbol{k}}(\boldsymbol{r}) = \exp(-\,\mathrm{i}\boldsymbol{k}\cdot\boldsymbol{t}_n)\,\phi_{\boldsymbol{k}}(\boldsymbol{r})\,, \tag{4.10}$$

but by the definition (3.3),

$$P(\{E \mid \boldsymbol{t}_n\})\,\phi_{\boldsymbol{k}}(\boldsymbol{r}) = \phi_{\boldsymbol{k}}(\{E \mid \boldsymbol{t}_n\}^{-1}\,\boldsymbol{r}) = \phi_{\boldsymbol{k}}(\boldsymbol{r} - \boldsymbol{t}_n)\,,$$

so that

$$\phi_{\boldsymbol{k}}(\boldsymbol{r} - \boldsymbol{t}_n) = \exp(-\,\mathrm{i}\boldsymbol{k}\cdot\boldsymbol{t}_n)\,\phi_{\boldsymbol{k}}(\boldsymbol{r})\,.$$

Thus

$$\phi_{\boldsymbol{k}}(\boldsymbol{r}) = \exp(\mathrm{i}\boldsymbol{k}\cdot\boldsymbol{r})\,u_{\boldsymbol{k}}(\boldsymbol{r})\,, \tag{4.11}$$

where $u_{\boldsymbol{k}}(\boldsymbol{r})$ is a function that has the periodicity of the lattice; that is, $u_{\boldsymbol{k}}(\boldsymbol{r}-\boldsymbol{t}_n) = u_{\boldsymbol{k}}(\boldsymbol{r})$ for any lattice vector $\boldsymbol{t}_n$.

Eq. (4.11) is the statement of the theorem of BLOCH [1928] in its usual form, for energy eigenfunctions must be basis functions of the irreducible representations $\boldsymbol{\Gamma}^{\boldsymbol{k}}$ of $\mathscr{T}$. A function of the form (4.11) is called a *Bloch function*. The corresponding energy eigenvalue is denoted by $E(\boldsymbol{k})$, so that

$$H(\boldsymbol{r})\,\phi_{\boldsymbol{k}}(\boldsymbol{r}) = E(\boldsymbol{k})\,\phi_{\boldsymbol{k}}(\boldsymbol{r})\,. \tag{4.12}$$

§ 4. THE RECIPROCAL LATTICE AND THE BRILLOUIN ZONE

The set of lattice vectors of the reciprocal lattice is defined by

$$\boldsymbol{K}_m = m_1\boldsymbol{b}_1 + m_2\boldsymbol{b}_2 + m_3\boldsymbol{b}_3\,, \tag{4.13}$$

where $\boldsymbol{m} = (m_1, m_2, m_3)$, m_1,m_2 and m_3 are integers, and $\boldsymbol{b}_1, \boldsymbol{b}_2$ and $\boldsymbol{b}_3$ are the basic lattice vectors of the reciprocal lattice defined in (4.6). They have the property that

$$\exp(\mathrm{i}\boldsymbol{K}_{\boldsymbol{m}} \cdot \boldsymbol{t}_{\boldsymbol{n}}) = 1 \tag{4.14}$$

for any $\boldsymbol{K}_{\boldsymbol{m}}$ and $\boldsymbol{t}_{\boldsymbol{n}}$. It is useful to note that

$$\sum_{\boldsymbol{t}_n} \exp(\mathrm{i}\boldsymbol{k} \cdot \boldsymbol{t}_{\boldsymbol{n}}) = \begin{cases} N, & \text{if} \quad \boldsymbol{k} = \boldsymbol{K}_{\boldsymbol{m}} \\ 0, & \text{if} \quad \boldsymbol{k} \neq \boldsymbol{K}_{\boldsymbol{m}}, \end{cases} \tag{4.15}$$

where the sum is over all the lattice vectors in the basic block of § 2, this result being a consequence of the fact that the left-hand side is a product of three simple geometric series.

In § 3 N irreducible representations of $\mathscr{T}$ were found and described by the allowed $\boldsymbol{k}$-vectors (4.8). These $\boldsymbol{k}$-vectors can be imagined as being plotted in the so-called '$\boldsymbol{k}$-*space*' or '*reciprocal space*' defined by the reciprocal lattice vectors (4.13). The allowed $\boldsymbol{k}$-vectors lie on a very fine lattice (defined by (4.8)) within and upon three faces of the parallelepiped having edges $\boldsymbol{b}_1, \boldsymbol{b}_2$ and $\boldsymbol{b}_3$ that is shown in fig. 4.1.

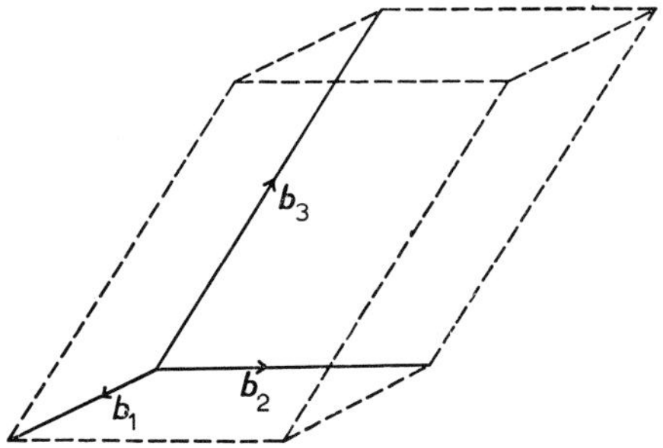

Fig. 4.1

It is, however, more convenient to re-plot the allowed $\boldsymbol{k}$-vectors into a more symmetrical region of $\boldsymbol{k}$-space surrounding the point $\boldsymbol{k} = \boldsymbol{0}$. To do this consider the equation

$$\boldsymbol{k}' = \boldsymbol{k} + \boldsymbol{K}_{\boldsymbol{m}}, \tag{4.16}$$

where $\boldsymbol{K}_m$ is a reciprocal lattice vector. A pair of vectors $\boldsymbol{k}$ and $\boldsymbol{k}'$ satisfying an equation of the form (4.16) are said to be *equivalent*, because $\exp(-i\boldsymbol{k}'\cdot\boldsymbol{t}_n)=\exp(-i\boldsymbol{k}\cdot\boldsymbol{t}_n)$ by (4.14), and hence

$$\Gamma^{\boldsymbol{k}'}(\{E \mid \boldsymbol{t}_n\}) = \Gamma^{\boldsymbol{k}}(\{E \mid \boldsymbol{t}_n\})$$

for every $\{E \mid \boldsymbol{t}_n\}$ of $\mathscr{T}$. Thus the irreducible representation of $\mathscr{T}$ described by $\boldsymbol{k}$ could equally well be described by $\boldsymbol{k}'$. The more symmetrical region of $\boldsymbol{k}$-space is called the *Brillouin zone* (or sometimes the 'first' Brillouin zone), and it is defined to consist of all those points of $\boldsymbol{k}$-space that lie closer to $\boldsymbol{k}=\boldsymbol{0}$ than to any other reciprocal lattice points. Its boundaries are therefore the planes that are the perpendicular bisectors of the lines joining the point $\boldsymbol{k}=\boldsymbol{0}$ to the nearer reciprocal lattice points, the plane bisecting the line from $\boldsymbol{k}=\boldsymbol{0}$ to $\boldsymbol{k}=\boldsymbol{K}_m$ having the equation

$$\boldsymbol{k}\cdot\boldsymbol{K}_m = \tfrac{1}{2}|\boldsymbol{K}_m|^2,$$

as is clear from fig. 4.2. For some lattices, such as the body-centred cubic lattice Γ_c^v, only *nearest*-neighbour reciprocal-lattice points are involved in the construction of the Brillouin zone, but for others, such as the face-centred cubic lattice Γ_c^f *next-nearest* neighbours are involved as well. The irreducible representations of $\mathscr{T}$ then

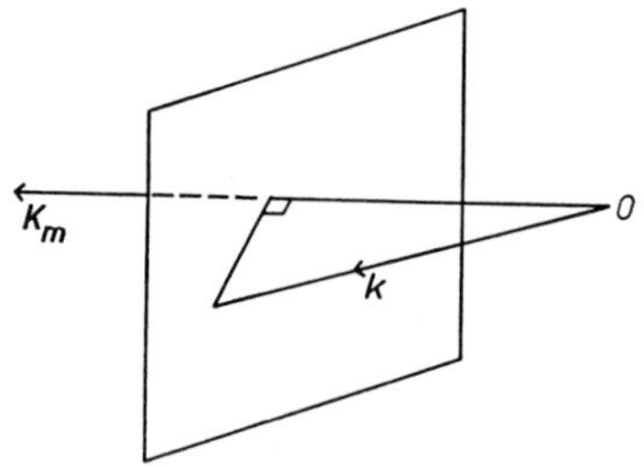

Fig. 4.2

correspond to a very fine lattice of points inside the Brillouin zone and on one half of its surface.

The mapping of the parallelepiped of fig. 4.1 into the Brillouin zone can be quite complicated because different regions of the parallelepiped are mapped using different reciprocal lattice vectors. The following *two*-dimensional example shown in fig. 4.3 of a

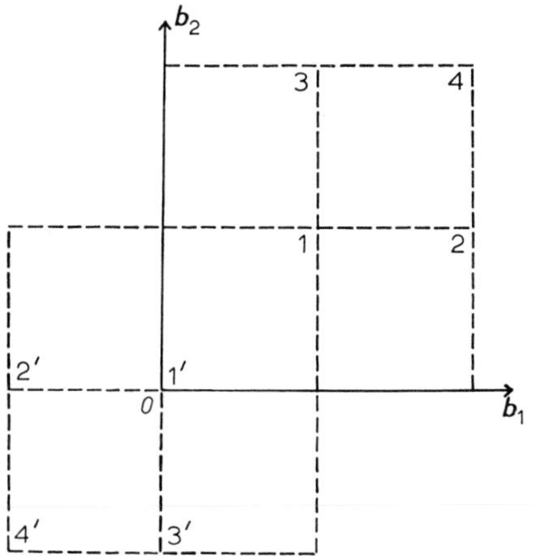

Fig. 4.3

square lattice demonstrates this clearly. In this example, the analogue of the three-dimensional parallelepiped of fig. 4.1 is the square with sides $\boldsymbol{b}_1$ and $\boldsymbol{b}_2$, which consists of four regions 1, 2, 3 and 4, and the analogue of the Brillouin zone is the square having $\boldsymbol{k}=\boldsymbol{0}$ at its centre, which consists of the four regions 1′, 2′, 3′ and 4′. The region 1 is mapped into 1′ by $\boldsymbol{K}_{(0,0,0)}=\boldsymbol{0}$, 2 is mapped into 2′ by $\boldsymbol{K}_{(-1,0,0)} = -\boldsymbol{b}_1$, 3 is mapped into 3′ by $\boldsymbol{K}_{(0,-1,0)}=-\boldsymbol{b}_2$, and 4 is mapped into 4′ by $\boldsymbol{K}_{(-1,-1,0)}=-\boldsymbol{b}_1-\boldsymbol{b}_2$.

By construction, the volume of the Brillouin zone is the same as that of the parallelepiped from which it is formed, namely $\boldsymbol{b}_1\cdot(\boldsymbol{b}_2\wedge\boldsymbol{b}_3)$. It follows from (4.7) that this is equal to

$$(2\pi)^3/\{\boldsymbol{a}_1\cdot(\boldsymbol{a}_2\wedge\boldsymbol{a}_3)\},$$

where $\boldsymbol{a}_1 \cdot (\boldsymbol{a}_2 \wedge \boldsymbol{a}_3)$ is the volume of the parallelepiped three of whose sides are $\boldsymbol{a}_1$, $\boldsymbol{a}_2$ and $\boldsymbol{a}_3$.

§ 5. THE RECIPROCAL LATTICES AND BRILLOUIN ZONES OF THE LATTICES Γ_c^v, Γ_c^f AND Γ_h

For the body-centred cubic lattice Γ_c^v, the basic lattice vectors of the reciprocal lattice obtained from eqs. (1.10) and (4.7) are

$$\left.\begin{aligned} \boldsymbol{b}_1 &= (2\pi/a)\,(1, 0, 1) \\ \boldsymbol{b}_2 &= (2\pi/a)\,(0, 1, -1) \\ \boldsymbol{b}_3 &= (2\pi/a)\,(1, -1, 0). \end{aligned}\right|$$

The Brillouin zone is shown in fig. 4.4. The position vectors of the 'symmetry points' marked are as follows: for Γ $\boldsymbol{k}=(0, 0, 0)$, for H $\boldsymbol{k}=(\pi/a)\,(0, 0, 2)$, for N $\boldsymbol{k}=(\pi/a)\,(0, 1, 1)$ and for P $\boldsymbol{k}=(\pi/a)(1, 1, 1)$. The significance of the term 'symmetry point' will be explained in ch. 5 § 1. The notation is that of Bouckaert *et al.* [1936].

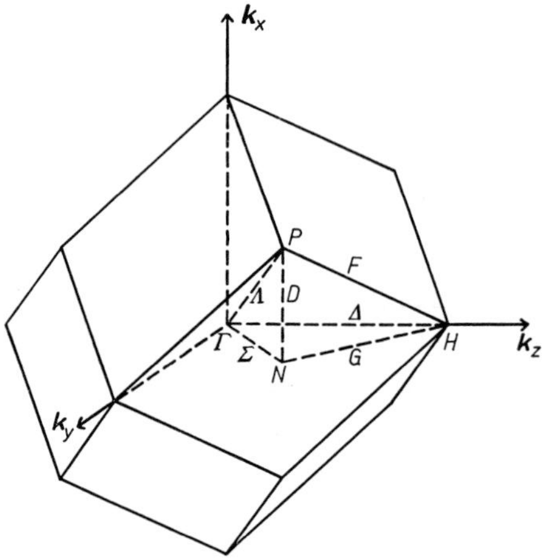

Fig. 4.4. The Brillouin zone corresponding to the body-centred cubic lattice Γ_c^v.

For the face-centred cubic lattice Γ_c^f the basic lattice vectors of the reciprocal lattice obtained from eqs. (1.11) and (4.7) are

$$\left.\begin{aligned} \boldsymbol{b}_1 &= (2\pi/a)\,(1, 1, -1) \\ \boldsymbol{b}_2 &= (2\pi/a)\,(-1, 1, 1) \\ \boldsymbol{b}_3 &= (2\pi/a)\,(1, -1, 1). \end{aligned}\right|$$

The Brillouin zone is shown in fig. 4.5, the position vectors of the symmetry points being as follows: for Γ $\boldsymbol{k}=(0, 0, 0)$, for K $\boldsymbol{k}=(\pi/a)\,(0, \frac{3}{2}, \frac{3}{2})$, for L $\boldsymbol{k}=(\pi/a)\,(1, 1, 1)$, for U $\boldsymbol{k}=(\pi/a)\,(\frac{1}{2}, 2, 2)$, for W $\boldsymbol{k}=(\pi/a)\,(0, 1, 2)$ and for X $\boldsymbol{k}=(\pi/a)\,(0, 0, 2)$. The notation is again that of BOUCKAERT *et al.* [1936].

For the simple hexagonal lattice Γ_h, which gives the translational symmetry of the hexagonal close-packed space group D_{6h}^4, the basic lattice vectors obtained from eqs. (1.12) and (4.7) are

$$\left.\begin{aligned} \boldsymbol{b}_1 &= (2\pi/c)\,(0, 0, 1) \\ \boldsymbol{b}_2 &= (2\pi/a\sqrt{3})\,(\sqrt{3}, -1, 0) \\ \boldsymbol{b}_3 &= (4\pi/a\sqrt{3})\,(0, -1, 0). \end{aligned}\right|$$

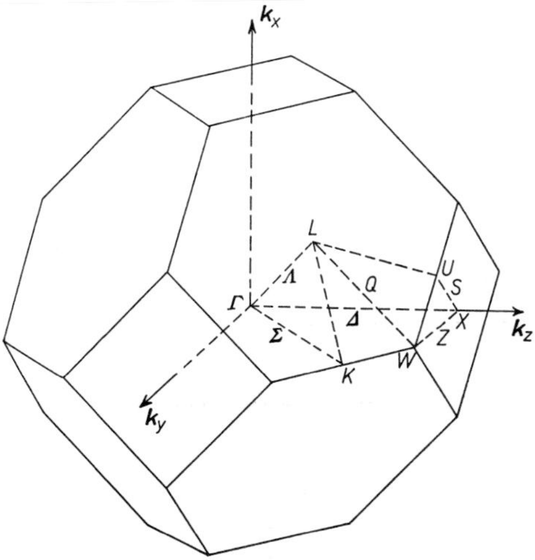

Fig. 4.5. The Brillouin zone corresponding to the face-centred cubic lattice Γ_c^f.

The Brillouin zone is shown in fig. 4.6, the symmetry points in the notation of HERRING [1942] being as follows: for Γ $\boldsymbol{k}=(0, 0, 0)$, for A $\boldsymbol{k}=(0, 0, \pi/c)$, for H $\boldsymbol{k}=(4\pi/3a, 0, \pi/c)$, for K $\boldsymbol{k}=(4\pi/3a, 0, 0)$, for L $\boldsymbol{k}=(0, 2\pi/a\sqrt{3}, \pi/c)$ and for M $\boldsymbol{k}=(0, 2\pi/a\sqrt{3}, 0)$.

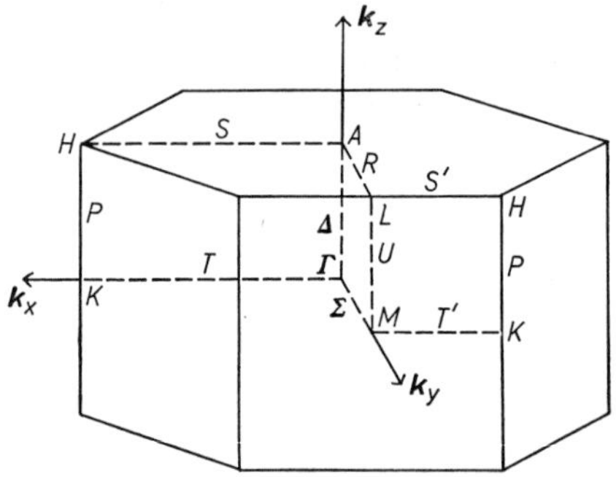

Fig. 4.6. The Brillouin zone corresponding to the simple hexagonal lattice Γ_h.

The Brillouin zones corresponding to the other eleven Bravais lattices may be found in the article by KOSTER [1957].

§ 6. ENERGY BANDS

An *infinite* set of energy eigenfunctions and eigenvalues correspond to each irreducible representation of the group of the Schrödinger equation and hence to each allowed $\boldsymbol{k}$-vector of the Brillouin zone. The members of the set of energy eigenvalues corresponding to wave vector $\boldsymbol{k}$ may be denoted by $E_1(\boldsymbol{k})$, $E_2(\boldsymbol{k})$,..., with the convention that

$$E_n(\boldsymbol{k}) \leqslant E_{n+1}(\boldsymbol{k}), \tag{4.17}$$

for all n. The set of energy eigenvalues $E_n(\boldsymbol{k})$ corresponding to a particular n are said to form the nth *energy band*, and the set of energy bands is said to constitute the energy band 'structure'.

To visualize the energy band structure, it is convenient to

consider one at a time the axes of the Brillouin zone that join the symmetry points, and for every allowed **k**-vector on each axis to plot the energy levels $E_n(\mathbf{k})$. A typical example of such a plot is shown in fig. 4.7, which gives the energy levels along the axis Δ

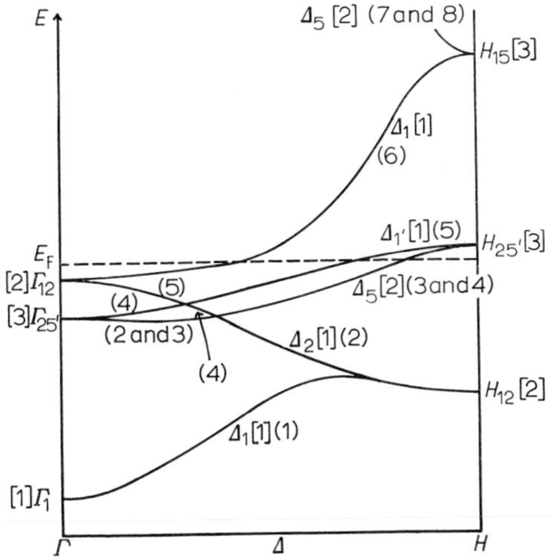

Fig. 4.7

between the symmetry points Γ and H for body-centred cubic iron, as calculated by Wood [1962]. The number in curved brackets gives the band index n, as defined in eq. (4.17), and the number in square brackets gives the degeneracy of the corresponding level. The occurrence of degenerate eigenvalues is a consequence of the rotational symmetry which has been neglected in this chapter but which will be exploited in ch. 5. The other symbols will also be explained in ch. 5.

The positive integers N_1, N_2 and N_3 introduced in eq. (4.1) are arbitrarily large, and it is frequently convenient to consider the limiting case when they tend to infinity. The allowed **k**-vectors can then take *all* values inside the Brillouin zone and on half its boundary, and the $E_n(\boldsymbol{k})$ are continuous functions of $\boldsymbol{k}$ for each n. Moreover, $\text{grad}_{\boldsymbol{k}}\, E_n(\boldsymbol{k})$ are also continuous functions of $\boldsymbol{k}$, except possibly at points where two bands touch. The plot in fig. 4.7 is made for this limiting case.

In the 'one-electron' approximation on which this whole theory is based, the Pauli exclusion principle implies that no two electrons can 'occupy' the same one-electron state, here specified by an allowed $\boldsymbol{k}$-vector, a band index n, and a spin quantum number that can take only one of two possible values. It follows that each energy level $E_n(\boldsymbol{k})$ can 'hold' *two* electrons, and hence each energy band can hold $2N$ electrons. If there are V valence (or conduction) electrons per atom, and A atoms per lattice point of the crystal lattice, there will be NVA valence electrons in the large basic block of the crystal of §2, which will therefore require the equivalent of $\frac{1}{2}VA$ bands to hold them.

In the ground state of the system, all the energy levels $E_n(\boldsymbol{k})$ will be doubly occupied up to a certain energy E_{F}, the *Fermi energy*, and all levels above this energy will be unoccupied. The surface in $\boldsymbol{k}$-space defined by

$$E_n(\boldsymbol{k}) = E_{\text{F}}$$

is called the *Fermi surface*. If one and only one band contains the Fermi energy, and all others are entirely above it or below it, then the Fermi surface merely consists of one sheet. If no band contains the Fermi energy, as happens for insulators and semi-conductors, there is no Fermi surface. In all other cases, the Fermi surface consists of several sheets, to visualize which one considers a number of identical Brillouin zones, with one zone for each band. A full band corresponds to a full Brillouin zone, but a partially occupied band corresponds to a partially occupied Brillouin zone and hence to a sheet of the Fermi surface in that Brillouin zone. As an example, consider body-centred cubic iron for which $A=1$ and $V=8$ so that the equivalent of 4 complete bands are needed to hold the valence electrons. The Fermi energy E_{F} is shown in fig. 4.7 and is clearly consistent with this. Bands 3, 4, 5 and 6 are partially occupied, giving rise to 4 sheets in the Fermi surface.

It is the distribution of energy levels near the Fermi energy that largely determines the electronic properties of a solid.

Occasionally it is convenient to work with the so-called *extended zone scheme*, in which the energy bands $E_n(\boldsymbol{k})$ are assumed to be periodic functions of $\boldsymbol{k}$ with the periodicity of the reciprocal lattice.

CHAPTER 5

SYMMORPHIC SPACE GROUPS

§ 1. IRREDUCIBLE REPRESENTATIONS OF A SYMMORPHIC SPACE GROUP

In ch. 4 § 3 it was shown that, as a consequence of the translational symmetry of the lattice *alone*, every energy eigenfunction can be written in the form of a Bloch function, that is, as

$$\phi_{\boldsymbol{k}}(\boldsymbol{r}) = \exp(\mathrm{i}\boldsymbol{k}\cdot\boldsymbol{r})\, u_{\boldsymbol{k}}(\boldsymbol{r}),$$

where $\boldsymbol{k}$ is an allowed $\boldsymbol{k}$-vector as defined in eq. (4.8). As the energy eigenfunctions form bases for the representations of the group of the Schrödinger equation, which in this context is the space group $\mathscr{G}$, the basis functions of the representations of $\mathscr{G}$ will be Bloch functions. As $\mathscr{G}$ contains more symmetry operations than its subgroup $\mathscr{T}$ of pure translations, it is to be expected that if a Bloch function $\phi_{\boldsymbol{k}}(\boldsymbol{r})$ is to be a basis function for an irreducible representation of $\mathscr{G}$, then in general some extra conditions will be imposed on $\phi_{\boldsymbol{k}}(\boldsymbol{r})$. This will be demonstrated explicitly in the theorem given below.

As noted in ch. 1 § 3.1, for a symmorphic space group $\mathscr{G}$, every symmetry operation is merely a product of a rotation of the point group $\mathscr{G}_o$ of the space group and a primitive translation of $\mathscr{T}$. The consequence of this neat separation is that the irreducible representations of $\mathscr{G}$ can be found fairly simply from subgroups of $\mathscr{G}_o$. For non-symmorphic space groups there is no such neat separation, and the corresponding theory is accordingly more complicated, and needs the development in ch. 6 of further abstract group theoretical concepts.

The whole theory for symmorphic space groups depends on a fundamental theorem. After stating and proving this theorem, the rest of this chapter is devoted to a study of its consequences. The body-centred and face-centred cubic space groups O_h^9 and O_h^5 will be treated as examples.

It is assumed throughout that the cyclic boundary conditions of ch. 4 §2 have been imposed on $\mathscr{T}$, but with $N_1 = N_2 = N_3$ so that every $P(T)$ of $\mathscr{G}$ is unitary. Then $\mathscr{G}$ is a *finite* group of order $g_o N$, g_o being the order of the point group $\mathscr{G}_o$ and N the order of $\mathscr{T}$.

Before stating the fundamental theorem, certain preliminary definitions have to be made.

Definition of $\mathscr{G}_o(\boldsymbol{k})$, the point group of the wave vector $\boldsymbol{k}$: The point group $\mathscr{G}_o(\boldsymbol{k})$ is the *subgroup* of the point group $\mathscr{G}_o$ of the space group that consists of all the rotations $\{\boldsymbol{R} \mid \boldsymbol{0}\}$ of $\mathscr{G}_o$ that rotate $\boldsymbol{k}$ into itself or an equivalent vector in the sense of eq. (4.16). That is, the rotation $\{\boldsymbol{R} \mid \boldsymbol{0}\}$ of $\mathscr{G}_o$ is a member of $\mathscr{G}_o(\boldsymbol{k})$ if there exists a lattice vector of the reciprocal lattice $\boldsymbol{K}_m$, as defined in eq. (4.13) and which may be zero, such that

$$\boldsymbol{R}\boldsymbol{k} = \boldsymbol{k} + \boldsymbol{K}_m .$$

Definitions of general points and symmetry points, symmetry axes, and symmetry planes of the Brillouin zone: If $\mathscr{G}_o(\boldsymbol{k})$ is the trivial group consisting only of the identity transformation $\{\boldsymbol{E} \mid \boldsymbol{0}\}$, then $\boldsymbol{k}$ is said to be a general point of the Brillouin zone. If $\boldsymbol{k}$ is such that $\mathscr{G}_o(\boldsymbol{k})$ is a larger group than those corresponding to *all* neighbouring points of the Brillouin zone, then $\boldsymbol{k}$ is known as a symmetry point. If all the points on a line or in a plane have the same *non-trivial* $\mathscr{G}_o(\boldsymbol{k})$, then this line or plane is said to be a symmetry axis or plane.

For example, for the body-centred cubic lattice Γ_c^v, the Brillouin zone is shown in fig. 4.4. For the corresponding space group O_h^9

the symmetry points are Γ, H, N and P, the symmetry axes are Δ, Σ, Λ, D, F and G, and the symmetry planes are those containing two symmetry axes.

Definition of the 'star' of $\boldsymbol{k}$*:* If $\{\boldsymbol{R} \mid \boldsymbol{0}\}$ is a member of $\mathscr{G}_o$ but not a member of $\mathscr{G}_o(\boldsymbol{k})$, then $\boldsymbol{Rk}$ is not equal or equivalent to $\boldsymbol{k}$. Let $\boldsymbol{k}_2, \boldsymbol{k}_3, \ldots$ be a set of distinct and non-equivalent vectors obtained in this way by working through all the rotations of $\mathscr{G}_o$ not in $\mathscr{G}_o(\boldsymbol{k})$. Let $\boldsymbol{k}_1 = \boldsymbol{k}$. (Any other sets obtained by this process are merely equivalent to the set $\boldsymbol{k}_1, \boldsymbol{k}_2, \ldots$.) The set of $M(\boldsymbol{k})$ vectors $\boldsymbol{k}_1, \boldsymbol{k}_2, \ldots$ (or any equivalent set) is called the star of $\boldsymbol{k}$.

One can associate with the star of $\boldsymbol{k}$ a set of $M(\boldsymbol{k})$ rotations $\{\boldsymbol{R}_1 \mid \boldsymbol{0}\}, \{\boldsymbol{R}_2 \mid \boldsymbol{0}\}, \ldots$ of $\mathscr{G}_o$ which are defined to be such that $\boldsymbol{R}_j\boldsymbol{k}$ is equivalent to $\boldsymbol{k}_j$. That is, for some $\boldsymbol{K}_m$, which may be zero,

$$\boldsymbol{R}_j\boldsymbol{k} = \boldsymbol{k}_j + \boldsymbol{K}_m, \qquad j = 1, 2, \ldots, M(\boldsymbol{k}). \tag{5.1}$$

Usually the choice of $\boldsymbol{R}_j$ is not unique, but this does not matter provided that, once a choice of $\boldsymbol{R}_j$ is made, it is adhered to. It is often convenient to take $\boldsymbol{R}_1$ to be $\boldsymbol{E}$. If g_o is the order of $\mathscr{G}_o$ and $g_o(\boldsymbol{k})$ is the order of $\mathscr{G}_o(\boldsymbol{k})$, then

$$g_o = g_o(\boldsymbol{k})\, M(\boldsymbol{k}). \tag{5.2}$$

This result is a consequence of the fact that $\mathscr{G}_o$ can be divided in the following way into $M(\boldsymbol{k})$ sets of rotations, each set containing $g_o(\boldsymbol{k})$ members. Let $\boldsymbol{R}_j\mathscr{G}_o(\boldsymbol{k})$ denote the set consisting of the $g_o(\boldsymbol{k})$ distinct rotations formed from the left product of $\boldsymbol{R}_j$ with each of the elements of $\mathscr{G}_o(\boldsymbol{k})$. Every element of $\boldsymbol{R}_j\mathscr{G}_o(\boldsymbol{k})$ acting on $\boldsymbol{k}$ produces $\boldsymbol{k}_j$ or a vector equivalent to $\boldsymbol{k}_j$, so by the definition of the star of $\boldsymbol{k}$ no rotation can be a member of the sets corresponding to two different rotations $\boldsymbol{R}_j$. However, every rotation of $\mathscr{G}_o$ must lie in one such set, as if $\boldsymbol{R}$ is a member of $\mathscr{G}_o$, then $\boldsymbol{Rk}$ must be equivalent to some member $\boldsymbol{k}_j$ of the star of $\boldsymbol{k}$, so that $\boldsymbol{R}_j^{-1}\boldsymbol{R}$ is a member of $\mathscr{G}_o(\boldsymbol{k})$, from which it follows that $\boldsymbol{R}$ is a member of $\boldsymbol{R}_j\mathscr{G}_o(\boldsymbol{k})$. [The process described here is actually an example of the decomposition of a group, which is here $\mathscr{G}_o$, into left cosets with respect to a subgroup, here $\mathscr{G}_o(\boldsymbol{k})$, that is described in detail in ch. 6 §2.]

The vital link between the irreducible representations of the space group $\mathscr{G}$ and those of $\mathscr{T}$ and the $\mathscr{G}_o(\boldsymbol{k})$ is provided by the following theorem.

Fundamental theorem on the basis functions of the irreducible representations of a symmorphic space group $\mathscr{G}$. Suppose that $\boldsymbol{k}$ is an allowed $\boldsymbol{k}$-vector, and $\phi^p_{\boldsymbol{k}s}(\boldsymbol{r})$ is a Bloch function with wave vector $\boldsymbol{k}$, which transforms as the sth row of the irreducible unitary representation $\boldsymbol{\Gamma}^p$ of $\mathscr{G}_o(\boldsymbol{k})$, this representation being of dimension l. Suppose that the star of $\boldsymbol{k}$ consists of the $M(\boldsymbol{k})$ vectors $\boldsymbol{k}_1(=\boldsymbol{k})$, $\boldsymbol{k}_2, \ldots$, associated with which are $M(\boldsymbol{k})$ rotations $\boldsymbol{R}_j$, as defined by eq. (5.1). Then the set of $lM(\boldsymbol{k})$ functions

$$P(\{\boldsymbol{R}_i \mid \boldsymbol{0}\})\, \phi^p_{\boldsymbol{k}s}(\boldsymbol{r}), \qquad i = 1, 2, \ldots, M(\boldsymbol{k}), \quad s = 1, 2, \ldots, l,$$

form a basis for a $lM(\boldsymbol{k})$-dimensional unitary irreducible representation of the space group $\mathscr{G}$. Moreover, *all* the irreducible representations of $\mathscr{G}$ may be obtained in this way, by working through all the irreducible representations of $\mathscr{G}_o(\boldsymbol{k})$ for all the allowed $\boldsymbol{k}$-vectors that are in different stars.

This theorem does away with the need for an explicit display of the character table for $\mathscr{G}$, which would anyhow be very difficult because of the vast number of irreducible representations that $\mathscr{G}$ possesses. All the information required about $\mathscr{G}$ on such things as basis functions and degeneracies is immediately provided by the theorem in terms of the corresponding quantities for point groups, which are very easily obtained.

The theorem implies that an irreducible representation of $\mathscr{G}$ can be specified by two quantities, namely an allowed $\boldsymbol{k}$-vector and a label, p, specifying the irreducible representation of $\mathscr{G}_o(\boldsymbol{k})$, and so may be conveniently denoted by $\boldsymbol{\Gamma}^{\boldsymbol{k}p}$.

It is very useful to note that as the matrices $\boldsymbol{R}_i$ are orthogonal,

$\boldsymbol{k}\cdot\boldsymbol{R}_i^{-1}\boldsymbol{r}=\boldsymbol{R}_i\boldsymbol{k}\cdot\boldsymbol{r}$, so that, for example, if

$$\phi^p_{\boldsymbol{k}s}(\boldsymbol{r}) = \exp(\mathrm{i}\boldsymbol{k}\cdot\boldsymbol{r})\, u^p_{\boldsymbol{k}s}(\boldsymbol{r}),$$

then

$$P(\{\boldsymbol{R}_i \mid \boldsymbol{0}\})\, \phi^p_{\boldsymbol{k}s}(\boldsymbol{r}) = \exp(\mathrm{i}\boldsymbol{R}_i\boldsymbol{k}\cdot\boldsymbol{r})\, u^p_{\boldsymbol{k}s}(\boldsymbol{R}_i^{-1}\boldsymbol{r}),$$

which is a Bloch function with wave vector $\boldsymbol{R}_i\boldsymbol{k}$.

An explicit expression for the matrix elements of $\boldsymbol{\Gamma}^{\boldsymbol{k}p}$ can be obtained from the theorem. It is convenient to distinguish each row and column of $\boldsymbol{\Gamma}^{\boldsymbol{k}p}$ by a pair of indices, so that, as

$$P(\{\boldsymbol{R}_i \mid \boldsymbol{0}\})\, \phi^p_{\boldsymbol{k}s}(\boldsymbol{r})$$

form a basis for $\boldsymbol{\Gamma}^{\boldsymbol{k}p}$, then for any $\{\boldsymbol{R} \mid \boldsymbol{t}_{\boldsymbol{n}}\}$ of $\mathscr{G}$,

$$P(\{\boldsymbol{R} \mid \boldsymbol{t}_{\boldsymbol{n}}\})\,[P(\{\boldsymbol{R}_i \mid \boldsymbol{0}\})\, \phi^p_{\boldsymbol{k}s}(\boldsymbol{r})] = \sum_{j=1}^{M(\boldsymbol{k})}\sum_{t=1}^{l} \Gamma^{\boldsymbol{k}p}(\{\boldsymbol{R} \mid \boldsymbol{t}_{\boldsymbol{n}}\})_{j,t;i,s}\,[P(\{\boldsymbol{R}_j \mid \boldsymbol{0}\})\, \phi^p_{\boldsymbol{k}t}(\boldsymbol{r})],$$

and hence by the unitary property (3.8) and the ortho-normality property of basis functions,

$$\Gamma^{\boldsymbol{k}p}(\{\boldsymbol{R} \mid \boldsymbol{t}_{\boldsymbol{n}}\})_{j,t;i,s} = (\phi^p_{\boldsymbol{k}t}, P(\{\boldsymbol{R}_j \mid \boldsymbol{0}\}^{-1}\{\boldsymbol{R} \mid \boldsymbol{t}_{\boldsymbol{n}}\}\{\boldsymbol{R}_i \mid \boldsymbol{0}\})\, \phi^p_{\boldsymbol{k}s}).$$

Thus, by (1.6) and (4.10),

$$\Gamma^{\boldsymbol{k}p}(\{\boldsymbol{R} \mid \boldsymbol{t}_{\boldsymbol{n}}\})_{j,t;i,s} = (\phi^p_{\boldsymbol{k}t}, P(\{\boldsymbol{R}_j^{-1}\boldsymbol{R}\boldsymbol{R}_i \mid \boldsymbol{0}\})\, \phi^p_{\boldsymbol{k}s})\exp(-\mathrm{i}\boldsymbol{R}\boldsymbol{R}_i\boldsymbol{k}\cdot\boldsymbol{t}_{\boldsymbol{n}}).$$

If $\boldsymbol{R}_j^{-1}\boldsymbol{R}\boldsymbol{R}_i\boldsymbol{k}$ is not equivalent to $\boldsymbol{k}$, the inner product involves basis functions of $\mathscr{T}$ belonging to different irreducible representations, and so is zero. Thus in terms of the irreducible representation $\boldsymbol{\Gamma}^p$ of $\mathscr{G}_o(\boldsymbol{k})$,

$$\Gamma^{\boldsymbol{k}p}(\{\boldsymbol{R} \mid \boldsymbol{t}_{\boldsymbol{n}}\})_{j,t;i,s} = \begin{cases} \Gamma^p(\{\boldsymbol{R}_j^{-1}\boldsymbol{R}\boldsymbol{R}_i \mid \boldsymbol{0}\})_{ts}\exp(-\mathrm{i}\boldsymbol{R}_j\boldsymbol{k}\cdot\boldsymbol{t}_{\boldsymbol{n}}), & \text{if } \boldsymbol{R}_j^{-1}\boldsymbol{R}\boldsymbol{R}_i \text{ is a member of } \mathscr{G}_o(\boldsymbol{k}), \\ 0, & \text{otherwise}. \end{cases} \quad (5.3)$$

The rest of this section is devoted to a proof of the fundamental theorem. The original proof by SEITZ [1936] applies to the more general theorem for non-symmorphic space groups that is given in ch. 7 § 1.

First it has to be established that the matrices of (5.3) do provide a representation. This can be shown as follows. Let $\{\boldsymbol{R} \mid \boldsymbol{t_n}\}$ and $\{\boldsymbol{R}' \mid \boldsymbol{t_{n'}}\}$ be any two elements of the space group $\mathscr{G}$. Then by (5.3) the matrix elements $\Gamma^{kp}(\{\boldsymbol{R} \mid \boldsymbol{t_n}\})_{j,t;m,u}$ and $\boldsymbol{\Gamma}^{kp}(\{\boldsymbol{R}' \mid \boldsymbol{t_{n'}}\})_{m,u;i,s}$ are non-zero only if $\boldsymbol{R}_j^{-1}\boldsymbol{R}\boldsymbol{R}_m\boldsymbol{k}$ and $\boldsymbol{R}_m^{-1}\boldsymbol{R}'\boldsymbol{R}_i\boldsymbol{k}$ are equivalent to $\boldsymbol{k}$ respectively. Thus

$$\sum_m \sum_u \Gamma^{kp}(\{\boldsymbol{R} \mid \boldsymbol{t_n}\})_{j,t;m,u}\, \Gamma^{kp}(\{\boldsymbol{R}' \mid \boldsymbol{t_{n'}}\})_{m,u;i,s}$$

$$= \begin{cases} 0, & \text{if } \boldsymbol{R}_j^{-1}\boldsymbol{R}\boldsymbol{R}'\boldsymbol{R}_i \text{ is not a member of } \mathscr{G}_o(\boldsymbol{k}), \\ \displaystyle\sum_{u=1}^{l} \Gamma^p(\{\boldsymbol{R}_j^{-1}\boldsymbol{R}\boldsymbol{R}_m \mid \boldsymbol{0}\})_{tu}\, \Gamma^p(\{\boldsymbol{R}_m^{-1}\boldsymbol{R}'\boldsymbol{R}_i \mid \boldsymbol{0}\})_{us} & \\ \qquad \exp(-\,\mathrm{i}\boldsymbol{R}_j\boldsymbol{k}\cdot\boldsymbol{t_n} - \mathrm{i}\boldsymbol{R}^{-1}\boldsymbol{R}_j\boldsymbol{k}\cdot\boldsymbol{t_{n'}}), & \text{otherwise}, \end{cases}$$

$$= \begin{cases} 0, & \text{if } \boldsymbol{R}_j^{-1}\boldsymbol{R}\boldsymbol{R}'\boldsymbol{R}_i \text{ is not a member of } \mathscr{G}_o(\boldsymbol{k}), \\ \Gamma^p(\{\boldsymbol{R}_j^{-1}\boldsymbol{R}\boldsymbol{R}'\boldsymbol{R}_i \mid \boldsymbol{0}\})_{ts} \exp\{-\,\mathrm{i}\boldsymbol{R}_j\boldsymbol{k}\cdot(\boldsymbol{t_n} + \boldsymbol{R}\boldsymbol{t_{n'}})\}, & \text{otherwise}, \end{cases}$$

$$= \Gamma^{kp}(\{\boldsymbol{R} \mid \boldsymbol{t_n}\}\,\{\boldsymbol{R}' \mid \boldsymbol{t_{n'}}\})_{j,t;i,s}\,.$$

It has next to be demonstrated that this representation is irreducible. This can be done by using the second theorem of ch. 2 § 6, for by eq. (5.3),

$$\sum_{\boldsymbol{R}} \sum_{\boldsymbol{t}_n} |\chi^{kp}(\{\boldsymbol{R} \mid \boldsymbol{t_n}\})|^2 = \sum_{\boldsymbol{R}} \sum_{\boldsymbol{t}_n} \Big|\sum_{\boldsymbol{R}_i} \chi^p(\{\boldsymbol{R}_i^{-1}\boldsymbol{R}\boldsymbol{R}_i \mid \boldsymbol{0}\}) \exp(-\,\mathrm{i}\boldsymbol{R}_i\boldsymbol{k}\cdot\boldsymbol{t_n})\Big|^2,$$

where the summations on the right-hand side are only over those $\boldsymbol{R}$ and $\boldsymbol{R}_i$ for which $\boldsymbol{R}_i^{-1}\boldsymbol{R}\boldsymbol{R}_i$ is a member of $\mathscr{G}_o(\boldsymbol{k})$, while the summation on the left-hand side is over all the members of the space group $\mathscr{G}$. The right-hand side can be rewritten as

$$\sum_{\boldsymbol{R}} \sum_{\boldsymbol{t}_n} \sum_{\boldsymbol{R}_i} \sum_{\boldsymbol{R}_j} \chi^p(\{\boldsymbol{R}_i^{-1}\boldsymbol{R}\boldsymbol{R}_i \mid \boldsymbol{0}\})^*\, \chi^p(\{\boldsymbol{R}_j^{-1}\boldsymbol{R}\boldsymbol{R}_j \mid \boldsymbol{0}\}) \exp\{\mathrm{i}(\boldsymbol{R}_i - \boldsymbol{R}_j)\,\boldsymbol{k}\cdot\boldsymbol{t_n}\}$$
$$= N \sum_{\boldsymbol{R}} \sum_{\boldsymbol{R}_i} |\chi^p(\{\boldsymbol{R}_i^{-1}\boldsymbol{R}\boldsymbol{R}_i \mid \boldsymbol{0}\})|^2,$$

by eq. (4.15), as either $\boldsymbol{k}=\boldsymbol{0}$, in which case the only $\boldsymbol{R}_i$ is $\boldsymbol{R}_1=\boldsymbol{E}$, or $\boldsymbol{k}\neq\boldsymbol{0}$, in which case $\boldsymbol{R}_i=\boldsymbol{R}_j$. But this is equal to $NM(\boldsymbol{k})\,g_o(\boldsymbol{k})$, which by eq. (5.2) is equal to Ng_o, which is the order of $\mathscr{G}$. Thus the conditions of the second theorem of ch. 2 § 6 are satisfied, and hence the representation $\boldsymbol{\Gamma}^{kp}$ is irreducible.

It is easily demonstrated that $\boldsymbol{\Gamma}^{kp}$ is unitary, for this merely requires that the basis functions should be ortho-normal, as was shown in ch. 3 §4.1. Here the inner product of two basis functions is $(P(\{\boldsymbol{R}_i \mid \boldsymbol{0}\})\, \phi^p_{\boldsymbol{k}t}, P(\{\boldsymbol{R}_j \mid \boldsymbol{0}\})\, \phi^p_{\boldsymbol{k}s})$. This is only non-zero if $\boldsymbol{k}_j = \boldsymbol{R}_j \boldsymbol{k}$ is equivalent to $\boldsymbol{k}_i = \boldsymbol{R}_i \boldsymbol{k}$, as otherwise the inner product involves Bloch functions with different wave vectors. However, the initial assumption was that $\boldsymbol{k}_i$ is equivalent to $\boldsymbol{k}_j$ only if $i=j$, so the inner product is non-zero only when $i=j$. It is then only non-zero when $s=t$, when it is unity, as the functions $\phi^p_{\boldsymbol{k}s}(\boldsymbol{r})$ are assumed ortho-normal in order that $\boldsymbol{\Gamma}^p$ should be a unitary representation of $\mathscr{G}_o(\boldsymbol{k})$.

It is clear from (5.3) that the character systems for irreducible representations belonging to different pairs of $\boldsymbol{k}$ and p are different, so that the corresponding representations are not equivalent. Therefore it remains only to prove that the prescription of the theorem gives *all* the irreducible representations of $\mathscr{G}$. This can be done very simply by applying the fifth theorem of ch. 2 §6 as follows. Consider a particular wave vector $\boldsymbol{k}$. The sum of the squares of the dimensions of the irreducible representations of $\mathscr{G}$ corresponding to $\boldsymbol{k}$ is $M(\boldsymbol{k})^2\, g_o(\boldsymbol{k})$, which is equal to $M(\boldsymbol{k})\, g_o$, by eq. (5.2). As there are $M(\boldsymbol{k})$ vectors in the star of $\boldsymbol{k}$, summing over all *different* stars gives $\sum M(\boldsymbol{k}) = N$. Thus the sum of the squares of the dimensions of all the irreducible representations of $\mathscr{G}$ that are given by the theorem is $g_o N$, which is the order of $\mathscr{G}$. Therefore there are *no* other non-equivalent irreducible representations of $\mathscr{G}$.

§ 2. SOME IMMEDIATE CONSEQUENCES OF THE FUNDAMENTAL THEOREM ON THE IRREDUCIBLE REPRESENTATIONS OF A SYMMORPHIC SPACE GROUP

§ 2.1. *Degeneracies of energy levels and the symmetry of* $E(\boldsymbol{k})$

Suppose first that $\boldsymbol{k}$ is a general point of the Brillouin zone. Then $\mathscr{G}_o(\boldsymbol{k})$ consists only of the identity transformation $\{\boldsymbol{E} \mid \boldsymbol{0}\}$, and so has only one irreducible representation, namely the one-dimensional representation for which $\boldsymbol{\Gamma}(\{\boldsymbol{E}\,|\,\boldsymbol{0}\})=(1)$. As $g_o(\boldsymbol{k})=1$, it follows from eq. (5.2) that $M(\boldsymbol{k})=g_o$. Therefore there is only one irreducible representation of $\mathscr{G}$ corresponding to this $\boldsymbol{k}$, and its dimension is $M(\boldsymbol{k})\,l=g_o\cdot 1=g_o$, so that the corresponding energy eigenvalue is g_o-fold degenerate.

Any Bloch function $\exp(i\boldsymbol{k}\cdot\boldsymbol{r})\,u_{\boldsymbol{k}}(\boldsymbol{r})$ with wave vector $\boldsymbol{k}$ is a basis function for the irreducible representation of this $\mathscr{G}_o(\boldsymbol{k})$, and the set of basis functions of the corresponding irreducible representation of $\mathscr{G}$ formed from this function are $\exp(i\boldsymbol{k}\cdot\boldsymbol{r})\,u_{\boldsymbol{k}}(\boldsymbol{r})$, $\exp(i\boldsymbol{k}_2\cdot\boldsymbol{r})\,u_{\boldsymbol{k}}(\boldsymbol{R}_2^{-1}\boldsymbol{r}),\ldots$. Now, suppose these Bloch functions are energy eigenfunctions. As they correspond to wave vectors $\boldsymbol{k}_1(=\boldsymbol{k})$, $\boldsymbol{k}_2,\ldots$, they correspond, according to the definition (4.12), to energy eigenvalues $E(\boldsymbol{k})$, $E(\boldsymbol{k}_2),\ldots$. As they are degenerate, being a basis for an irreducible representation of $\mathscr{G}$, it follows that

$$E(\boldsymbol{R}_i\boldsymbol{k})\equiv E(\boldsymbol{k}_i)=E(\boldsymbol{k})$$

for $i=1,2,\ldots,g_o$, the $\boldsymbol{R}_i$ being here the set of rotations of $\mathscr{G}_o$. Thus the g_o wave vectors in the star of $\boldsymbol{k}$ have the same energies, and $E(\boldsymbol{k})$ has the symmetry of the point group $\mathscr{G}_o$ of the space group $\mathscr{G}$. This means that if the band structure is known in one basic section of the Brillouin zone containing only $(1/g_o)$ of the volume of the Brillouin zone and no two wave vectors in the same star, then it can be obtained immediately throughout the *whole* Brillouin zone. For example, for the body-centred cubic space group O_h^9, whose Brillouin zone is shown in fig. 4.4, $g_o=48$ and the basic section is the wedge-shaped region ΓHNP (or more precisely the region bounded by the planes containing three of the four points Γ, H, N and P).

The symmetry of $E(\boldsymbol{k})$ is widely exploited in calculations of energy band structures and in determinations of the Fermi surface from experimental measurements, such as those of the de Haas–van Alphen effect. However, for a general point of the Brillouin zone, the inclusion of the rotational parts of $\mathscr{G}$ in addition to the translational parts of $\mathscr{T}$ already taken into account in Bloch's theorem is of *no* assistance in simplifying the numerical task of actually finding the energy eigenvalues by the technique described in ch. 3 §6.2. For this reason, few accurate calculations of $E(\boldsymbol{k})$ have been performed for general points of the Brillouin zone.

The second case that will be considered is at the other extreme. For the point $\boldsymbol{k}=\boldsymbol{0}$, $\boldsymbol{Rk}=\boldsymbol{0}$ for every $\{\boldsymbol{R}\,|\,\boldsymbol{0}\}$ of $\mathcal{G}_o$, so that $\mathcal{G}_o(\boldsymbol{0})=\mathcal{G}_o$. The star of $\boldsymbol{k}$ then consists only of $\boldsymbol{k}=\boldsymbol{0}$ itself, and so $M(\boldsymbol{0})=1$. The basis functions of $\mathcal{G}$ corresponding to $\boldsymbol{k}=\boldsymbol{0}$ are merely the periodic basis functions of $\mathcal{G}_o$, and the corresponding degeneracies of energy eigenvalues are those of the dimensions of the irreducible representations of $\mathcal{G}_o$. In this case, the technique of ch. 3 §6.2 allows an appreciable simplification of the numerical work involved in finding the energy eigenvalues even beyond that already brought about by the consideration of $\mathcal{T}$ alone.

Some space groups possess other symmetry points for which $\mathcal{G}_o(\boldsymbol{k})=\mathcal{G}_o$. These do not require further examination, as the comments made about the point $\boldsymbol{k}=\boldsymbol{0}$ also apply to these points. The point H of the Brillouin zone of the body-centred cubic lattice is an example.

The third and final case is that of the intermediate situation in which $\mathcal{G}_o(\boldsymbol{k})$ is not trivial but is a proper subgroup of $\mathcal{G}_o$. Included in this case are all symmetry points other than those of the second case, and all points on symmetry axes and planes. For such a point $\mathcal{G}_o(\boldsymbol{k})$ will have more than one irreducible representation, some of these possibly being of more than one dimension. As $g_o(\boldsymbol{k})<g_o$ it follows from eq. (5.2) that $M(\boldsymbol{k})>1$.

Consider the energy eigenvalue corresponding to a l-dimensional irreducible representation of $G_o(\boldsymbol{k})$. This will be $lM(\boldsymbol{k})$-fold degenerate by the theorem, and this degeneracy is made up as follows. (a) By a similar argument to that used in the case of a general point, it follows that

$$E(\boldsymbol{k}_i)=E(\boldsymbol{k}), \qquad i=1,2,\ldots,M(\boldsymbol{k}),$$

so that again $E(\boldsymbol{k})$ exhibits the symmetry of the point group $\mathcal{G}_o$.

(b) In addition, each $E(\boldsymbol{k}_i)$ is 'l-fold degenerate', in the sense that there are l linearly independent Bloch eigenfunctions of $H(\boldsymbol{r})$ corresponding to this eigenvalue *and* to this particular wave vector $\boldsymbol{k}_i$. The degree of simplification of numerical work depends on the order $g_o(\boldsymbol{k})$.

These arguments show that, although the concept of a star appears in the fundamental theorem, it is in some contexts possible and convenient to revert to the *description* that appeared in ch. 4 with Bloch's theorem, in which there corresponds a set of energy levels to *every* allowed $\boldsymbol{k}$-vector of the Brillouin zone and not merely to those lying in different stars. In this description an l-fold degeneracy of $E(\boldsymbol{k})$ means, as above, that there are l linearly independent Bloch eigenfunctions of $H(\boldsymbol{r})$ corresponding to this eigenvalue and to this particular wave vector $\boldsymbol{k}$. An l-fold degeneracy of $E(\boldsymbol{k})$ then corresponds to an l-dimensional irreducible representation of $\mathscr{G}_o(\boldsymbol{k})$. This degeneracy was indicated in fig. 4.7 by $[l]$. Furthermore, each energy band then has the symmetry of $\mathscr{G}_o$. This is the description that is used in the energy band literature.

It is worthwhile pointing out here that even if $\mathscr{G}_o$ does not contain the inversion operator I, the symmetry $E(\boldsymbol{k})=E(-\boldsymbol{k})$ remains because of time-reversal symmetry, as will be shown in § 7.1.

§ 2.2. *A matrix element theorem for Bloch basis functions*

Theorem: Suppose that $\phi^p_{\boldsymbol{k}i}(\boldsymbol{r})$ and $\phi^q_{\boldsymbol{k}'j}(\boldsymbol{r})$ are two Bloch functions transforming as the ith row of the pth irreducible representation of $\mathscr{G}_o(\boldsymbol{k})$ and the jth row of the qth irreducible representation of $\mathscr{G}_o(\boldsymbol{k}')$ respectively. Then the matrix elements $(\phi^p_{\boldsymbol{k}i}(\boldsymbol{r}), \phi^q_{\boldsymbol{k}'j}(\boldsymbol{r}))$ and $(\phi^p_{\boldsymbol{k}i}, H\phi^q_{\boldsymbol{k}'j})$ are both zero unless $\boldsymbol{k}=\boldsymbol{k}'$, $p=q$ and $i=j$. Moreover, $(\phi^p_{\boldsymbol{k}i}, \phi^p_{\boldsymbol{k}i})$ and $(\phi^p_{\boldsymbol{k}i}, H\phi^p_{\boldsymbol{k}i})$ are independent of i.

The proof of this theorem follows almost immediately from the matrix element theorems of ch. 3 § 6.1 and the fundamental theorem of § 1. If $\boldsymbol{k}$ and $\boldsymbol{k}'$ are not in the same star, or if $\boldsymbol{k}$ and $\boldsymbol{k}'$ are in the same star but $p\neq q$, then $\phi^p_{\boldsymbol{k}i}(\boldsymbol{r})$ and $\phi^q_{\boldsymbol{k}'j}(\boldsymbol{r})$ transform according to different irreducible representations of the space group $\mathscr{G}$. If $\boldsymbol{k}$ and $\boldsymbol{k}'$ are in the same star and $p=q$ but $\boldsymbol{k}\neq\boldsymbol{k}'$, or if $\boldsymbol{k}=\boldsymbol{k}'$ and $p=q$ but $i\neq j$, then $\phi^{p}_{\boldsymbol{k}i}(\boldsymbol{r})$ and $\phi^{q}_{\boldsymbol{k}'j}(\boldsymbol{r})$ transform according to different rows of the same irreducible representation of $\mathscr{G}$. Thus only if $\boldsymbol{k}=\boldsymbol{k}'$, $p=q$ and $i=j$ do the functions transform according to the same row of the same irreducible representation of $\mathscr{G}$.

The theorem shows that if an energy eigenvalue corresponding to the irreducible representation $\boldsymbol{\Gamma}^{\boldsymbol{k}p}$ of the space group $\mathscr{G}$ is found by the method of ch. 3 § 6.2, the only functions that need be included in the expansion (3.27) of the eigenfunction are Bloch functions with wave vector $\boldsymbol{k}$ that transform according to a particular arbitrarily chosen row of the corresponding irreducible representation $\boldsymbol{\Gamma}^{p}$ of $\mathscr{G}_{o}(\boldsymbol{k})$.

§ 3. IRREDUCIBLE REPRESENTATIONS FOR THE BODY-CENTRED CUBIC SPACE GROUP O_h^9 AND THE FACE-CENTRED CUBIC SPACE GROUP O_h^5

The body-centred and face-centred cubic space groups O_h^9 and O_h^5 are typical and common examples of symmorphic space groups. The following description of them will serve to introduce the standard notations and conventions that are used for all symmorphic space groups. (A list is given in appendix 3 of papers containing character tables relating to other space groups).

As shown by the fundamental theorem of § 1, an irreducible representation of the space group $\mathscr{G}$ is specified by an allowed $\boldsymbol{k}$-vector and a label, p, of the irreducible representation of $\mathscr{G}_{o}(\boldsymbol{k})$ to which it corresponds. The convention is that the symmetry points and axes of the Brillouin zone are denoted by capital letters, as for example in figs. 4.4 and 4.5. The irreducible representations of the corresponding point groups $\mathscr{G}_{o}(\boldsymbol{k})$ are then labelled by assigning a subscript or set of subscripts to the appropriate capital letter. For example, the centre of the Brillouin zone for the space groups O_h^9 and O_h^5 is the point Γ, so that the 10 irreducible representations of the point group $\mathscr{G}_o(\boldsymbol{k})$ for Γ (which is actually O_h) are called $\Gamma_1, \Gamma_2, \Gamma_{12}, \Gamma_{15}, \Gamma_{25}, \Gamma_{1'}, \Gamma_{2'}, \Gamma_{12'}, \Gamma_{15'}$ and $\Gamma_{25'}$ in the most commonly used notation of BOUCKAERT *et al.* [1936]. (This assignment of subscripts is almost entirely arbitrary, and unfortunately conveys no direct information about the nature of the corresponding irreducible representation. More informative notations have been proposed by HOWARTH and JONES [1952] and by BELL [1954], but the notation of BOUCKAERT *et al.* [1936] is now so widely used that to employ any other notation now would simply cause confusion.)

The point groups $\mathscr{G}_o(\boldsymbol{k})$ for the symmetry points, axes and planes of the space groups O_h^9 and O_h^5 are given in tables 5.1 and 5.2 respectively. The notation for the point groups is that of SCHÖNFLIESS [1923]. This notation is used again in appendix 1, where there are listed the character tables for all the 32 crystallographic point groups, together with explicit sets of matrices for the irreducible representations of more than one dimension. The tables of appendix 1 also give, when appropriate, the notation of BOUCKAERT *et al.* [1936] for the irreducible representations of $\mathscr{G}_o(\boldsymbol{k})$. The group elements of a $\mathscr{G}_o(\boldsymbol{k})$ listed in appendix 1 are for the particular value of $\boldsymbol{k}$ listed in tables 5.1 or 5.2.

It is possible for two or more points in different stars to have point groups $\mathscr{G}_o(\boldsymbol{k})$ that are isomorphic. In such a situation the actual group elements of the $\mathscr{G}_o(\boldsymbol{k})$ may be different, the isomorphic groups merely differing in the orientation of their defining axes. For example, for the body-centred cubic space group O_h^9, the points on the axes Σ and D correspond to a $\mathscr{G}_o(\boldsymbol{k})$ that is C_{2v}. However, with the coordinates of Σ and D given in table 5.1, the group elements of $\mathscr{G}_o(\boldsymbol{k})$ for Σ (arranged in classes) are

$$\mathscr{C}_1 = E, \qquad \mathscr{C}_2 = C_{2e}, \qquad \mathscr{C}_3 = IC_{2x}, \qquad \mathscr{C}_4 = IC_{2f},$$

whereas the group elements of $\mathscr{G}_o(\boldsymbol{k})$ for D are

$$\mathscr{C}_1 = E, \qquad \mathscr{C}_2 = C_{2x}, \qquad \mathscr{C}_3 = IC_{2e}, \qquad \mathscr{C}_4 = IC_{2f}.$$

TABLE 5.1

The point groups $\mathscr{G}_o(\boldsymbol{k})$ for the symmetry points, axes and planes of the body-centred cubic space group O_h^9

Point	Coordinate	$\mathscr{G}_o(\boldsymbol{k})$
Γ	$(0, 0, 0)$	O_h
H	$(\pi/a)\,(0, 0, 2)$	O_h
N	$(\pi/a)\,(0, 1, 1)$	D_{2h}
P	$(\pi/a)\,(1, 1, 1)$	T_d
Axis	**Coordinates, $0<\kappa<1$**	**$\mathscr{G}_o(\boldsymbol{k})$**
Δ	$(\pi/a)\,(0, 0, 2\kappa)$	C_{4v}
Λ	$(\pi/a)\,(\kappa, \kappa, \kappa)$	C_{3v}
Σ	$(\pi/a)\,(0, \kappa, \kappa)$	C_{2v}
D	$(\pi/a)\,(\kappa, 1, 1)$	C_{2v}
F	$(\pi/a)\,(1-\kappa, 1-\kappa, 1+\kappa)$	C_{3v}
G	$(\pi/a)\,(0, 1-\kappa, 1+\kappa)$	C_{2v}
Plane	**Equation**	**$\mathscr{G}_o(\boldsymbol{k})$**
ΓHN	$k_x=0$	$C_s(IC_{2x})$
ΓNP	$k_y=k_z>k_x$	$C_s(IC_{2f})$
ΓHP	$k_x=k_y<k_z$	$C_s(IC_{2b})$
HNP	$k_y+k_z=2\pi/a$	$C_s(IC_{2e})$

TABLE 5.2

The point groups $\mathscr{G}_o(\boldsymbol{k})$ for the symmetry points, axes and planes of the face-centred cubic space group O_h^5

Point	Coordinate	$\mathscr{G}_o(\boldsymbol{k})$
Γ	$(0, 0, 0)$	O_h
K	$(\pi/a)\ (0, \frac{3}{2}, \frac{3}{2})$	C_{2v}
L	$(\pi/a)\ (1, 1, 1)$	D_{3d}
U	$(\pi/a)\ (\frac{1}{2}, \frac{1}{2}, 2)$	C_{2v}
W	$(\pi/a)\ (0, 1, 2)$	D_{2d}
X	$(\pi/a)\ (0, 0, 2)$	D_{4h}
Axis	**Coordinates, $0<\kappa<1$**	$\mathscr{G}_o(\boldsymbol{k})$
Δ	$(\pi/a)\ (0, 0, 2\kappa)$	C_{4v}
Λ	$(\pi/a)\ (\kappa, \kappa, \kappa)$	C_{3v}
Σ	$(\pi/a)\ (0, \frac{3}{2}\kappa, \frac{3}{2}\kappa)$	C_{2v}
Q	$(\pi/a)\ (1-\kappa, 1, 1+\kappa)$	C_2
S	$(\pi/a)\ (\frac{1}{2}\kappa, \frac{1}{2}\kappa, 2)$	C_{2v}
Z	$(\pi/a)\ (0, \kappa, 2)$	C_{2v}
Plane	**Equation**	$\mathscr{G}_o(\boldsymbol{k})$
ΓKWX	$k_x = 0$	$C_s(IC_{2x})$
ΓKL	$k_y = k_z > k_x$	$C_s(IC_{2f})$
ΓLUX	$k_x = k_y < k_z$	$C_s(IC_{2b})$
UWX	$k_z = 2\pi/a$	$C_s(IC_{2z})$

The irreducible representations of $\mathscr{G}_o(\boldsymbol{k})$ for Σ are denoted by Σ_1, Σ_2, Σ_3 and Σ_4, while those for D are denoted by D_1, D_2, D_3 and D_4. The different sets of group elements occurring in this way are all listed in the description of the corresponding point group in appendix 1.

For every point $\boldsymbol{k}$ on a symmetry plane, the group $\mathscr{G}_o(\boldsymbol{k})$ is C_s, which contains just the identity transformation and a reflection. The appropriate reflection for each plane is indicated in parentheses in the third column of tables 5.1 and 5.2. The irreducible representations of C_s for which the character of the reflection is $+1$ is described as being 'even', and is denoted by a $+$, and the other irreducible representation is described as being 'odd', and is denoted by a $-$. It should be noted that the plane $KLUW$ for the face-centred cubic space group O_h^5 is not a symmetry plane.

Fig. 4.7 is an example of an energy band diagram employing the notation described in this section.

§ 4. SELECTION RULES FOR SYMMORPHIC SPACE GROUPS

§ 4.1. *Formulation in terms of the point groups $\mathscr{G}_o(\boldsymbol{k})$*

The theory of selection rules given in ch. 3 §7.2 can be readily applied to the case when the group of the Schrödinger equation is a symmorphic space group $\mathscr{G}$. If the initial state eigenfunction $\psi_i(\boldsymbol{r})$, the final state eigenfunction $\psi_f(\boldsymbol{r})$ and the perturbation $V'(\boldsymbol{r})$ transform as the irreducible representations $\boldsymbol{\Gamma}^{\boldsymbol{k}p}$, $\boldsymbol{\Gamma}^{\boldsymbol{k}''p''}$ and $\boldsymbol{\Gamma}^{\boldsymbol{k}'p'}$ respectively, and

$$n_{\boldsymbol{k}p,\,\boldsymbol{k}'p'}^{\boldsymbol{k}''p''} = (1/g_o N) \sum \chi^{\boldsymbol{k}p}(\{\boldsymbol{R} \mid \boldsymbol{t}_n\})\, \chi^{\boldsymbol{k}'p'}(\{\boldsymbol{R} \mid \boldsymbol{t}_n\})\, \chi^{\boldsymbol{k}''p''}(\{\boldsymbol{R} \mid \boldsymbol{t}_n\})^*, \tag{5.4}$$

where the summation is over all the transformations $\{\boldsymbol{R} \mid \boldsymbol{t}_n\}$ of $\mathscr{G}$, which has order $g_o N$, then $(\psi_f, V'(\boldsymbol{r})\,\psi_i)=0$ if $n_{\boldsymbol{k}p,\,\boldsymbol{k}'p'}^{\boldsymbol{k}''p''}=0$. This is true whether $V'(\boldsymbol{r})$ is an operator or merely a function of $\boldsymbol{r}$. In general $V'(\boldsymbol{r})$ will contain parts transforming as several irreducible representations of $\mathscr{G}$, the theory given here being easily generalized to this situation.

Using eq. (5.3), the formula (5.4) can be expressed in terms of the characters of the irreducible representations of the point groups $\mathscr{G}_o(\boldsymbol{k})$, $\mathscr{G}_o(\boldsymbol{k}')$ and $\mathscr{G}_o(\boldsymbol{k}'')$, and can be much simplified. The first step is to note that (5.3) and (4.15) imply that

$$n_{\boldsymbol{k}p,\,\boldsymbol{k}'p'}^{\boldsymbol{k}''p''} = (1/g_o) \sum_{\boldsymbol{R}} \sum_{\boldsymbol{R}_i} \sum_{\boldsymbol{R}'_j} \sum_{\boldsymbol{R}''_l} \chi^p(\{\boldsymbol{R}_i^{-1}\boldsymbol{R}\boldsymbol{R}_i \mid \boldsymbol{0}\})\, \chi^{p'}(\{\boldsymbol{R}_j'^{-1}\boldsymbol{R}\boldsymbol{R}'_j \mid \boldsymbol{0}\})$$
$$\times \chi^{p''}(\{\boldsymbol{R}_l''^{-1}\boldsymbol{R}\boldsymbol{R}''_l \mid \boldsymbol{0}\})^* \, J_{\boldsymbol{k}}(\boldsymbol{R}_i^{-1}\boldsymbol{R}\boldsymbol{R}_i)\, J_{\boldsymbol{k}'}(\boldsymbol{R}_j'^{-1}\boldsymbol{R}\boldsymbol{R}'_j)\, J_{\boldsymbol{k}''}(\boldsymbol{R}_l''^{-1}\boldsymbol{R}\boldsymbol{R}''_l)$$
$$\times \delta(\boldsymbol{R}_i\boldsymbol{k} + \boldsymbol{R}'_j\boldsymbol{k}' - \boldsymbol{R}''_l\boldsymbol{k}'' - \boldsymbol{K}_{\boldsymbol{m}}), \tag{5.5}$$

where, by definition,

$$J_{\boldsymbol{k}}(\boldsymbol{R}_i^{-1}\boldsymbol{R}\boldsymbol{R}_i) = \begin{cases} 1, & \text{if } \boldsymbol{R}_i^{-1}\boldsymbol{R}\boldsymbol{R}_i\boldsymbol{k} \text{ is equivalent to } \boldsymbol{k}, \\ 0, & \text{otherwise}. \end{cases} \tag{5.6}$$

Here χ^p, $\chi^{p'}{}_j$ and $\chi^{p''}_l$ are characters of the irreducible representations of $\mathscr{G}_o(\boldsymbol{k})$, $\mathscr{G}_o(\boldsymbol{k}')$ and $\mathscr{G}_o(\boldsymbol{k}'')$ corresponding to $\boldsymbol{\Gamma}^{\boldsymbol{k}p}$, $\boldsymbol{\Gamma}^{\boldsymbol{k}'p'}$ and $\boldsymbol{\Gamma}^{\boldsymbol{k}''p''}$ respectively, and the rotations $\boldsymbol{R}_i$, $\boldsymbol{R}'_i$ and $\boldsymbol{R}''_j$ generate the stars of $\boldsymbol{k}$, $\boldsymbol{k}'$ and $\boldsymbol{k}''$ respectively. The Kronecker delta factor indicates that the only non-zero contributions to the process come from terms where $\boldsymbol{R}_i\boldsymbol{k}+\boldsymbol{R}'_j\boldsymbol{k}'-\boldsymbol{R}''_l\boldsymbol{k}''$ is equal to a lattice vector $\boldsymbol{K}_{\boldsymbol{m}}$ of the reciprocal lattice, which may be zero. If at least one non-zero contribution exists, the labelling of vectors in the stars may be chosen so that

$$\boldsymbol{k} + \boldsymbol{k}' - \boldsymbol{k}'' = \boldsymbol{K}_{\boldsymbol{m}}, \tag{5.7}$$

where $\boldsymbol{K}_{\boldsymbol{m}}$ may be zero. An example of such a relabelling will be given shortly. If $\mathscr{G}_o(\boldsymbol{k}, \boldsymbol{k}', \boldsymbol{k}'')$ is defined as the group of rotations that are common to $\mathscr{G}_o(\boldsymbol{k})$, $\mathscr{G}_o(\boldsymbol{k}')$ and $\mathscr{G}_o(\boldsymbol{k}'')$, then

$$n_{\boldsymbol{k}p,\,\boldsymbol{k}'p'}^{\boldsymbol{k}''p''} = (1/g') \sum_{\boldsymbol{R}} \chi^p(\{\boldsymbol{R} \mid \boldsymbol{0}\})\, \chi^{p'}(\{\boldsymbol{R} \mid \boldsymbol{0}\})\, \chi^{p''}(\{\boldsymbol{R} \mid \boldsymbol{0}\})^*, \tag{5.8}$$

where the sum is over the elements $\{\boldsymbol{R} \mid \boldsymbol{0}\}$ of $\mathscr{G}_o(\boldsymbol{k}, \boldsymbol{k}', \boldsymbol{k}'')$ and g' is the order of $\mathscr{G}_o(\boldsymbol{k}, \boldsymbol{k}', \boldsymbol{k}'')$.

The proof of (5.8) is as follows. Consider the sum in (5.5) corresponding to a particular $\boldsymbol{R}_i$. As the corresponding $\boldsymbol{R}'_j$ and $\boldsymbol{R}''_l$ giving non-zero contributions are such that $\boldsymbol{R}_i\boldsymbol{k}$, $\boldsymbol{R}'_j\boldsymbol{k}'$ and $\boldsymbol{R}''_l\boldsymbol{k}''$ are produced from $\boldsymbol{k}$, $\boldsymbol{k}'$ and $\boldsymbol{k}''$ respectively by the *same* rotation, then there exists rotations $\boldsymbol{S}$ and $\boldsymbol{S}''$ that are members of $\mathscr{G}_o(\boldsymbol{R}_i\boldsymbol{k})$ and $\mathscr{G}_o(\boldsymbol{R}''_l\boldsymbol{k}'')$ respectively such that $\boldsymbol{R}'_j=\boldsymbol{S}\boldsymbol{R}_i=\boldsymbol{S}''\boldsymbol{R}''_l$. The summation in (5.5) over $\boldsymbol{R}$ is over all the elements of $\mathscr{G}_o(\boldsymbol{R}_i\boldsymbol{k}, \boldsymbol{R}'_j\boldsymbol{k}', \boldsymbol{R}''_l\boldsymbol{k}'')$. Let $\boldsymbol{R}'$ be a typical element of $\mathscr{G}_o(\boldsymbol{R}_i\boldsymbol{k}, \boldsymbol{R}'_j\boldsymbol{k}', \boldsymbol{R}''_l\boldsymbol{k}'')$. Then $\boldsymbol{R}'^{-1}_j\boldsymbol{R}'\boldsymbol{R}'_j$ is a member of $\mathscr{G}_o(\boldsymbol{k}, \boldsymbol{k}', \boldsymbol{k}'')$, as this rotation transforms $\boldsymbol{k}$, $\boldsymbol{k}'$ and $\boldsymbol{k}''$ into equivalent wave vectors. Let $\boldsymbol{R}$ be this element, so that $\boldsymbol{R}'=\boldsymbol{R}_j\boldsymbol{R}\boldsymbol{R}_j^{-1}$. Then necessarily $\chi^{p'}(\boldsymbol{R}'^{-1}_j\boldsymbol{R}'\boldsymbol{R}'_j)=\chi^{p'}(\boldsymbol{R})$. Moreover

$$\boldsymbol{R}_i^{-1}\boldsymbol{R}'\boldsymbol{R}_i=(\boldsymbol{R}_i^{-1}\boldsymbol{S}\boldsymbol{R}_i)^{-1}\,\boldsymbol{R}(\boldsymbol{R}_i^{-1}\boldsymbol{S}\boldsymbol{R}_i),$$

and as $\boldsymbol{R}_i^{-1}\boldsymbol{S}\boldsymbol{R}_i$ is a member of $\mathscr{G}_o(\boldsymbol{k})$ then $\boldsymbol{R}_i^{-1}\boldsymbol{R}'\boldsymbol{R}_i$ is a member of $\mathscr{G}_o(\boldsymbol{k})$ and is in the same class as $\boldsymbol{R}$, so that $\chi^p(\boldsymbol{R}_i^{-1}\boldsymbol{R}'\boldsymbol{R}_i)=\chi^p(\boldsymbol{R})$. Similarly $\chi^{p''}(\boldsymbol{R}''^{-1}_l\boldsymbol{R}'\boldsymbol{R}''_l)=\chi^{p''}(\boldsymbol{R})$. Thus this sum over the elements of $\mathscr{G}_o(\boldsymbol{R}_i\boldsymbol{k}, \boldsymbol{R}'_j\boldsymbol{k}', \boldsymbol{R}''_l\boldsymbol{k}'')$ produces (g'/g_o) times the right-hand side of (5.8). However, there are (g_o/g') distinct sets of $\boldsymbol{R}_i$, $\boldsymbol{R}'_j$ and $\boldsymbol{R}''_l$ produced by common rotation that give non-zero contributions to (5.5), so that summing over these sets just produces (5.8).

As an example of the application of (5.8), consider the body-centred cubic space group O_h^9 and the case where $\boldsymbol{k}=(0, 0, k)$, $\boldsymbol{k}'=(0, k, 0)$ and $\boldsymbol{k}''=(0, k, k)$. Clearly here $\boldsymbol{k}+\boldsymbol{k}'=\boldsymbol{k}''$. The point $\boldsymbol{k}''$ is on a Σ axis, while $\boldsymbol{k}$ and $\boldsymbol{k}'$ are on two different Δ axes. In fact $\boldsymbol{k}'$ is in the star of $\boldsymbol{k}$, and $\boldsymbol{k}'=\boldsymbol{R}(C_{4x})\,\boldsymbol{k}$. This is an example of the relabelling that is referred to above. The groups $\mathscr{G}_o(\boldsymbol{k})$ and $\mathscr{G}_o(\boldsymbol{k}')$ are here isomorphic, and if $\boldsymbol{R}(T)$ is a rotation of $\mathscr{G}_o(\boldsymbol{k})$, then the corresponding rotation of $\mathscr{G}_o(\boldsymbol{k}')$ is $\boldsymbol{R}(C_{4x})\;\boldsymbol{R}(T)\;\boldsymbol{R}(C_{4x}^{-1})$. The character table of $\mathscr{G}_o(\boldsymbol{k}')$ is then easily constructed from that given for $\mathscr{G}_o(\boldsymbol{k})$ in appendix 1 in table A.15, for the classes of $\mathscr{G}_o(\boldsymbol{k}')$ are $\mathscr{C}_1=E$; $\mathscr{C}_2=C_{2y}$; $\mathscr{C}_3=C_{4y}, C_{4y}^{-1}$; $\mathscr{C}_4=IC_{2x}, IC_{2z}$; $\mathscr{C}_5=IC_{2d}, IC_{2c}$. (The above correspondence is a consequence of the following argument. If $\phi^p_{\boldsymbol{k}s}(\boldsymbol{r})$ are a set of basis functions of the pth irreducible representation of $\mathscr{G}_o(\boldsymbol{k})$, then $P(\{\boldsymbol{R}(C_{4x}) \mid \boldsymbol{0}\})\, \phi^p_{\boldsymbol{k}s}(\boldsymbol{r})$ have wave vector $\boldsymbol{k}' = \boldsymbol{R}(C_{4x})\, \boldsymbol{k}$, and therefore transform as a set of basis functions for an irreducible representation of $\mathscr{G}_o(\boldsymbol{k}')$, which may naturally be called the pth irreducible representation. Then if $\{\boldsymbol{R}' \mid \boldsymbol{0}\}$ is a member of $\mathscr{G}_o(\boldsymbol{k}')$,

$$\begin{aligned} P(\{\boldsymbol{R}' \mid \boldsymbol{0}\})\, [P(\{\boldsymbol{R}(C_{4x}) \mid \boldsymbol{0}\})\, \phi^p_{\boldsymbol{k}s}(\boldsymbol{r})] &\\ = P(\{\boldsymbol{R}(C_{4x}) \mid \boldsymbol{0}\})\, P(\{\boldsymbol{R}(C_{4x}^{-1})\, \boldsymbol{R}'\boldsymbol{R}(C_{4x}) \mid \boldsymbol{0}\})\, \phi^p_{\boldsymbol{k}s}(\boldsymbol{r}) &\\ = \sum_t \Gamma^p(\boldsymbol{R}(C_{4x}^{-1})\, \boldsymbol{R}'\boldsymbol{R}(C_{4x}))_{ts}\, [P(\{\boldsymbol{R}(C_{4x}) \mid \boldsymbol{0}\})\, \phi^p_{\boldsymbol{k}t}(\boldsymbol{r})]\,, & \end{aligned}$$

which demonstrates the correspondence.) In this particular example the only elements of $\mathscr{G}_o(\boldsymbol{k}, \boldsymbol{k}', \boldsymbol{k}'')$ are E and IC_{2x}. Then, for instance, if $V'(\boldsymbol{r})$ transforms as Δ_2 and $\psi_i(\boldsymbol{r})$ as $\Delta_{2'}$, the only allowed transitions are to a final state $\psi_f(\boldsymbol{r})$ transforming as Σ_2 or Σ_3. In the literature this is sometimes written compactly but misleadingly as

$$\Delta_2 \otimes \Delta_{2'} = \Sigma_2 \oplus \Sigma_3\,, \tag{5.9}$$

but this equation should not be interpreted literally, for the irreducible representations involved belong to different groups. Indeed, the dimension $l_p l_{p'}$ of the 'direct-product representation' on the left-hand side of (5.9) is in general not equal to the sum of the dimensions $\sum l_{p''}$ of the representations on the right-hand side. The reason for this is, of course, that this result is true for the corresponding irreducible representations of the space group $\mathscr{G}$, which have dimensions $M(\boldsymbol{k})\, l_p$, $M(\boldsymbol{k}')\, l_{p'}$ and $M(\boldsymbol{k}'')\, l_{p''}$ respectively, $M(\boldsymbol{k})$, $M(\boldsymbol{k}')$ and $M(\boldsymbol{k}'')$ being the number of wave vectors in the stars of $\boldsymbol{k}$, $\boldsymbol{k}'$ and $\boldsymbol{k}''$ respectively, so that $M(\boldsymbol{k})\, M(\boldsymbol{k}')\, l_p l_{p'} = M(\boldsymbol{k}'') \sum l_{p''}$, and in general $M(\boldsymbol{k})\, M(\boldsymbol{k}') \neq M(\boldsymbol{k}'')$. In fact, in the above example $M(\boldsymbol{k}) = 6$, $M(\boldsymbol{k}') = 6$ and $M(\boldsymbol{k}'') = 12$.

Selection rules for the zinc blende space group T_d^2 have been tabulated by BIRMAN [1962, 1963] and MITRA [1964]. Alternative formulations of the method of calculating selection rules are mentioned at the end of ch. 7 §5.1.

§ 4.2. *Direct optical transitions*

The direct optical transitions between electronic states provide an important example of the theory developed in §4.1. As shown for example by STERN [1963], the relevant matrix elements involve the operator $\boldsymbol{a}_o \cdot (\boldsymbol{p} + \frac{1}{2}\hbar \boldsymbol{k}) \exp(\mathrm{i}\boldsymbol{k} \cdot \boldsymbol{r})$, where $p_x = (\hbar/\mathrm{i})\, \partial/\partial x$, $p_y = (\hbar/\mathrm{i})\, \partial/\partial y$, and $p_z = (\hbar/\mathrm{i})\, \partial/\partial z$, $\boldsymbol{a}_o$ is a unit vector specifying the polarisation and $\hbar\boldsymbol{k}$ the momentum of the photon involved. This latter momentum for visible and infrared light is so small compared with all other momenta involved that $\boldsymbol{k}$ may be taken to be zero leaving the perturbing operator as $V'(\boldsymbol{r}) = \boldsymbol{a}_o \cdot \boldsymbol{p}$.

To apply the theory of §4.1, it has first to be established how $\boldsymbol{a}_o \cdot \boldsymbol{p}$ transforms. Now

$$P(T)\,(\partial/\partial x)\,P(T^{-1}) = R(T)_{11}\,\partial/\partial x + R(T)_{21}\,\partial/\partial y + R(T)_{31}\,\partial/\partial z\,,$$

with similar expressions for $\partial/\partial y$ and $\partial/\partial z$, where $R(T)_{ij}$ are the matrix elements of the rotational part of T. [These relations may be verified by operating with both sides on any function $f(\boldsymbol{r})$.] Thus $\partial/\partial x$, $\partial/\partial y$ and $\partial/\partial z$ transform as the rows of the representation $\boldsymbol{\Gamma}^{\boldsymbol{k}'p'}$ of the space group $\mathcal{G}$ for which $\Gamma^{\boldsymbol{k}'p'}(T)_{ij} = R(T)_{ij}$. As $R(T)_{ij} = \delta_{ij}$ for a pure translation T, this representation corresponds to $\boldsymbol{k}' = \boldsymbol{0}$. Thus $\partial/\partial x$, $\partial/\partial y$ and $\partial/\partial z$ transform as rows of some representation of $\mathcal{G}_o(\boldsymbol{0})$, which may be reducible.

In the case of the body-centred and face-centred cubic space groups O_h^9 and O_h^5, $\mathcal{G}_o(\boldsymbol{0})$ is $\boldsymbol{O}_h$, and, as noted in appendix 1, the matrices $\boldsymbol{R}$ belong to the three-dimensional irreducible representation Γ_{15}. $V'(\boldsymbol{r})$ therefore transforms as Γ_{15} for any direction of the polarisation vector $\boldsymbol{a}_o$.

The condition (5.7) here implies that $\boldsymbol{k}'' = \boldsymbol{k}$, so that, in an energy band diagram of $E(\boldsymbol{k})$ plotted against $\boldsymbol{k}$, the transitions are 'vertical'. As $\mathcal{G}_o(\boldsymbol{0})$ is equal to $\mathcal{G}_o$ and contains $\mathcal{G}_o(\boldsymbol{k}) = \mathcal{G}_o(\boldsymbol{k}'')$ as a subgroup, the group $\mathcal{G}_o(\boldsymbol{k}, \boldsymbol{k}', \boldsymbol{k}'')$ of eq. (5.8) is in this case merely $\mathcal{G}_o(\boldsymbol{k})$.

§ 5. PROPERTIES OF ENERGY BANDS

§ 5.1. *Continuity and compatibility of the irreducible representations of $\mathscr{G}_o(\boldsymbol{k})$*

A typical section of an energy band diagram for a symmetry axis is shown in fig. 5.1. The axis displayed there is the Δ axis of the body-centred or face-centred cubic space groups. The numbers in parentheses are the band labels, as defined in (4.17), so that the bands 1 and 2 'touch' at one point $\boldsymbol{k}_o$. This figure exhibits two general characteristics of energy bands.

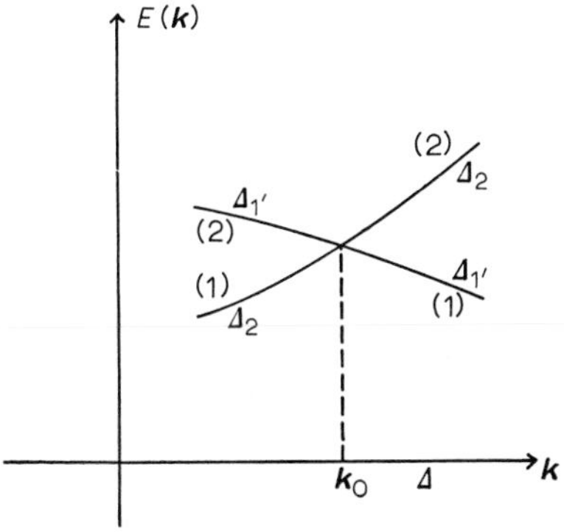

Fig. 5.1

The first characteristic is that, along any part of a band which does not touch another band, the corresponding irreducible representation of $\mathscr{G}_o(\boldsymbol{k})$ remains the same. Thus, for example, the whole of the left-hand part of band 1 corresponds to Δ_2, and the whole of the right-hand part corresponds to $\Delta_{1'}$. It is therefore possible to talk of *the* symmetry of a band, or part of a band, along an axis. The reason for this behaviour is essentially that the energy eigenvalues corresponding to a particular irreducible representation of $\mathscr{G}_o(\boldsymbol{k})$ can be obtained from a secular equation involving only basis functions of that representation, as was noted in ch. 3 §6.2. A small change in $\boldsymbol{k}$ will then only produce a small change in the energy eigenvalues produced from this secular equation.

The second general characteristic is that the symmetries of the bands are *interchanged* at a point where the bands touch. For example, in fig. 5.1 band 1 changes from Δ_2 to $\Delta_{1'}$ on moving from left to right, while band 2 changes from $\Delta_{1'}$ to Δ_2. The reason for this is that the secular equation corresponding to an irreducible representation of $\mathscr{G}_o(\boldsymbol{k})$ produces energy eigenvalues that are analytic functions of $\boldsymbol{k}$, the degeneracy corresponding to the touching of two bands having no effect on this as it is 'accidental'. This also implies that $\text{grad}_{\boldsymbol{k}} E(\boldsymbol{k})$ is continuous for bands, or parts of bands, belonging to the *same* irreducible representation, and this continuity is not affected when band labels are interchanged when bands touch. Thus, for example, in fig. 5.1 the limit as $\boldsymbol{k} \rightarrow \boldsymbol{k}_o$ from the left of $\text{grad}_{\boldsymbol{k}} E(\boldsymbol{k})$ for band 1 is equal to the limit as $\boldsymbol{k} \rightarrow \boldsymbol{k}_o$ from the right of $\text{grad}_{\boldsymbol{k}} E(\boldsymbol{k})$ for band 2.

The touching of energy bands has been investigated in detail by HERRING [1937b]. Amongst other things, Herring showed that, for the type of potential $V(\boldsymbol{r})$ corresponding to a real crystal, two bands corresponding to the *same* irreducible representation of $\mathscr{G}_o(\boldsymbol{k})$ only touch in the following circumstances.

(a) For a crystal with an inversion centre, two such bands can touch only at points lying on an endless curve in $\boldsymbol{k}$-space, or on a number of such curves. These curves do not lie in planes of symmetry, although they can cross symmetry axes at points where bands corresponding to *different* irreducible representations of $\mathscr{G}_o(\boldsymbol{k})$ touch. (The crystals having O_h^9 or O_h^5 as their space group possess of course an inversion centre.)

(b) For a crystal with no inversion centre, two such bands may touch at isolated points which may lie in a symmetry plane or in a plane perpendicular to a two-fold axis.

There is nothing to prevent two bands corresponding to *different* irreducible representations of $G_o(\boldsymbol{k})$ from touching, and many such degeneracies do occur. The conclusions (a) and (b) do not apply to exceptional potentials, such as that corresponding to the model in which the electrons are free, for which $V(\boldsymbol{r})=0$ for all $\boldsymbol{r}$. This model will be examined in more detail in §6.3, and fig. 5.11 in particular shows that many such degeneracies do occur in this case. However, the insertion of even a small lattice potential is sufficient to break most of the degeneracies in accordance with the above conclusions for general potentials. This is illustrated in figs. 5.2a and 5.2b. The curves of fig. 5.2a correspond to $V(\boldsymbol{r})\equiv 0$,

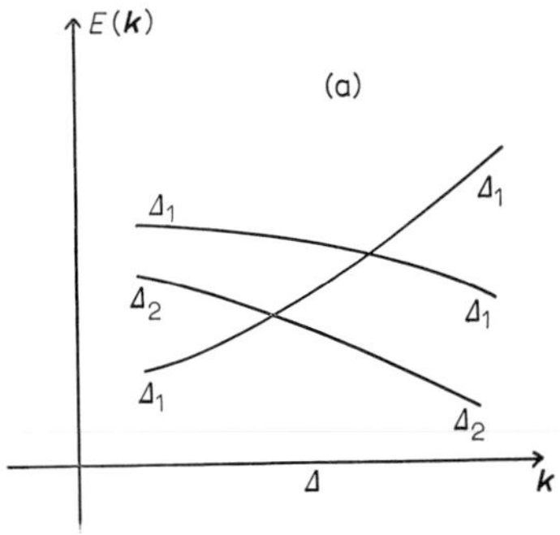

Fig. 5.2a

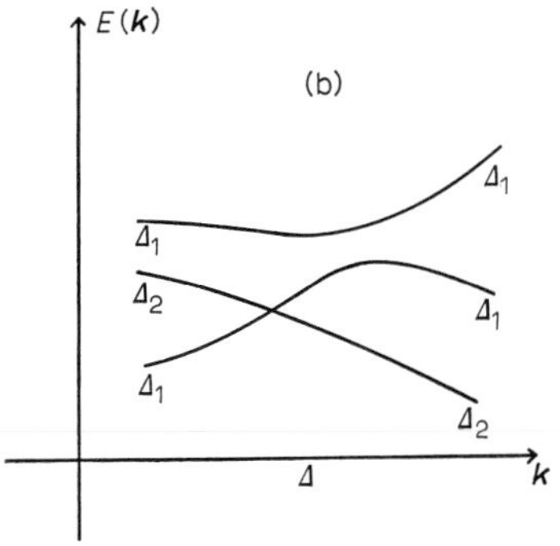

Fig. 5.2b

while the curves of fig. 5.2b correspond to a very small lattice potential. This behaviour is quite useful in sketching the energy bands of the so-called 'nearly free electron' metals.

The concept of *compatibility* is best described by examining an example. Consider, therefore, a Γ_{12} energy level at the point Γ of the Brillouin zone for the body-centred cubic space group O_h^9. This level, being two-fold degenerate, belongs to two energy bands. What then are the symmetries of these bands near Γ along the symmetry axis Δ? That is, what irreducible representations of the group $\mathscr{G}_o(\boldsymbol{k})$ for Δ are 'compatible' with Γ_{12}?

The investigation proceeds as follows. Let $\mathscr{G}_o(\Gamma)$ and $\mathscr{G}_o(\Delta)$

be the point groups $\mathscr{G}_o(\boldsymbol{k})$ for $\boldsymbol{k}$ at Γ and on Δ respectively. Then $\mathscr{G}_o(\Delta)$ is a subgroup of $\mathscr{G}_o(\Gamma)$, and so the irreducible representations of $\mathscr{G}_o(\Gamma)$ are representations of $\mathscr{G}_o(\Delta)$ that are in general reducible. The actual reduction can be determined immediately from the characters, for if $\boldsymbol{\Gamma}$ and $\boldsymbol{\Gamma}^p$ are irreducible representations of $\mathscr{G}_o(\Gamma)$ and $\mathscr{G}_o(\Delta)$ respectively, with characters denoted by χ and χ^p, then the number of times n_p that $\boldsymbol{\Gamma}^p$ appears in the reduction of $\boldsymbol{\Gamma}$ is given by

$$n_p = (1/g_o(\Delta)) \sum_{\boldsymbol{R}} \chi(\{\boldsymbol{R} \mid \boldsymbol{0}\})\, \chi^p(\{\boldsymbol{R} \mid \boldsymbol{0}\})^*,$$

where $g_o(\Delta)$ is the order of $\mathscr{G}_o(\Delta)$, and the summation is over all the rotations $\{\boldsymbol{R} \mid \boldsymbol{0}\}$ of $\mathscr{G}_o(\Delta)$. This is just an immediate application of the third theorem of ch. 2 §6. Thus, for example, the reduction of Γ_{12} in this context is given by

$$\Gamma_{12} = \Delta_1 \oplus \Delta_2 .$$

The point Γ could be considered as an ordinary point of Δ by ignoring the elements of $\mathscr{G}_o(\Gamma)$ that are not in $\mathscr{G}_o(\Delta)$, and then the energy levels at Γ could be classified in terms of the irreducible representations of $\mathscr{G}_o(\Delta)$. The Γ_{12} level would then be regarded as a non-degenerate Δ_1 level and a non-degenerate Δ_2 level that happen to have the same value. Because of the continuity of irreducible representations along a part of an axis that was men-

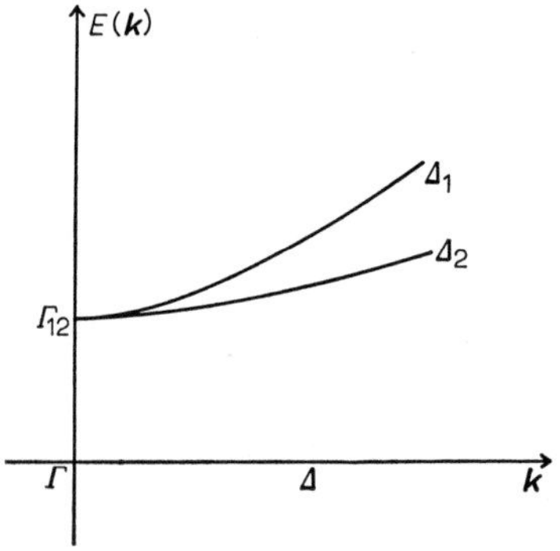

Fig. 5.3

tioned above, the two bands that touch at Γ in the Γ_{12} level will have symmetries Δ_1 and Δ_2 along the Δ axis near Γ. The degeneracy that exists at Γ is split on moving away from Γ. This situation is shown in fig. 5.3.

TABLE 5.3

Compatibility relations between symmetry points and symmetry axes for the body-centred cubic space group O_h^9

Γ_1	Γ_2	Γ_{12}	Γ_{15}	Γ_{25}	$\Gamma_{1'}$	$\Gamma_{2'}$	$\Gamma_{12'}$	$\Gamma_{15'}$	$\Gamma_{25'}$
Δ_1	Δ_2	$\Delta_1\Delta_2$	$\Delta_1\Delta_5$	$\Delta_2\Delta_5$	$\Delta_{1'}$	$\Delta_{2'}$	$\Delta_{1'}\Delta_{2'}$	$\Delta_{1'}\Delta_5$	$\Delta_{2'}\Delta_5$
Σ_1	Σ_4	$\Sigma_1\Sigma_4$	$\Sigma_1\Sigma_3\Sigma_4$	$\Sigma_1\Sigma_2\Sigma_4$	Σ_2	Σ_3	$\Sigma_2\Sigma_3$	$\Sigma_2\Sigma_3\Sigma_4$	$\Sigma_1\Sigma_2\Sigma_3$
Λ_1	Λ_2	Λ_3	$\Lambda_1\Lambda_3$	$\Lambda_2\Lambda_3$	Λ_2	Λ_1	Λ_3	$\Lambda_2\Lambda_3$	$\Lambda_1\Lambda_3$

H_1	H_2	H_{12}	H_{15}	H_{25}	$H_{1'}$	$H_{2'}$	$H_{12'}$	$H_{15'}$	$H_{25'}$
Δ_1	Δ_2	$\Delta_1\Delta_2$	$\Delta_1\Delta_5$	$\Delta_2\Delta_5$	$\Delta_{1'}$	$\Delta_{2'}$	$\Delta_{1'}\Delta_{2'}$	$\Delta_{1'}\Delta_5$	$\Delta_{2'}\Delta_5$
F_1	F_2	F_3	F_1F_3	F_2F_3	F_2	F_1	F_3	F_2F_3	F_1F_3
G_1	G_4	G_1G_4	$G_1G_3G_4$	$G_1G_2G_4$	G_2	G_3	G_2G_3	$G_2G_3G_4$	$G_1G_2G_3$

P_1	P_2	P_3	P_4	P_5
Λ_1	Λ_2	Λ_3	$\Lambda_1\Lambda_3$	$\Lambda_2\Lambda_3$
F_1	F_2	F_3	F_1F_3	F_2F_3
D_1	D_2	D_1D_2	$D_1D_3D_4$	$D_2D_3D_4$

N_1	N_2	N_3	N_4	$N_{1'}$	$N_{2'}$	$N_{3'}$	$N_{4'}$
Σ_1	Σ_2	Σ_3	Σ_4	Σ_1	Σ_2	Σ_3	Σ_4
D_1	D_4	D_3	D_2	D_3	D_2	D_1	D_4
G_1	G_3	G_2	G_4	G_4	G_2	G_3	G_1

Although this argument shows that the Δ_1 and Δ_2 irreducible representations are compatible with the Γ_{12} level, it cannot predict whether the band having Δ_1 symmetry lies lower or higher than the band having Δ_2 symmetry. This can only be determined by direct calculation.

TABLE 5.4

Compatibility relations between symmetry axes and symmetry planes for the body-centred cubic space group O_h^9

	Δ_1	Δ_2	$\Delta_{1'}$	$\Delta_{2'}$	Δ_5
ΓHN	+	+	−	−	+−
ΓHP	+	−	−	+	+−

	Λ_1	Λ_2	Λ_3
ΓHP	+	−	+−
ΓNP	+	−	+−

	Σ_1	Σ_2	Σ_3	Σ_4
ΓHN	+	−	−	+
ΓNP	+	−	+	−

	F_1	F_2	F_3
ΓHP	+	−	+−
HNP	+	−	+−

	D_1	D_2	D_3	D_4
ΓNP	+	−	+	−
HNP	+	−	−	+

	G_1	G_2	G_3	G_4
ΓHN	+	−	−	+
HNP	+	−	+	−

The same analysis can clearly be applied to all irreducible representations at every symmetry point for all the symmetry axes going through that point. A similar analysis can also be used to

TABLE 5.5

Compatibility relations between symmetry points and symmetry axes for the face-centred cubic space group O_h^5

K_1	K_2	K_3	K_4
Σ_1	Σ_2	Σ_3	Σ_4

L_1	L_2	L_3	$L_{1'}$	$L_{2'}$	$L_{3'}$
Λ_1	Λ_2	Λ_3	Λ_2	Λ_1	Λ_3
Q_1	Q_2	Q_1Q_2	Q_1	Q_2	Q_1Q_2

U_1	U_2	U_3	U_4
S_1	S_2	S_3	S_4

W_1	W_2	$W_{1'}$	$W_{2'}$	W_3
Q_1	Q_2	Q_1	Q_2	Q_1Q_2
Z_1	Z_2	Z_2	Z_1	Z_3Z_4

X_1	X_2	X_3	X_4	X_5	$X_{1'}$	$X_{2'}$	$X_{3'}$	$X_{4'}$	$X_{5'}$
S_1	S_4	S_1	S_4	S_2S_3	S_2	S_3	S_2	S_3	S_1S_4
Z_1	Z_1	Z_4	Z_4	Z_2Z_3	Z_2	Z_2	Z_3	Z_3	Z_1Z_4
Δ_1	Δ_2	$\Delta_{2'}$	$\Delta_{1'}$	Δ_5	$\Delta_{1'}$	$\Delta_{2'}$	Δ_2	Δ_1	Δ_5

The compatibility relations between the point Γ and the axes Δ, Σ, and Λ are as given in table 5.3.

determine the compatibility of the irreducible representations corresponding to a symmetry axis with those of the symmetry planes containing the axis. These results can be expressed in 'compatibility tables'. The compatibility tables for the body-centred cubic space group O_h^9 are given in tables 5.3 and 5.4 and those for the face-centred cubic space group O_h^5 are given in tables 5.5 and 5.6.

§ 5.2. *Fine detail of the symmetry of energy bands*

In §2.1 it was shown that the energy bands are invariant under all the rotations of $\mathscr{G}_o$. It is now possible to investigate certain finer

TABLE 5.6

Compatibility relations between symmetry axes and symmetry planes for the face-centred cubic space group O_h^5

	Δ_1	Δ_2	$\Delta_{1'}$	$\Delta_{2'}$	Δ_5
ΓKWX	+	+	−	−	+ −
ΓLUX	+	−	−	+	+ −

	Λ_1	Λ_2	Λ_3
ΓLUX	+	−	+ −
ΓKL	+	−	+ −

	Σ_1	Σ_2	Σ_3	Σ_4
ΓKWX	+	−	−	+
ΓKL	+	−	+	−

	S_1	S_2	S_3	S_4
ΓLUX	+	−	+	−
UWX	+	−	−	+

	Z_1	Z_2	Z_3	Z_4
ΓKWX	+	−	+	−
UWX	+	−	−	+

features of the symmetry of energy bands, particularly the vanishing of components of $\mathrm{grad}_{\boldsymbol{k}} E(\boldsymbol{k})$ and the form of the intersection of constant energy contours with symmetry axes.

It is necessary to begin by studying the reflection of $E(\boldsymbol{k})$ in symmetry planes. In the case of a symmetry plane contained within the Brillouin zone, the discussion of §2.1 shows immediately that the energy band structure is symmetrical with respect to reflection in the symmetry plane. That is, to each energy level corresponding to a particular $\boldsymbol{k}$ and a particular irreducible representation of $\mathscr{G}_0(\boldsymbol{k})$, there is another energy level having the same value and corresponding to the same irreducible representation of $\mathscr{G}_0(\boldsymbol{k})$, whose wave vector is $\boldsymbol{k}$ reflected in the symmetry plane. The same result is true of symmetry planes on the surface of a Brillouin zone in the extended zone scheme, but the reasoning is slightly more complicated. Fig. 5.4 illustrates the argument. The point $\boldsymbol{k}$

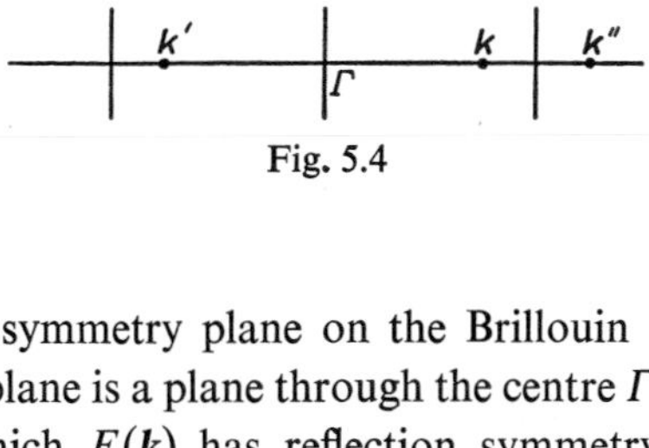

Fig. 5.4

is just inside a symmetry plane on the Brillouin zone boundary. Parallel to this plane is a plane through the centre Γ of the Brillouin zone, about which $E(\boldsymbol{k})$ has reflection symmetry. The point $\boldsymbol{k}'$ is the reflection of point $\boldsymbol{k}$ in this latter plane. The point $\boldsymbol{k}''$ is then obtained by a translation through a lattice vector of the reciprocal lattice in the direction of the line from $\boldsymbol{k}'$ to $\boldsymbol{k}$, so that it is the reflection of the point $\boldsymbol{k}$ in the symmetry plane on the Brillouin zone surface. As noted just above, there is an energy level at $\boldsymbol{k}'$ having the same value and corresponding to the same irreducible representation as that at $\boldsymbol{k}$. However, the points $\boldsymbol{k}'$ and $\boldsymbol{k}''$ are equivalent, so in the extended zone scheme there is an energy level at $\boldsymbol{k}''$ having the same value as those at $\boldsymbol{k}$ and $\boldsymbol{k}'$, and corresponding to the same irreducible representation.

It is now possible to investigate the component of $\mathrm{grad}_{\boldsymbol{k}} E(\boldsymbol{k})$ in the direction normal to a symmetry plane at a point on the symmetry plane. There are essentially four possibilities, as illustrated in figs. 5.5a, b, c and d. If the band n under consideration does not

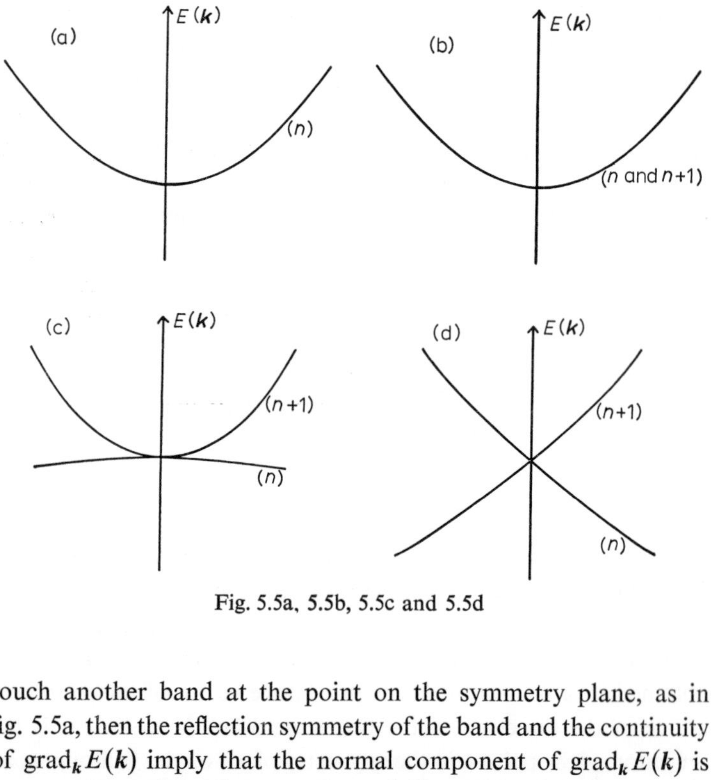

Fig. 5.5a, 5.5b, 5.5c and 5.5d

touch another band at the point on the symmetry plane, as in fig. 5.5a, then the reflection symmetry of the band and the continuity of $\mathrm{grad}_{\boldsymbol{k}} E(\boldsymbol{k})$ imply that the normal component of $\mathrm{grad}_{\boldsymbol{k}} E(\boldsymbol{k})$ is zero at that point. The second possibility, which is shown in fig. 5.5b, is that two bands n and $n+1$ correspond to a two-fold degenerate irreducible representation of $\mathscr{G}_o(\boldsymbol{k})$ for all $\boldsymbol{k}$ along the normal through the point under consideration, but do not touch any other band at that point. The argument and conclusion are then the same as in the first case. The third possibility is that two bands n and $n+1$ touch *only* at the point on the symmetry plane, but correspond to *different* irreducible representations of $\mathscr{G}_o(\boldsymbol{k})$ for $\boldsymbol{k}$ along the normal. In this case, the normal components of $\text{grad}_{\boldsymbol{k}} E(\boldsymbol{k})$ vanish for both bands, as $\text{grad}_{\boldsymbol{k}} E(\boldsymbol{k})$ is continuous for each irreducible representation and hence for each band. This case is illustrated in fig. 5.5c. The last possibility, which is shown in fig. 5.5d, is that the two bands n and $n+1$ touch only at the point on the symmetry plane and correspond to the *same* irreducible representation of $\mathscr{G}_o(\boldsymbol{k})$ for $\boldsymbol{k}$ along the normal. The vanishing of the normal components of $\text{grad}_{\boldsymbol{k}} E(\boldsymbol{k})$ is *not* implied in this case, because $\text{grad}_{\boldsymbol{k}} E(\boldsymbol{k})$ is not continuous in each band.

It should be noted that, whereas *every* plane of the boundary of the Brillouin zone for the body-centred cubic space group O_h^9 is a symmetry plane, the same is not true for the face-centred cubic space group O_h^5. The plane $LKUW$ of O_h^5 is not a symmetry plane, so the above results do not apply to it.

These results can have interesting implications for the form of the $E(\boldsymbol{k})$ curves at the symmetry points. For example, for the body-centred cubic space group O_h^9, each of the symmetry points Γ, H, N and P lie at the intersection of *three* symmetry planes. If an energy level at one of these symmetry points is non-degenerate, or is degenerate and compatible along each symmetry axis through the point with bands that belong to a set of irreducible representations in which none appears more than once, then all three components of $\text{grad}_{\boldsymbol{k}} E(\boldsymbol{k})$ are zero at the symmetry point. Table 5.3 shows that these conditions are always satisfied for the space group O_h^9, provided only that there is not an accidental degeneracy at the symmetry point itself. Hence, for the body-centred cubic space group O_h^9, all three components of $\text{grad}_{\boldsymbol{k}} E(\boldsymbol{k})$ are zero at each of the symmetry points, provided that there is not an accidental degeneracy at the symmetry point.

For the face-centred cubic space group O_h^5, the same result is true for the symmetry points lying in three symmetry planes, namely for Γ and X. The other symmetry points, K, L, U and W

each lie in only two symmetry planes, so that some of the components of $\text{grad}_k E(\boldsymbol{k})$ vanish at these points, but not all components.

Another interesting question concerns the intersection of contours of constant energy drawn in a symmetry plane with the symmetry axes that lie in the plane. For example, consider the plane ΓHN of the body-centred cubic space group O_h^9, which contains the symmetry axes Δ, G and Σ. Every line in this plane that is normal to Δ is also normal to another symmetry plane, namely the plane $k_y=0$, so that the theory described above can be applied to each such line. In particular, it follows that every contour is symmetrical with respect to Δ. There are, therefore, only two possible types of intersection of a contour with Δ; either it intersects normally, as shown in fig. 5.6a, or it intersects in a series of cusps, as shown in fig. 5.6b. However, it will be shown that the

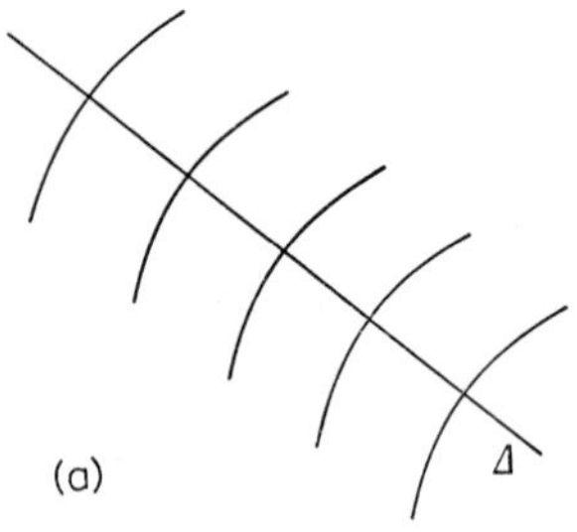

Fig. 5.6a

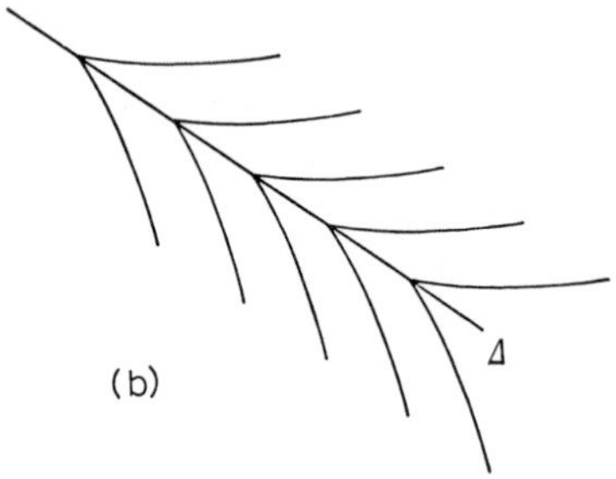

Fig. 5.6b

latter possibility only occurs if there is an *accidental* degeneracy between two bands along a *whole section* of the Δ axis, and not merely at isolated points. In real solids the constant energy contours will therefore intersect Δ normally, but this may not happen in special models, of which the free-electron model is an example. The reason for the exclusion of the second possibility is that the cusp behaviour shown in fig. 5.6b implies that the $E(\boldsymbol{k})$ curve plotted along a normal to Δ in a symmetry plane has the form shown in fig. 5.5.d That is, two bands must be involved and they must touch along Δ in such a way that $\mathrm{grad}_{\boldsymbol{k}} E(\boldsymbol{k})$ in the direction normal to Δ is not zero for points on Δ. The two bands must therefore correspond to the *same* irreducible representations of $\mathscr{G}_o(\boldsymbol{k})$ for $\boldsymbol{k}$ along the normal to Δ in the symmetry plane. Inspection of table 5.1 and the corresponding character table A.15 shows that Δ_5 is the only irreducible representation of $\mathscr{G}_o(\boldsymbol{k})$ for $\boldsymbol{k}$ on Δ of more than one dimension. Table 5.4 then shows that Δ_5 is compatible with a pair of *different* irreducible representations of $\mathscr{G}_o(\boldsymbol{k})$ for $\boldsymbol{k}$ in the symmetry plane ΓHN. The possibility of obtaining cusps therefore cannot occur for Δ, unless an accidental degeneracy occurs along a whole section of Δ.

An identical analysis leading to the same conclusion can be applied to any symmetry axis in a symmetry plane, provided that there exists another symmetry plane perpendicular to the one under consideration which also contains the symmetry axis. Thus, for the body-centred cubic space group O_h^9, in addition to the case of the axis Δ in the plane ΓHN, the analysis also applies to Σ and G in ΓHN, Δ in ΓHP, Σ and D in ΓNP, and D and G in HNP. For the face-centred cubic space group O_h^5, the analysis applies to Δ and Σ in ΓKWX, Σ in ΓKL, Δ and S in ΓLUX, and S and Z in UWX.

Further analysis along these lines for other space groups has been described by RASHBA [1959], SHEKA [1960] and KUDRYAVTSEVA [1967].

§ 5.3. *Critical points*

It is possible to use some of the results of §5.2 to deduce *directly* certain features of the density of states. The density of states per atom in the nth energy band $N_n(E)$ is defined to be such that the number of electronic states per atom in the nth energy band having energies between E and $E+\mathrm{d}E$ is $N_n(E)\,\mathrm{d}E$. It is fairly easily shown that

$$N_n(E) = \frac{2\{\boldsymbol{a}_1 \cdot (\boldsymbol{a}_2 \wedge \boldsymbol{a}_3)\}}{(2\pi)^3 A} \iint \frac{\mathrm{d}S}{|\mathrm{grad}_{\boldsymbol{k}} E_n(\boldsymbol{k})|},$$

where A is the number of atoms per lattice point of the crystal and the integral is over the surface of constant energy $E_n(\boldsymbol{k}) = E$. The total density of states per atom is then given by $\sum_n N_n(E)$.

A point $\boldsymbol{k}_o$ at which $\mathrm{grad}_{\boldsymbol{k}} E_n(\boldsymbol{k}) = 0$ is called a *critical point*, and if the expansion

$$E_n(\boldsymbol{k}) = E_n(\boldsymbol{k}_o) + \alpha_1 (k_x - k_{ox})^2 + \alpha_2 (k_y - k_{oy})^2 + \alpha_3 (k_z - k_{oz})^2$$

is valid for $\boldsymbol{k}$ near $\boldsymbol{k}_o$, the critical point is called *ordinary* or *analytic*. VAN HOVE [1953] has shown that such a point produces an infinite discontinuity of $\mathrm{d}N_n(E)/\mathrm{d}E$ at $E = E_n(\boldsymbol{k}_o)$, although $N_n(E)$ remains continuous. The exact behaviour of $N_n(E)$ for E near $E_n(\boldsymbol{k}_o)$ depends on the signs of α_1, α_2 and α_3. The four possible cases are as follows.

(a) α_1, α_2 and α_3 all are negative. This corresponds to a local maximum of the energy band, and

$$N_n(E) = \begin{cases} \alpha + \beta\{E - E_n(\boldsymbol{k}_o)\}^{\frac{1}{2}} + O\{E - E_n(\boldsymbol{k}_o)\}, & E < E_n(\boldsymbol{k}_o), \\ \alpha + O\{E - E_n(\boldsymbol{k}_o)\}, & E > E_n(\boldsymbol{k}_o). \end{cases}$$

Here α and β are constants.

(b) Two of the set α_1, α_2, α_3 are negative and one is positive. This corresponds to a 'saddle point' of the energy band, and

$$N_n(E) = \begin{cases} \alpha + O\{E - E_n(\boldsymbol{k}_o)\}, & E < E_n(\boldsymbol{k}_o), \\ \alpha + \beta\{E - E_n(\boldsymbol{k}_o)\}^{\frac{1}{2}} + O\{E - E_n(\boldsymbol{k}_o)\}, & E > E_n(\boldsymbol{k}_o). \end{cases}$$

(c) One of set α_1, α_2, α_3 is negative and two are positive. This also corresponds to a saddle point of the energy band, and

$$N_n(E) = \begin{cases} \alpha + \beta\{E - E_n(\boldsymbol{k}_o)\}^{\frac{1}{2}} + O\{E - E_n(\boldsymbol{k}_o)\}, & E < E_n(\boldsymbol{k}_o), \\ \alpha + O\{E - E_n(\boldsymbol{k}_o)\}, & E > E_n(\boldsymbol{k}_o). \end{cases}$$

(d) α_1, α_2 and α_3 are all positive. This corresponds to a local minimum of the energy band, and

$$N_n(E) = \begin{cases} \alpha + O\{E - E_n(\boldsymbol{k}_o)\}, & E < E_n(\boldsymbol{k}_o), \\ \alpha + \beta\{E - E_n(\boldsymbol{k}_o)\}^{\frac{1}{2}} + O\{E - E_n(\boldsymbol{k}_o)\}, & E > E_n(\boldsymbol{k}_o). \end{cases}$$

The development given in §5.2 allows the ordinary critical points to be located very easily. For example, it was shown in §5.2 that all three components of $\text{grad}_{\boldsymbol{k}} E_n(\boldsymbol{k})$ vanish at each of the symmetry points of the space group O_h^9, provided that there is not an accidental degeneracy at the symmetry point. Therefore, with this proviso, all the symmetry points of O_h^9 are ordinary critical points. Critical points may also be easily located along symmetry axes, for it was shown in §5.2 that, in the absence of accidental degeneracies along the axis, if the axis lies in two intersecting symmetry planes, then both components of $\text{grad}_{\boldsymbol{k}} E_n(\boldsymbol{k})$ normal to the axis vanish. An ordinary critical point on such a symmetry axis will then occur at points where the component of $\text{grad}_{\boldsymbol{k}} E_n(\boldsymbol{k})$ *along* the axis vanishes. Such points can be detected immediately from a plot of $E_n(\boldsymbol{k})$ against $\boldsymbol{k}$ along the axis. Moreover, the signs of α_1, α_2 and α_3 may easily be found from this plot and the curvature of the constant energy contours drawn in the two intersecting symmetry planes, which may be sketched in a way described in §5.4. Although it is not impossible for ordinary critical points to occur at other points of the Brillouin zone, it is unlikely for this to happen, for this requires that two or more components of $\text{grad}_{\boldsymbol{k}} E_n(\boldsymbol{k})$ must simultaneously vanish for reasons other than those of symmetry.

As noted in §5.1 and illustrated in fig. 5.1, at a point $\boldsymbol{k}_o$ where two bands touch there is at least one component of $\text{grad}_{\boldsymbol{k}} E_n(\boldsymbol{k})$ that is discontinuous. PHILLIPS [1956] has shown that such a point $\boldsymbol{k}_o$ must also be regarded as a critical point, but as the above Taylor series expansion for $E_n(\boldsymbol{k})$ is no longer valid, it is described as a *singular* critical point. Singular critical points do not produce discontinuities in $N_n(E)$. Only a singular local maximum or minimum having only one discontinuous component of $\text{grad}_{\boldsymbol{k}} E_n(\boldsymbol{k})$ can produce an infinite discontinuity in $\mathrm{d}N_n(E)/\mathrm{d}E$. All other singular critical points produce no discontinuity in $\mathrm{d}N_n(E)/\mathrm{d}E$. Singular critical points on symmetry axes may be located and classified in exactly the same way as described above for ordinary singular points.

Because of the periodicity of $E_n(\boldsymbol{k})$ in the extended zone scheme, VAN HOVE [1953] and PHILLIPS [1956] have shown by topological arguments that a minimal set of critical points must occur for each energy band. For complicated energy band structures, such as those belonging to the transition metals for example, this minimal set is considerably exceeded. The results of a calculation of the density of states for body-centred cubic iron in which the above considerations were used has been described by WOHLFARTH and CORNWELL [1961].

§ 5.4. *Determination of the whole energy band structure from a knowledge of the energy levels at the symmetry points of the Brillouin zone*

As mentioned in §2.1, accurate calculations of energy levels are usually performed only for the symmetry points of the Brillouin zone. However, by using the results on continuity, compatibility and symmetry that have been described in §§5.1, 5.2, it is possible to get a rough picture of the *whole* band structure without further calculation.

The first stage is to sketch the energy bands along every symmetry axis by joining up the known energy levels at the symmetry points at each end of the axis in the simplest way allowed by the rules of compatibility and continuity. The second stage is to sketch for each band separately the contours of constant energy in each of the symmetry planes of the Brillouin zone. It is important that the definition (4.17) of the ordering of energy bands should be invoked here. As an example, consider the energy band plotted for the axes Σ, G and Δ of the body-centred cubic space group O_h^9 that is given in fig. 5.7. The dashed curves represent other energy

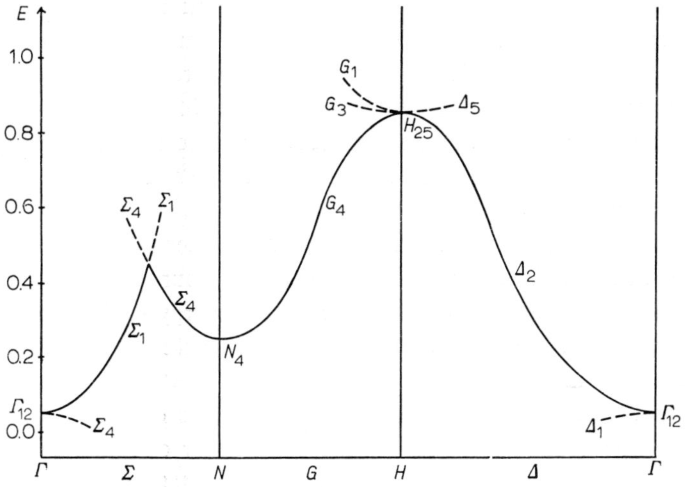
E
1.0
0.8
0.6
0.4
0.2
0.0
Γ_{12}
Σ_4
Σ_1
Σ_4
Σ_1
Σ_4
N_4
G_1
G_3
Δ_5
H_{25}
G_4
Δ_2
Δ_1
Γ_{12}
Γ
Σ
N
G
H
Δ
Γ

Fig. 5.7

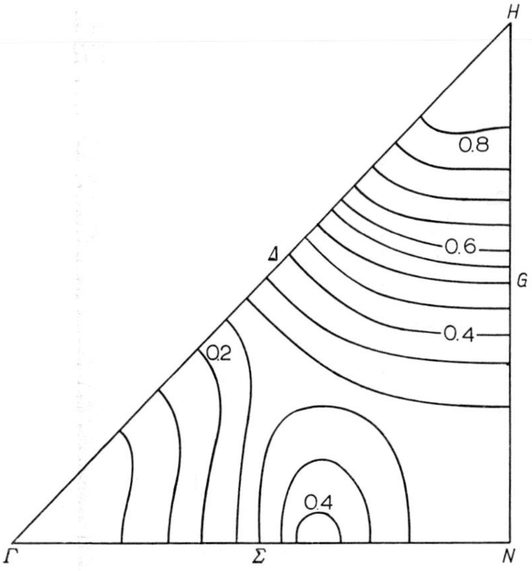
H
0.8
0.6
G
0.4
Δ
0.2
0.4
Γ
Σ
N

Fig. 5.8

bands that touch the band under consideration. For the symmetry plane ΓHN, the points where the constant energy contours intersect the axes Σ, G and Δ can be plotted by reference to fig. 5.7. A picture of the energy contours can then be obtained by smoothly joining up points with the same energy, as for example in fig. 5.8, bearing in mind that the contours intersect Σ, G and Δ normally, as was shown in §5.2.

This construction may be repeated for the other symmetry planes ΓNP, ΓHP and HNP, and the set of four contour diagrams would then give a rough picture of the constant energy surfaces of this energy band. The other energy bands may be studied in the same way.

In fig. 5.7 *all* the irreducible representations of the $\mathscr{G}_o(\boldsymbol{k})$ for $\boldsymbol{k}$ on the axes Σ, G and Δ for the band under consideration are compatible with the 'even' irreducible representation of $\mathscr{G}_o(\boldsymbol{k})$ for $\boldsymbol{k}$ in the symmetry plane ΓHN. A more complex situation arises if this is not so. Fig. 5.9 shows such a band, with fig. 5.10 giving the corresponding contours of constant energy. The dashed curve of fig. 5.10 marks the boundary between regions of ΓHN corresponding

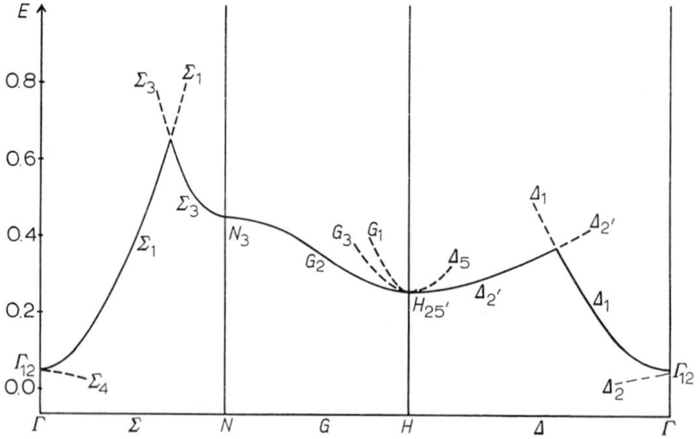

Fig. 5.9

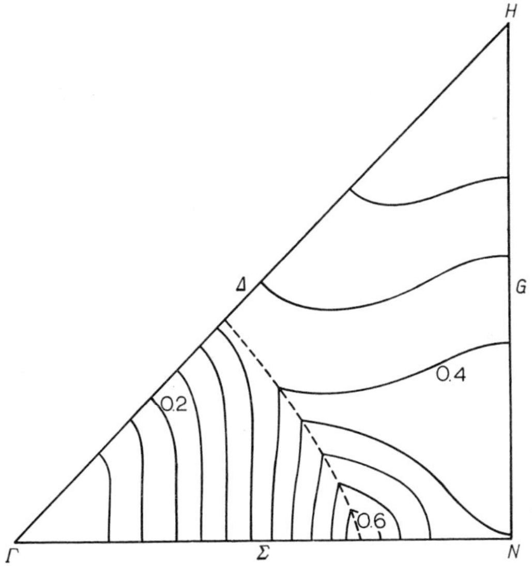

Fig. 5.10

to the 'even' and 'odd' irreducible representations. The constant energy contours are continuous across this boundary, but their derivatives are not continuous.

A more accurate determination of the whole energy band structure from a knowledge of the energy levels at the symmetry points of the Brillouin zone requires an interpolation scheme. Applying such a scheme would give, first, a more accurate set of $E(\boldsymbol{k})$ curves along the symmetry axes, and then, better constant energy contours. The situation at general points of the Brillouin zone could also be investigated, and the volumes in $\boldsymbol{k}$ space contained by surfaces of constant energy could be computed numerically. From these it is possible to deduce the Fermi energy and the density of electronic states as a function of energy. An interpolation scheme based on the tight-binding method has been proposed by SLATER and KOSTER [1954], and investigated in detail for body-centred cubic iron by CORNWELL *et al.* [1968]. A plane wave method has been studied by CORNWELL [1961], and a combination of the two methods has been tried by HODGES *et al.* [1966] and MUELLER [1967].

§ 6. SYMMETRIZED WAVE FUNCTIONS

§ 6.1. *Introduction*

The application of the fundamental theorem of § 1 on the irreducible representations of a symmorphic space group to the simplification of approximate calculations of energy eigenvalues was described in § 2.2. It was shown there that, in finding energy eigenfunctions and eigenvalues corresponding to a wave vector $\boldsymbol{k}$ and the irreducible representation $\boldsymbol{\Gamma}^p$ of $\mathscr{G}_o(\boldsymbol{k})$, the only functions that appear in the expansion of the required energy eigenfunction are products of the fixed factor $\exp(\mathrm{i}\boldsymbol{k}\cdot\boldsymbol{r})$ with a set of different periodic functions such that each product transforms according to a fixed row of $\boldsymbol{\Gamma}^p$.

The various methods of calculation differ essentially in the choice of basis functions. For example, in the cellular method the basis functions are formed from spherical harmonics, in the tight-binding method they are formed from linear combinations of atomic orbitals, while in the orthogonalized plane wave method both plane waves and linear combinations of atomic orbitals are used. In all cases the basis functions can be found by the projection operator method of ch. 3 § 5.2, all the required information on the point groups $\mathscr{G}_o(\boldsymbol{k})$ being given in appendix 1.

The following subsections deal with the application of the projection operators to spherical harmonics and plane waves, and then with the use of the symmetrized combinations so formed in electronic energy band calculations. The various methods of calculation will not be described in great detail here as good accounts exist in the reviews of CALLAWAY [1958, 1964], PINCHERLE [1960] and REITZ [1955].

§ 6.2. *Symmetrized spherical harmonics*

The spherical harmonic $Y_l^m(\theta, \phi)$ is defined for $|m| \leqslant l$ by

$$Y_l^m(\theta, \phi) = P_l^m(\cos\theta)\exp(im\phi),$$

θ and ϕ being two of the set of spherical polar coordinates and P_l^m being the associated Legendre polynomial. Frequently it is convenient to use the fact that, for low values of l, the functions $r^l Y_l^m(\theta, \phi)$ are linear combinations of the following simple Cartesian coordinate functions:

$$\left.\begin{array}{ll} l=0: & 1; \\ l=1: & x, y, z; \\ l=2: & xy, yz, zx, x^2-y^2, 3z^2-r^2. \end{array}\right|$$

The basis functions formed from these functions for all the rows of all the irreducible representations of the point groups $\mathscr{G}_o(\boldsymbol{k})$ for each symmetry point, axis and plane of the body-centred cubic space group O_h^9 and the face-centred cubic space group O_h^5 are given in tables 5.7 to 5.12. These combinations were obtained using the projection operator technique of ch. 3 § 5.2 and the explicit form of the matrix representations that are given in appendix 1. The coordinates of the symmetry points, axes and planes are those given in tables 5.1 and 5.2. The factor $1/\sqrt{3}$ appearing with $3z^2-r^2$ is to give it the same normalizing constant as the function x^2-y^2, the normalization being performed as in ch. 3 § 5.3.

If a spherically symmetric function transforms as a basis function for an irreducible representation, then that representation is said to be ‘s-like’. If linear combinations of x, y and z transform as basis functions of a representation that is not s-like, then that representation is said to be ‘p-like’. The other irreducible representations may similarly be described as ‘d-like’, ‘f-like’, and so on, by analogy with the corresponding description of atomic states.

Symmetrized linear combinations of spherical harmonics for the cubic space groups O_h^1, O_h^5 and O_h^9 were called ‘kubic harmonics’

TABLE 5.7

Symmetrized linear combinations of spherical harmonics for the symmetry points of the Brillouin zone for the body-centred cubic space group O_h^9

Representation	Row	Symmetrized linear combinations
Γ_1, H_1	1	1
Γ_{12}, H_{12}	1	(x^2-y^2)
	2	$(3z^2-r^2)/\sqrt{3}$
$\Gamma_{25'}, H_{25'}$	1	xy
	2	yz
	3	zx
Γ_{15}, H_{15}	1	x
	2	y
	3	z

table 5.7 (continued)

P_1	1	1
P_3	1	(x^2-y^2)
	2	$(3z^2-r^2)/\sqrt{3}$
P_4	1	x; yz
	2	y; zx
	3	z; xy
N_1	1	$1;\quad yz;\quad \frac{1}{2}\{\sqrt{3}(x^2-y^2)-(3z^2-r^2)/\sqrt{3}\}$
N_2	1	$(xy-zx)/\sqrt{2}$
N_3	1	$(xy+zx)/\sqrt{2}$
N_4	1	$\frac{1}{2}\{(x^2-y^2)+(3z^2-r^2)\}$
$N_{1'}$	1	$(y+z)/\sqrt{2}$
$N_{3'}$	1	x
$N_{4'}$	1	$(y-z)/\sqrt{2}$

TABLE 5.8

Symmetrized linear combinations of spherical harmonics for the symmetry axes of the Brillouin zone for the body-centred cubic space group O_h^9

Representation	Row	Symmetrized linear combinations
Δ_1	1	$1;\ z;\ (3z^2-r^2)/\sqrt{3}$
Δ_2	1	(x^2-y^2)
$\Delta_{2'}$	1	xy
Δ_5	1	x ; zx
	2	y ; yz
Σ_1	1	$1;\ (y+z)/\sqrt{2};\ yz;\ \frac{1}{2}\{\sqrt{3}(x^2-y^2)-(3z^2-r^2)/\sqrt{3}\}$
Σ_2	1	$(xy-zx)/\sqrt{2}$
Σ_3	1	$x;\ (xy+zx)/\sqrt{2}$
Σ_4	1	$(y-z)/\sqrt{2};\ \frac{1}{2}\{(x^2-y^2)+(3z^2-r^2)\}$
Λ_1	1	$1;\ (x+y+z)/\sqrt{3};\ (xy+yz+zx)/\sqrt{3}$
Λ_3	1	$(x-y)/\sqrt{2}$; $(yz-zx)/\sqrt{2}$; (x^2-y^2)
	2	$(2z-x-y)/\sqrt{6}$; $(2xy-yz-zx)/\sqrt{6}$; $(3z^2-r^2)/\sqrt{3}$

table 5.8 (continued)

D_1	1	$1;\quad x;\quad yz;\quad \frac{1}{2}\{\sqrt{3}(x^2-y^2)-(3z^2-r^2)/\sqrt{3}\}$
D_2	1	$\frac{1}{2}\{(x^2-y^2)+(3z^2-r^2)\}$
D_3	1	$(y+z)/\sqrt{2};\quad (xy+zx)/\sqrt{2}$
D_4	1	$(y-z)/\sqrt{2};\quad (xy-zx)/\sqrt{2}$
F_1	1	$1;\quad (x+y-z)/\sqrt{3};\quad (xy-yz-zx)/\sqrt{3}$
F_3	1 2	$\left.\begin{matrix}(x-y)/\sqrt{2}\\(x+y+2z)/\sqrt{6}\end{matrix}\right\};\quad \left.\begin{matrix}(zx-yz)/\sqrt{2}\\(2xy+yz+zx)/\sqrt{6}\end{matrix}\right\};\quad \left.\begin{matrix}(x^2-y^2)\\(3z^2-r^2)/\sqrt{3}\end{matrix}\right\}$
G_1	1	$1;\quad (y-z)/\sqrt{2};\quad yz;\quad \frac{1}{2}\{\sqrt{3}(x^2-y^2)-(3z^2-r^2)/\sqrt{3}\}$
G_2	1	$(xy+zx)/\sqrt{2}$
G_3	1	$x;\quad (xy-zx)/\sqrt{2}$
G_4	1	$(y+z)/\sqrt{2};\quad \frac{1}{2}\{(x^2-y^2)+(3z^2-r^2)\}$

TABLE 5.9

Symmetrized linear combinations of spherical harmonics for the symmetry planes of the Brillouin zone for the body-centred cubic space group O_h^9

Representation	Symmetrized linear combinations
$(\Gamma HN)_+$	$1;\ y;\ z;\ yz;\ (x^2-y^2);\ (3z^2-r^2)/\sqrt{3}$
$(\Gamma HN)_-$	$x;\ xy;\ zx$
$(\Gamma HP)_+$	$1;\ (x+y)/\sqrt{2};\ z;\ xy;\ (yz+zx)/\sqrt{2};\ (3z^2-r^2)/\sqrt{3}$
$(\Gamma HP)_-$	$(x-y)/\sqrt{2};\ (yz-zx)/\sqrt{2};\ (x^2-y^2)$
$(\Gamma NP)_+$	$1;\ x;\ (y+z)/\sqrt{2};\ yz;\ (xy+zx)/\sqrt{2};\ \frac{1}{2}\{\sqrt{3}(x^2-y^2)-(3z^2-r^2)/\sqrt{3}\}$
$(\Gamma NP)_-$	$(y-z)/\sqrt{2};\ (xy-zx)/\sqrt{2};\ \frac{1}{2}\{(x^2-y^2)+(3z^2-r^2)\}$
$(HNP)_+$	$1;\ x;\ (y-z)/\sqrt{2};\ yz;\ (xy-zx)/\sqrt{2};\ \frac{1}{2}\{\sqrt{3}(x^2-y^2)-(3z^2-r^2)/\sqrt{3}\}$
$(HNP)_-$	$(y+z)/\sqrt{2};\ (xy+zx)/\sqrt{2};\ \frac{1}{2}\{(x^2-y^2)+(3z^2-r^2)\}$

TABLE 5.10

Symmetrized linear combinations of spherical harmonics for the symmetry points of the Brillouin zone for the face-centred cubic space group O_h^5

Representation	Row	Symmetrized linear combinations
K_1	1	$1;\ (y+z)/\sqrt{2};\ yz;\ \frac{1}{2}\{\sqrt{3}(x^2-y^2)-(3z^2-r^2)/\sqrt{3}\}$
K_2	1	$(xy-zx)/\sqrt{2}$
K_3	1	$x;\ (xy+zx)/\sqrt{2}$
K_4	1	$(y-z)/\sqrt{2};\ \frac{1}{2}\{(x^2-y^2)+(3z^2-r^2)\}$
L_1	1	$1;\ (xy+yz+zx)/\sqrt{3}$
L_3	1	$\left.\begin{matrix}(yz-zx)/\sqrt{2}\\(2xy-yz-zx)/\sqrt{6}\end{matrix}\right\};\ \left.\begin{matrix}(x^2-y^2)\\(3z^2-r^2)/\sqrt{3}\end{matrix}\right\}$
	2	
$L_{2'}$	1	$(x+y+z)/\sqrt{3}$
$L_{3'}$	1	$\left.\begin{matrix}(x+y-2z)/\sqrt{6}\\(x-y)/\sqrt{2}\end{matrix}\right\}$
	2	

table 5.10 (continued)

U_1	1	$1;\quad (x+y)/\sqrt{2};\quad xy;\quad (3z^2-r^2)/\sqrt{3}$
U_2	1	$(yz-zx)/\sqrt{2}$
U_3	1	$z;\quad (yz+zx)/\sqrt{2}$
U_4	1	$(x-y)/\sqrt{2};\quad (x^2-y^2)$
W_1	1	$1;\quad \frac{1}{2}\{\sqrt{3}(x^2-y^2)+(3z^2-r^2)/\sqrt{3}\}$
$W_{1'}$	1	zx
$W_{2'}$	1	$y;\quad \frac{1}{2}\{(x^2-y^2)-(3z^2-r^2)\}$
W_3	1	$x \quad \Big\}; \quad xy \quad \Big\}$
	2	$z \quad \Big\} \quad -yz \quad \Big\}$
X_1	1	$1;\quad (3z^2-r^2)/\sqrt{3}$
X_2	1	(x^2-y^2)
X_3	1	xy
X_5	1	$yz \quad \Big\}$
	2	$-zx \quad \Big\}$
$X_{4'}$	1	z
$X_{5'}$	1	$x \quad \Big\}$
	2	$y \quad \Big\}$

The symmetrized linear combinations for Γ are given in table 5.7.

TABLE 5.11

Symmetrized linear combinations of spherical harmonics for the symmetry axes of the Brillouin zone for the face-centred cubic space group O_h^5

Representation	Symmetrized linear combination
Q_1	$1;\ (x-z)/\sqrt{2};\ (xy+yz)/\sqrt{2};\ zx;\ \frac{1}{2}\{\sqrt{3}(x^2-y^2)+(3z^2-r^2)/\sqrt{3}\}$
Q_2	$(x+z)/\sqrt{2};\ y;\ (xy-yz)/\sqrt{2};\ \frac{1}{2}\{(x^2-y^2)-(3z^2-r^2)\}$
S_1	$1;\ (x+y)/\sqrt{2};\ xy;\ (3z^2-r^2)/\sqrt{3}$
S_2	$(yz-zx)/\sqrt{2}$
S_3	$z;\ (yz+zx)/\sqrt{2}$
S_4	$(x-y)/\sqrt{2};\ (x^2-y^2)$
Z_1	$1;\ y;\ (x^2-y^2);\ (3z^2-r^2)/\sqrt{3}$
Z_2	zx
Z_3	$z;\ yz$
Z_4	$x;\ xy$

The symmetrized linear combinations for Δ, Σ and Λ are given in table 5.8.

TABLE 5.12

Symmetrized linear combinations of spherical harmonics for the symmetry planes of the Brillouin zone for the face-centred cubic space group O_h^5

Representation	Symmetrized linear combinations
$(\Gamma KWX)_+$	$1;\ y;\ z;\ yz;\ (x^2-y^2);\ (3z^2-r^2)/\sqrt{3}$
$(\Gamma KWX)_-$	$x;\ xy;\ zx$
$(\Gamma LUX)_+$	$1;\ (x+y)/\sqrt{2};\ z;\ xy;\ (yz+zx)/\sqrt{2};\ (3z^2-r^2)/\sqrt{3}$
$(\Gamma LUX)_-$	$(x-y)/\sqrt{2};\ (yz-zx)/\sqrt{2};\ (x^2-y^2)$
$(\Gamma KL)_+$	$1;\ x;\ (y+z)/\sqrt{2};\ yz;\ (xy+zx)/\sqrt{2};\ \frac{1}{2}\{\sqrt{3}(x^2-y^2)-(3z^2-r^2)/\sqrt{3}\}$
$(\Gamma KL)_-$	$(y-z)/\sqrt{2};\ (xy-zx)/\sqrt{2};\ \frac{1}{2}\{(x^2-y^2)+(3z^2-r^2)\}$
$(UWX)_+$	$1;\ x;\ y;\ xy;\ (x^2-y^2);\ (3z^2-r^2)/\sqrt{3}$
$(UWX)_-$	$z;\ yz;\ zx$

by VON DER LAGE and BETHE [1947], who tabulated those for $l \leqslant 6$ for a small number of symmetry points. Additional kubic harmonics have been given by HOWARTH and JONES [1952], BELL [1954], ALTMANN [1957] and ALTMANN and CRACKNELL [1965]. This last paper lists the kubic harmonics for $l \leqslant 12$ for all the symmetry points, axes and planes of the cubic space groups O_h^1, O_h^5 and O_h^9.

These symmetrized spherical harmonics are of direct use in the cellular method and the augmented plane wave (or A.P.W.) method. (Good accounts of both methods appear in the review article by REITZ [1955], and the latter method is the subject of a book by LOUCKS [1967].) They are also needed in constructing symmetrized linear combinations of atomic orbitals for use in the tight-binding method, as will be described in §6.4.

One difficulty with the cellular method, that does not occur with such methods as the orthogonalized plane wave method and the tight-binding method, is that while the basis functions are chosen to be exact basis functions of the irreducible representations of $\mathscr{G}_o(\boldsymbol{k})$ (in the form of symmetrized spherical harmonics), they are *not exact* Bloch functions. The condition (4.11) is merely taken as a boundary condition to be satisfied at a finite number of points on the boundary between two atomic cells of the crystal. Thus the approximate eigenfunctions of the cellular method are not exact basis functions of the irreducible representations of the space group $\mathscr{G}$.

§ 6.3. *Symmetrized plane waves*

Basis functions formed from plane waves are very convenient to use. The members of the set of plane wave Bloch functions corresponding to wave vector $\boldsymbol{k}$ have the form

$$\phi_{\boldsymbol{k}\boldsymbol{m}}(\boldsymbol{r}) = \exp\{i(\boldsymbol{k} + \boldsymbol{K}_{\boldsymbol{m}})\cdot\boldsymbol{r}\}, \tag{5.10}$$

where $\boldsymbol{K}_{\boldsymbol{m}}$ is a lattice vector of the reciprocal lattice. (As shown by eq. (4.14), the factor $\exp(\mathrm{i}\boldsymbol{K}_m \cdot \boldsymbol{r})$ has the periodicity of the crystal lattice, so that $\phi_{\boldsymbol{k}m}(\boldsymbol{r})$ has the form (4.11) of a Bloch function.) Unfortunately, an expansion of the form

$$\sum_m C_m \phi_{\boldsymbol{k}m}(\boldsymbol{r}) \tag{5.11}$$

alone converges much too slowly to be of any practical use in accurate energy band calculations, even when the symmetry of the crystal is exploited to the full. However, the nearly-free electron approximation, which is based on a severely truncated form of this expansion, can provide useful qualitative information, as was notably demonstrated by the work of HARRISON [1960a, 1960b, 1966]. Moreover, the orthogonalized plane-wave method, which is often more briefly called the O.P.W. method, and which involves an expansion of the form (5.11) together with additional terms, does provide a practicable method for calculating energy levels at the symmetry points of the Brillouin zone. The O.P.W. method will be described in § 6.5. The symmetrized plane waves that are introduced in this section may be used, therefore, in both the nearly-free electron approximation and in the O.P.W. method.

The basis function formed from the function $\phi_{\boldsymbol{k}m}(\boldsymbol{r})$ of eq. (5.10) that transforms according to the sth row of the irreducible representation $\boldsymbol{\Gamma}^p$ of $\mathscr{G}_o(\boldsymbol{k})$ is obtained simply by operating on $\phi_{\boldsymbol{k}m}(\boldsymbol{r})$ with the projection operator $\mathscr{P}^p_{ss}$ of (3.22), where $\mathscr{P}^p_{st}$ is here given by

$$\mathscr{P}^p_{st} = \{l_p/g_o(\boldsymbol{k})\} \sum \Gamma^p(\{\boldsymbol{R} \mid \boldsymbol{0}\})^*_{st}\, P(\{\boldsymbol{R} \mid \boldsymbol{0}\}), \tag{5.12}$$

where the summation is over all the rotations $\{\boldsymbol{R} \mid \boldsymbol{0}\}$ of $\mathscr{G}_o(\boldsymbol{k})$, $g_o(\boldsymbol{k})$ is the order of $\mathscr{G}_o(\boldsymbol{k})$, and l_p is the dimension of $\boldsymbol{\Gamma}^p$. As noted in § 1, because $\boldsymbol{R}$ is an orthogonal matrix,

$$P(\{\boldsymbol{R} \mid \boldsymbol{0}\})\, \phi_{\boldsymbol{k}m}(\boldsymbol{r}) = \exp\{\mathrm{i}\boldsymbol{R}(\boldsymbol{k} + \boldsymbol{K}_m) \cdot \boldsymbol{r}\}.$$

Because $\{\boldsymbol{R} \mid \boldsymbol{0}\}$ is a member of $\mathscr{G}_o(\boldsymbol{k})$, $\boldsymbol{R}\boldsymbol{k}$ is equivalent to $\boldsymbol{k}$. As $\boldsymbol{R}\boldsymbol{K}_m$ is also a lattice vector of the reciprocal lattice, then there exists a lattice vector of the reciprocal lattice $\boldsymbol{K}_{m'}$ such that

$$\boldsymbol{R}(\boldsymbol{k} + \boldsymbol{K}_m) = \boldsymbol{k} + \boldsymbol{K}_{m'}\,.$$

Moreover, $\boldsymbol{R}$ is orthogonal,

$$|\boldsymbol{R}(\boldsymbol{k} + \boldsymbol{K}_m)| = |\boldsymbol{k} + \boldsymbol{K}_m|\,. \tag{5.13}$$

As an example, consider the point H of the Brillouin zone of the body-centred cubic space group O_h^9, for which $\boldsymbol{k} = (\pi/a)\,(0, 0, 2)$, by table 5.1. In the basis functions of $\mathscr{G}_o(\boldsymbol{k})$ formed from $\phi_{\boldsymbol{k}0}(\boldsymbol{r}) = \exp(\mathrm{i}\boldsymbol{k}\cdot\boldsymbol{r})$, six plane waves appear with wave vectors $\pm\boldsymbol{k}, \pm\boldsymbol{k}', \pm\boldsymbol{k}''$, where $\boldsymbol{k}' = (\pi/a)\,(0, 2, 0)$ and $\boldsymbol{k}'' = (\pi/a)\,(2, 0, 0)$. Table 1.5 gives that

$$\begin{aligned}
\boldsymbol{R}(T)\,\boldsymbol{k} &= \boldsymbol{k} && \text{for } T = E,\; C_{2z},\; C_{4z},\; C_{4z}^{-1},\; IC_{2x},\; IC_{2y},\; IC_{2a},\; IC_{2b}\,,\\
\boldsymbol{R}(T)\,\boldsymbol{k} &= \boldsymbol{k}' && \text{for } T = C_{3\gamma},\; C_{3\delta},\; C_{4x},\; C_{2e},\; IC_{3\alpha},\; IC_{3\beta},\; IC_{4x}^{-1},\; IC_{2f}\,,\\
\boldsymbol{R}(T)\,\boldsymbol{k} &= \boldsymbol{k}'' && \text{for } T = C_{3\beta}^{-1},\; C_{3\delta}^{-1},\; C_{4y}^{-1},\; C_{2c},\; IC_{3\alpha}^{-1},\; IC_{3\gamma}^{-1},\; IC_{4y},\; IC_{2d}\,,
\end{aligned}$$

and, of course, $\boldsymbol{R}(IT)\,\boldsymbol{k} = -\boldsymbol{R}(T)\,\boldsymbol{k}$. Application to $\exp(\mathrm{i}\boldsymbol{k}\cdot\boldsymbol{r})$ of the character projection operator (3.25), which here has the form

$$\mathscr{P}^p = \{l_p/g_o(\boldsymbol{k})\} \sum \chi^p(\{\boldsymbol{R}\,|\,\boldsymbol{0}\})^*\, P(\{\boldsymbol{R}\,|\,\boldsymbol{0}\}),$$

shows that the irreducible representations H_1, H_{12} and H_{15} all possess basis functions containing $\exp(\mathrm{i}\boldsymbol{k}\cdot\boldsymbol{r})$. On using the matrices for the irreducible representations of $\mathscr{G}_o(\boldsymbol{k})$ that are given in appendix 1 in the application of the projection operator $\mathscr{P}_{ss}^p$ of (5.12) to $\exp(\mathrm{i}\boldsymbol{k}\cdot\boldsymbol{r})$, one finds only the following non-zero combinations:

$$\begin{aligned}
&H_1 \text{ (first row):}\\
&\quad \phi_1^1 = (\psi_{\boldsymbol{k}} + \psi_{\boldsymbol{k}'} + \psi_{\boldsymbol{k}''} + \psi_{-\boldsymbol{k}} + \psi_{-\boldsymbol{k}'} + \psi_{-\boldsymbol{k}''})\,(6\Omega)^{-\frac{1}{2}}\\
&H_{12} \text{ (second row):}\\
&\quad \phi_1^{12} = (\psi_{\boldsymbol{k}} - \tfrac{1}{2}\psi_{\boldsymbol{k}'} - \tfrac{1}{2}\psi_{\boldsymbol{k}''} + \psi_{-\boldsymbol{k}} - \tfrac{1}{2}\psi_{-\boldsymbol{k}'} - \tfrac{1}{2}\psi_{-\boldsymbol{k}''})\,(3\Omega)^{-\frac{1}{2}}\\
&H_{15} \text{ (third row):}\\
&\quad \phi_3^{15} = (\psi_{\boldsymbol{k}} - \psi_{-\boldsymbol{k}})\,(2\Omega)^{-\frac{1}{2}}
\end{aligned} \tag{5.14}$$

Here $\psi_{\boldsymbol{k}}$ denotes $\exp(\mathrm{i}\boldsymbol{k}\cdot\boldsymbol{r})$, and so on. These combinations are normalized in the basic block of the crystal of ch. 4 §2, which has volume $\Omega = N(\boldsymbol{a}_1 \wedge \boldsymbol{a}_2)\cdot\boldsymbol{a}_3$, using the fact that

$$(\exp(\mathrm{i}\boldsymbol{k}\cdot\boldsymbol{r}), \exp(\mathrm{i}\boldsymbol{k}'\cdot\boldsymbol{r})) = \begin{cases} \Omega, & \text{if} \quad \boldsymbol{k} = \boldsymbol{k}' \\ 0, & \text{if} \quad \boldsymbol{k} = \boldsymbol{k}' + \boldsymbol{K}_m, \quad \text{but} \quad \boldsymbol{K}_m \neq \boldsymbol{0}. \end{cases}$$

The fact that the projection operator $\mathscr{P}^p_{ss}$ of (5.12) applied to $\exp(\mathrm{i}\boldsymbol{k}\cdot\boldsymbol{r})$ gives zero for the first row of H_{12} and the first and second rows of H_{15} implies that basis functions transforming according to these rows do not contain $\psi_{\boldsymbol{k}}$, although they will contain $\psi_{\boldsymbol{k}'}$ and $\psi_{\boldsymbol{k}''}$. In fact, the partners of the basis functions of (5.14) obtained using the projection operators $\mathscr{P}^{12}_{12}$, $\mathscr{P}^{15}_{13}$ and $\mathscr{P}^{15}_{23}$ are

$$\begin{aligned} &H_{12}\ \text{(first row)}: && \phi^{12}_1 = (\psi_{\boldsymbol{k}'} - \psi_{\boldsymbol{k}''} + \psi_{-\boldsymbol{k}'} - \psi_{-\boldsymbol{k}''})(4\Omega)^{-\frac{1}{2}}, \\ &H_{15}\ \text{(first row)}: && \phi^{15}_1 = (\psi_{\boldsymbol{k}''} - \psi_{-\boldsymbol{k}''})(2\Omega)^{-\frac{1}{2}}, \\ &H_{15}\ \text{(second row)}: && \phi^{15}_2 = (\psi_{\boldsymbol{k}'} - \psi_{-\boldsymbol{k}'})(2\Omega)^{-\frac{1}{2}}. \end{aligned}$$

In the free-electron model the potential energy $V(\boldsymbol{r})$ is zero everywhere, so that the plane waves are themselves energy eigenfunctions, the energy eigenvalue corresponding to $\phi_{\boldsymbol{km}}(\boldsymbol{r})$ of (5.10) being $\hbar^2(\boldsymbol{k}+\boldsymbol{K}_m)^2/2m$, m being the mass of the electron. By virtue of eq. (5.13), all the basis functions formed from $\phi_{\boldsymbol{km}}(\boldsymbol{r})$ correspond to the same energy. For example, at the point H considered above, there will be a six-fold degeneracy at the energy $2\hbar^2\pi^2/ma^2$, consisting of a non-degenerate H_1 level, a two-fold degenerate H_{12} level and a three-fold degenerate H_{15} level. The lower lying energy eigenvalues $E(\boldsymbol{k})$ are plotted in fig. 5.11 for $\boldsymbol{k}$ at the symmetry points Γ and H and the symmetry axis Δ for the body-centred cubic space group O^9_h. The irreducible representations of $\mathscr{G}_o(\boldsymbol{k})$ corresponding to each level is shown. It will be noted that, in this model, several accidental degeneracies occur between levels corresponding to the *same* irreducible representation.

In the nearly-free electron approximation, the effect of the periodic potential $V(\boldsymbol{r})$ is taken into account using first-order perturbation theory. For a symmetry point of the Brillouin zone, this is equivalent to taking the energy eigenfunctions to be linear

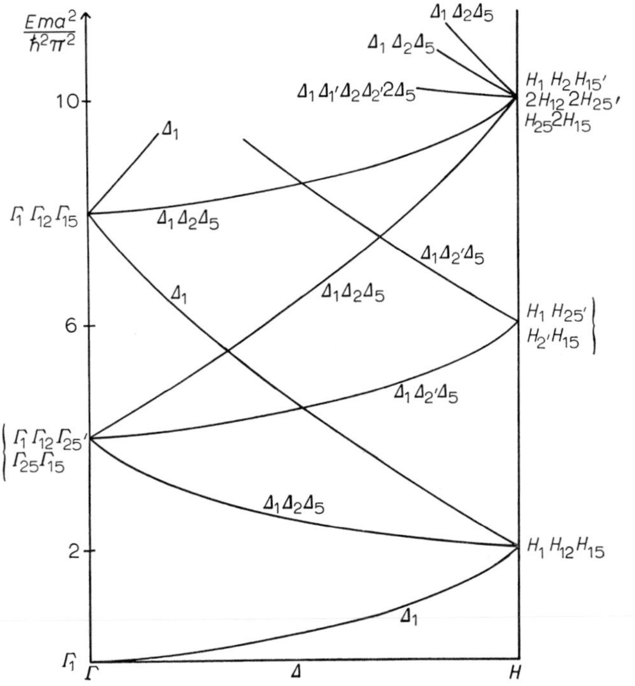

Fig. 5.11

combinations of plane waves corresponding to a particular free-electron energy, and then diagonalizing the Hamiltonian matrix. The Hamiltonian matrix is automatically diagonal if these linear combinations correspond to basis functions of different rows of different irreducible representations of $\mathscr{G}_o(\boldsymbol{k})$. For example, considering the nearly-free electron levels coming from the free-electron energy $2\hbar^2\pi^2/ma^2$ at H, the six-fold degeneracy would be split to give the following energy levels. Firstly, there is a non-degenerate H_1 level at

$$(\phi_1^1, H\phi_1^1) = V(0,0,0) + 4V(1,0,1) + V(2,0,0) + 2\hbar^2\pi^2/ma^2 \,. \tag{5.15}$$

There is a two-fold degenerate H_{12} level at

$$(\phi_1^{12}, H\phi_1^{12}) = (\phi_2^{12}, H\phi_2^{12})$$
$$= V(0,0,0) - 2V(1,0,1) + V(2,0,0) + 2\hbar^2\pi^2/ma^2, \tag{5.16}$$

and finally, there is a three-fold degenerate H_{15} level at

$$(\phi_1^{15}, H\phi_1^{15}) = (\phi_2^{15}, H\phi_2^{15}) = (\phi_3^{15}, H\phi_3^{15})$$
$$= V(0,0,0) - V(2,0,0) + 2\hbar^2\pi^2/ma^2. \tag{5.17}$$

Here

$$V(l,m,n) = \Omega^{-1} \iiint \exp\{(2\pi i/a)(l,m,n)\cdot \boldsymbol{r}\} \, V(\boldsymbol{r}) \, dx \, dy \, dz,$$

the integration being over the basic block of the crystal of volume Ω that was introduced in ch. 4 § 2. In obtaining these expressions, use has been made of the result that $V(l',m',n') = V(l,m,n)$ if

$$\begin{pmatrix} l' \\ m' \\ n' \end{pmatrix} = \boldsymbol{R} \begin{pmatrix} l \\ m \\ n \end{pmatrix},$$

where $\boldsymbol{R}$ is a rotation of $\mathscr{G}_o$, which is here O_h. This result follows immediately from the symmetry of $V(\boldsymbol{r})$. As a consequence of the second part of the second matrix element theorem of ch. 3 § 6.1, of course it is necessary to work out basis functions transforming according to *only one* row of each irreducible representation of $\mathscr{G}_o(\boldsymbol{k})$ in order to obtain expressions like (5.15), (5.16) and (5.17).

The energy levels at the symmetry points of the Brillouin zone for sodium, as calculated by Howarth and Jones [1952], using the cellular method, are given in table 5.13. The free-electron levels corresponding to the observed value of the parameter a for sodium are also shown. The values calculated by Howarth and Jones that correspond to a set of *degenerate* free-electron levels are bracketed together. The closeness of the two sets of levels provides striking justification for the description of sodium as being a 'nearly-free-electron-like metal'.

TABLE 5.13

Energy levels for sodium. All energies are in Rydbergs. The zero of energy has been chosen so that the free-electron Γ_1 level coincides with the level calculated by HOWARTH and JONES [1952]

Irreducible representation	Energy level calculated by Howarth and Jones	Free-electron energy level
Γ_1	-0.608	-0.608
N_1	-0.315	-0.303
$N_{1'}$	-0.268	
P_1	-0.110	-0.152
P_4	0.020	
H_{15}	-0.014	-0.012
H_{12}	0.017	
H_1	0.100	

The construction of symmetrized plane waves for all 73 symmorphic space groups may be simplified by use of the tables given by LUEHRMANN [1968], which supplement those given previously for the space groups O_h^1, O_h^9 and T_d^2 by SCHLOSSER [1962].

§ 6.4. *Symmetrized linear combinations of atomic orbitals*

In the tight-binding method proposed by BLOCH [1928], the energy eigenfunctions for an electron in a crystal are taken to be linear combinations of atomic orbitals, and for this reason the method is sometimes called the L.C.A.O. method. For simplicity it will be assumed that the space group of the crystal is symmorphic and that only one type of atomic nucleus is present. The construction of energy eigenfunctions then proceeds as follows. Suppose that $\psi_m(\boldsymbol{r})$ is an atomic energy eigenfunction. Then the '*Bloch sum*' $\psi_{\boldsymbol{k}m}(\boldsymbol{r})$ formed from $\psi_m(\boldsymbol{r})$ is defined by

$$\psi_{\boldsymbol{k}m}(\boldsymbol{r}) = M^{-\frac{1}{2}} \sum_{\boldsymbol{t}_n} \exp(\mathrm{i}\boldsymbol{k}\cdot\boldsymbol{t}_n)\, \psi_m(\boldsymbol{r} - \boldsymbol{t}_n), \tag{5.18}$$

where the summation is over all the lattice vectors of the crystal and M is a normalizing factor. The function $\psi_{\boldsymbol{k}m}(\boldsymbol{r})$ satisfies the defining condition (4.11) of a Bloch function. Each energy eigenfunction $\psi_{\boldsymbol{k}}(\boldsymbol{r})$ may then be taken as a linear combination of Bloch sums formed from different atomic orbitals, that is,

$$\psi_{\boldsymbol{k}}(\boldsymbol{r}) = \sum_{m} a_m \psi_{\boldsymbol{k}m}(\boldsymbol{r}). \tag{5.19}$$

As usual, the calculations are simplified by forming basis functions from linear combinations of Bloch sums. If $\{\boldsymbol{R} \mid \boldsymbol{0}\}$ is a rotation of $\mathscr{G}_o$, then

$$P(\{\boldsymbol{R} \mid \boldsymbol{0}\})\, \psi_{\boldsymbol{k}m}(\boldsymbol{r}) = M^{-\frac{1}{2}} \sum_{\boldsymbol{t}_n} \exp(\mathrm{i}\boldsymbol{k}\cdot\boldsymbol{t}_n)\, \psi_m(\boldsymbol{R}^{-1}\boldsymbol{r} - \boldsymbol{t}_n). \tag{5.20}$$

Suppose that $\boldsymbol{t}_{n'}$ is the lattice vector such that $\boldsymbol{R}^{-1}\boldsymbol{t}_{n'} = \boldsymbol{t}_n$. Then

$$\begin{aligned} \exp(\mathrm{i}\boldsymbol{k}\cdot\boldsymbol{t}_n)\, \psi_m(\boldsymbol{R}^{-1}\boldsymbol{r} - \boldsymbol{t}_n) &= \exp(\mathrm{i}\boldsymbol{k}\cdot\boldsymbol{R}^{-1}\boldsymbol{t}_{n'})\, \psi_m(\boldsymbol{R}^{-1}(\boldsymbol{r} - \boldsymbol{t}_{n'})) \\ &= \exp(\mathrm{i}\boldsymbol{R}\boldsymbol{k}\cdot\boldsymbol{t}_{n'})\, P(\{\boldsymbol{R} \mid \boldsymbol{0}\})\, \psi_m(\boldsymbol{r}_{n'}), \end{aligned}$$

where $\boldsymbol{r}_{n'}$ is replaced by $\boldsymbol{r} - \boldsymbol{t}_{n'}$ after the $P(\{\boldsymbol{R} \mid \boldsymbol{0}\})$ operation has taken place. If $\{\boldsymbol{R} \mid \boldsymbol{0}\}$ is a member of $\mathscr{G}_o(\boldsymbol{k})$ then $\exp(\mathrm{i}\boldsymbol{R}\boldsymbol{k}\cdot\boldsymbol{t}_{n'}) = \exp(\mathrm{i}\boldsymbol{k}\cdot\boldsymbol{t}_{n'})$. As the summation in (5.20) is over all lattice sites, it follows that

$$P(\{\boldsymbol{R} \mid \boldsymbol{0}\})\, \psi_{\boldsymbol{k}m}(\boldsymbol{r}) = M^{-\frac{1}{2}} \sum_{\boldsymbol{t}_n} \exp(\mathrm{i}\boldsymbol{k}\cdot\boldsymbol{t}_n)\, P(\{\boldsymbol{R} \mid \boldsymbol{0}\})\, \psi_m(\boldsymbol{r}_n),$$

where again $\boldsymbol{r}_n$ is replaced by $\boldsymbol{r} - \boldsymbol{t}_n$ after the $P(\{\boldsymbol{R} \mid \boldsymbol{0}\})$ operation has taken place. This function is a Bloch sum formed from the atomic orbital $P(\{\boldsymbol{R} \mid \boldsymbol{0}\})\psi_m(\boldsymbol{r})$. The Bloch sums, therefore, have the symmetry properties of the atomic orbitals from which they are formed. As linear combinations of the atomic orbitals may be arranged to transform as the spherical harmonics, x, yz, x^2-y^2, and so on, the corresponding linear combinations of Bloch sums also transform as these spherical harmonics, and so they may be labelled appropriately. Thus, for example, $\psi_{kxy}(\boldsymbol{r})$ has the property that if

$$P(\{\boldsymbol{R} \mid \boldsymbol{0}\})\, xy = b_1 xy + b_2 yz + b_3 zx\,,$$

then

$$P(\{\boldsymbol{R} \mid \boldsymbol{0}\})\, \psi_{kxy}(\boldsymbol{r}) = b_1 \psi_{kxy}(\boldsymbol{r}) + b_2 \psi_{kyz}(\boldsymbol{r}) + b_3 \psi_{kzx}(\boldsymbol{r})\,.$$

The functions $\psi_{kn}(\boldsymbol{r})$ do not form an ortho-normal set, but LÖWDIN [1950] has shown that it is possible, by taking linear combinations of atomic orbitals, to construct *orthogonalized* atomic orbitals, denoted by $\phi_j(\boldsymbol{r})$. Bloch sums may be formed from these, and Löwdin has shown that the coefficients are the same as in the ordinary Bloch sum (5.18), and that

$$\phi_{km}(\boldsymbol{r}) = M^{-\frac{1}{2}} \sum_{t_n} \exp(\mathrm{i}\boldsymbol{k}\cdot\boldsymbol{t}_n)\, \phi_m(\boldsymbol{r} - \boldsymbol{t}_n) \qquad (5.21)$$

is normalized if $M = N$, where N is the number of atoms in the basic block of the crystal. These Bloch sums have the property that

$$(\phi_{km}, \phi_{k'm'}) = \delta_{kk'}\delta_{mm'}\,.$$

If the matrix elements of the Hamiltonian H are denoted by

$$(n/m)_k = (\phi_{kn}, H\phi_{km})\,,$$

it follows that

$$(n/m)_k = \sum_{t_n} \exp(\mathrm{i}\boldsymbol{k}\cdot\boldsymbol{t}_n)\, E_{nm}(\boldsymbol{t}_n)\,, \qquad (5.22)$$

where

$$E_{nm}(\boldsymbol{t}_n) = \iiint \phi_n(\boldsymbol{r})^*\, H\phi_m(\boldsymbol{r} - \boldsymbol{t}_n)\, \mathrm{d}x\, \mathrm{d}y\, \mathrm{d}z\,. \qquad (5.23)$$

SLATER and KOSTER [1954] have shown that the Bloch sums (5.21) have the same symmetry properties as the Bloch sums (5.18). Therefore they transform in the same way as spherical harmonics, and again may be labelled by spherical harmonics. The linear combinations of spherical harmonics for $l \leqslant 2$ that form basis functions for the irreducible representations of the point groups $\mathscr{G}_o(\boldsymbol{k})$ for the body-centred cubic and face-centred cubic space groups O_h^9 and O_h^5 have been listed in tables 5.7 to 5.12. The basis functions formed from Bloch sums (5.21) can then be immediately obtained by the 1:1 correspondences:

$$\left.\begin{array}{ll}\phi_{\boldsymbol{k}s} \leftrightarrow 1\,, & \phi_{\boldsymbol{k}xy} \leftrightarrow xy\,, \\ \phi_{\boldsymbol{k}x} \leftrightarrow x\,, & \phi_{\boldsymbol{k}yz} \leftrightarrow yz\,, \\ \phi_{\boldsymbol{k}y} \leftrightarrow y\,, & \phi_{\boldsymbol{k}zx} \leftrightarrow zx\,, \\ \phi_{\boldsymbol{k}z} \leftrightarrow z\,, & \phi_{\boldsymbol{k}(x^2-y^2)} \leftrightarrow x^2 - y^2\,, \\ & \phi_{\boldsymbol{k}(3z^2-r^2)} \leftrightarrow (3z^2 - r^2)/\sqrt{3}\,. \end{array}\right| \qquad (5.24)$$

The factor $1/\sqrt{3}$ in the last correspondence appears because (x^2-y^2) and $(3z^2-r^2)/\sqrt{3}$ have the same normalizing constant.

The matrix elements of the Hamiltonian between these basis functions may be denoted by $(s/s)_{\boldsymbol{k}}$, $(s/x)_{\boldsymbol{k}}$, and so on. The corresponding integrals (5.23) may similarly be denoted by $E_{s,s}(\boldsymbol{t}_n)$, $E_{s,x}(\boldsymbol{t}_n)$, and so on. There exist many relationships between the energy integrals $E_{nm}(\boldsymbol{t}_n)$, which may be obtained as follows. By the definition (5.23),

$$E_{nm}(\boldsymbol{t}_n) = (\phi_n, HP(\{\boldsymbol{E} \mid \boldsymbol{t}_n\})\, \phi_m)\,.$$

The unitary property (3.8), the commutation property (3.12), and the relations (1.6) and (3.13) then show that if $\{\boldsymbol{R} \mid \boldsymbol{0}\}$ is a member of the space group $\mathscr{G}$, then

$$\begin{aligned} E_{nm}(\boldsymbol{t}_n) &= (P(\{\boldsymbol{R} \mid \boldsymbol{0}\})\, \phi_n, HP(\{\boldsymbol{E} \mid \boldsymbol{R}\boldsymbol{t}_n\})\, P(\{\boldsymbol{R} \mid \boldsymbol{0}\})\, \phi_m) \\ &= \sum_i \sum_j \boldsymbol{\Gamma}(\boldsymbol{R})^*_{jn}\, \boldsymbol{\Gamma}'(\boldsymbol{R})_{im}\, E_{ji}(\boldsymbol{R}\boldsymbol{t}_n)\,, \end{aligned} \qquad (5.25)$$

where $\boldsymbol{\Gamma}(\boldsymbol{R})$ and $\boldsymbol{\Gamma}'(\boldsymbol{R})$ are such that

$$P(\{\boldsymbol{R} \mid \boldsymbol{0}\})\, \phi_n(\boldsymbol{r}) = \sum_j \boldsymbol{\Gamma}(\boldsymbol{R})_{jn}\, \phi_j(\boldsymbol{r})\,,$$

and

$$P(\{\boldsymbol{R} \mid \boldsymbol{0}\})\, \phi_m(\boldsymbol{r}) = \sum_i \boldsymbol{\Gamma}'(\boldsymbol{R})_{im}\, \phi_i(\boldsymbol{r})\,.$$

[The matrices $\boldsymbol{\Gamma}(\boldsymbol{R})$ and $\boldsymbol{\Gamma}'(\boldsymbol{R})$ will be different if the atomic functions $\phi_n(\boldsymbol{r})$ and $\phi_m(\boldsymbol{r})$ correspond to different sets of kubic harmonics.]

As an example, consider the body-centred cubic space group O_h^9 and the energy integrals $E_{sx}((\pm a, 0, 0))$, $E_{sx}((0, \pm a, 0))$, $E_{sx}((0, 0, \pm a))$, and the similar integrals obtained by replacing x by y and then z. In this case

$$\begin{aligned} E_{sx}((a, 0, 0)) & \\ &= -E_{sx}((-a, 0, 0)) = E_{sy}((0, a, 0)) = -E_{sy}((0, -a, 0)) \\ &= E_{sz}((0, 0, a)) = -E_{sz}((0, 0, -a)), \end{aligned}$$

and all the other 12 integrals under consideration are equal to zero. The relationship between $E_{sx}((a, 0, 0))$ and $E_{sx}((-a, 0, 0))$ follows from (5.25) when $\boldsymbol{R}$ is taken to be $\boldsymbol{R}(I)$, $n=s$ and $m=x$, as

$$\left.\begin{aligned} P(\{\boldsymbol{R}(I) \mid \boldsymbol{0}\})\, \phi_s(\boldsymbol{r}) &= \phi_s(\boldsymbol{r}), \\ P(\{\boldsymbol{R}(I) \mid \boldsymbol{0}\})\, \phi_x(\boldsymbol{r}) &= -\phi_x(\boldsymbol{r}), \end{aligned}\right|$$

and $\boldsymbol{R}(I)(a, 0, 0) = (-a, 0, 0)$. Similarly the relationship between $E_{sx}((a, 0, 0))$ and $E_{sy}((0, a, 0))$ follows from (5.25) when $\boldsymbol{R}$ is taken to be $\boldsymbol{R}(C_{3\alpha})$, $n=s$ and $m=y$, as

$$\left.\begin{aligned} P(\{\boldsymbol{R}(C_{3\alpha}) \mid \boldsymbol{0}\})\, \phi_s(\boldsymbol{r}) &= \phi_s(\boldsymbol{r}), \\ P(\{\boldsymbol{R}(C_{3\alpha}) \mid \boldsymbol{0}\})\, \phi_y(\boldsymbol{r}) &= \phi_x(\boldsymbol{r}), \end{aligned}\right|$$

and $\boldsymbol{R}(C_{3\alpha})(0, a, 0) = (a, 0, 0)$. The vanishing of $E_{sx}((0, a, 0))$ follows from (5.25) when $\boldsymbol{R}$ is taken to be $\boldsymbol{R}(C_{2y})$, $n=s$ and $m=x$, as then (5.25) gives $E_{sx}((0, a, 0)) = -E_{sx}((0, a, 0))$. The other relations may be obtained in a similar way. The sum of the terms in the expression (5.22) for $(s/x)_{\boldsymbol{k}}$ corresponding to these integrals is then simply $2\mathrm{i} \sin(ak_x)\, E_{sx}((a, 0, 0))$. Similar expressions for all the other matrix elements (5.22) for the cubic space groups may be found in table II of the paper by SLATER and KOSTER [1954].

The information given in tables 5.7 to 5.12 together with the correspondences (5.24) is sufficient for a complete description of the energy bands of transition metals of the iron series, using the tight-binding method. Thus, if the expansion (5.19) contains Bloch sums based on atomic s, p and d orbitals, then the resulting secular equations for the energy eigenvalues at general points of the Brillouin zone are of ninth order. However, those corresponding to symmetry points, axes and planes will be partially or completely diagonalized by the use of the above basis functions. For example, for the space group O_h^9, as there is only one symmetrized spherical harmonic for $l \leqslant 2$ transforming as Γ_1, the Γ_1 energy eigenvalue is simply equal to $(s/s)_{\boldsymbol{k}}$, evaluated at $\boldsymbol{k}=0$. Similarly, as there is only one symmetrized spherical harmonic for $l \leqslant 2$ transforming as N_2, namely $(xy-zx)/\sqrt{2}$, the N_2 energy eigenvalue is equal to

$$\{\tfrac{1}{2}(xy/xy)_{\boldsymbol{k}} - (xy/zx)_{\boldsymbol{k}} + \tfrac{1}{2}(zx/zx)_{\boldsymbol{k}}\},$$

evaluated at $\boldsymbol{k}=(\pi/a)\,(0, 1, 1)$. On the other hand, there are two symmetrized spherical harmonics with $l \leqslant 2$ transforming according to each row of P_4, so that the P_4 energy eigenvalues are given by the secular equation

$$\begin{vmatrix} \{(x/x)_{\boldsymbol{k}} - E\} & (x/yz)_{\boldsymbol{k}} \\ (yz/x)_{\boldsymbol{k}} & \{(yz/yz)_{\boldsymbol{k}} - E\} \end{vmatrix} = 0,$$

each element being evaluated at $\boldsymbol{k}=(\pi/a)\,(1, 1, 1)$. Both of the eigenvalues produced by this equation are 3-fold degenerate.

The great difficulty in an *ab initio* calculation using the tight-binding method is the evaluation of the energy integrals (5.23). However, this method does provide a very convenient energy band interpolation scheme, in which the energy integrals (5.23) are taken as parameters, whose values are obtained by fitting to the known energy levels at the symmetry points and axes of the Brillouin zone. This interpolation role was proposed by SLATER and KOSTER [1954].

§ 6.5 *Symmetrized orthogonalized plane waves*

One reason why the plane wave expansion (5.11) does not converge rapidly is that the energy eigenfunction for a valence electron has to be orthogonal to the ion-core energy eigenfunctions, and consequently it has an oscillatory behaviour near each nucleus. A method for dealing with this difficulty was proposed by HERRING [1940]. This involves adding to each plane wave a set of Bloch sums formed from each of the ion-core atomic orbitals, giving the orthogonalized plane wave, or O.P.W.,

$$Y_{\boldsymbol{km}}(\boldsymbol{r}) = \exp\{\mathrm{i}(\boldsymbol{k} + \boldsymbol{K}_{\boldsymbol{m}})\cdot\boldsymbol{r}\} - \sum_{n} \mu_{\boldsymbol{m}n}\phi_{(\boldsymbol{k}+\boldsymbol{K}_{\boldsymbol{m}})n}(\boldsymbol{r}),$$

where $\phi_{\boldsymbol{k}n}(\boldsymbol{r})$ is the Bloch sum (5.21), and the summation is over all the ion-core atomic orbitals. Here $\mu_{\boldsymbol{m}n}$ is a constant chosen so that $Y_{\boldsymbol{km}}(\boldsymbol{r})$ is orthogonal to all the ion-core Bloch sums, that is, so that $(Y_{\boldsymbol{km}}, \phi_{\boldsymbol{k}n})=0$ for all ion-core states n.

Basis functions can be formed from O.P.W.s in the usual way. They consist simply of symmetrized plane waves, as found in §6.3, together with symmetrized Bloch sums, as found in §6.4. In some cases, the Bloch sums disappear. For example, in applying the method to lithium, which contains only 1s states in the ion-core, only the Bloch sums $\phi_{\boldsymbol{k}s}(\boldsymbol{r})$ appear in O.P.W.s. If the symmetrized O.P.W. transforms according to a non-s-like irreducible representation, such as Γ_{12} for example, as the plane wave part is automatically orthogonal to $\phi_{\boldsymbol{k}s}(\boldsymbol{r})$ by the first matrix element theorem of ch. 3 §6.1, then $\phi_{\boldsymbol{k}s}(\boldsymbol{r})$ does not appear.

§ 7. TIME-REVERSAL SYMMETRY

§ 7.1. *Criteria for extra degeneracies*

The Hamiltonian $H(\boldsymbol{r})$ of ch. 3 §3.2 is real, and so are all its eigenvalues. Thus, when the complex conjugate is taken of the Schrödinger equation in the form (4.12), the result is

$$H(\boldsymbol{r})\,\phi_{\boldsymbol{k}}(\boldsymbol{r})^* = E(\boldsymbol{k})\,\phi_{\boldsymbol{k}}(\boldsymbol{r})^*.$$

However, $\phi_{\boldsymbol{k}}(\boldsymbol{r})^*$ is a Bloch function corresponding to wave vector $-\boldsymbol{k}$, and so $E(\boldsymbol{k})=E(-\boldsymbol{k})$ for all $\boldsymbol{k}$. Thus $E(\boldsymbol{k})$ has inversion symmetry, *even* when $\mathscr{G}_o$ does not contain I.

The name 'time-reversal symmetry' for this phenomenon occurs because, if the process of complex conjugation is applied to the time-dependent Schrödinger equation $H(\boldsymbol{r})\,\psi = \mathrm{i}\hbar\partial\psi/\partial t$, the result is $H(\boldsymbol{r})\,\psi^* = \mathrm{i}\hbar\partial\psi^*/\partial(-t)$. That is, the sign of the time t is reversed.

The equality $E(\boldsymbol{k}) = E(-\boldsymbol{k})$ represents a degeneracy that is additional to those caused by rotational symmetry if $\mathscr{G}_o$ does not contain I. However, time-reversal symmetry can also cause further degeneracies. These can be predicted very simply from the character tables of the groups $\mathscr{G}_o(\boldsymbol{k})$, as the following analysis will show.

The original study of the effects of time-reversal symmetry on a general group of the Schrödinger equation was made by WIGNER [1932]. (A translated version is given in the book by WIGNER [1959].) The specific application to space groups was investigated by HERRING [1937a]. The present analysis up to eq. (5.27) is merely Herring's theory, modified to the special case of symmorphic space groups.

Wigner showed that there are three cases to be considered, and these depend on the nature of the irreducible representation $\boldsymbol{\Gamma}$ of the group of the Schrödinger equation to which the energy eigenvalue under consideration belongs. These are

(a) $\boldsymbol{\Gamma}$ equivalent to a representation by real matrices,

(b) $\boldsymbol{\Gamma}$ not equivalent to $\boldsymbol{\Gamma}^*$, the complex conjugate representation, and

(c) $\boldsymbol{\Gamma}$ equivalent to $\boldsymbol{\Gamma}^*$, but not equivalent to a representation by real matrices.

In case (a), the energy eigenvalue corresponding to $\boldsymbol{\Gamma}$ is different, in general, from all other energy eigenvalues, so that no additional degeneracies occur. In cases (b) and (c), the energy eigenvalues corresponding to $\boldsymbol{\Gamma}$ and $\boldsymbol{\Gamma}^*$ are equal, the eigenfunctions corresponding to $\boldsymbol{\Gamma}^*$ are the complex conjugates of those corresponding to $\boldsymbol{\Gamma}$, and the two sets of eigenfunctions are linearly independent. Therefore, in cases (b) and (c), there is an additional degeneracy.

Fortunately, there exists a simple criterion, formulated by

FROBENIUS and SCHUR [1906], for deciding to which case $\boldsymbol{\Gamma}$ belongs, namely

$$\sum_T \chi(T^2) = \begin{cases} g\,, & \text{case } (a)\,, \\ 0\,, & \text{case } (b)\,, \\ -g\,, & \text{case } (c)\,. \end{cases} \tag{5.26}$$

Here the summation is over all transformations T of the group, $\chi(T)$ is the character of T, and g is the order of the group. For electrons in a crystal, the group of the Schrödinger equation is the space group $\mathscr{G}$. By the fundamental theorem of § 1, every irreducible representation of $\mathscr{G}$ is specified by a wave vector $\boldsymbol{k}$ and an irreducible representation $\boldsymbol{\Gamma}^p$ of the point group $\mathscr{G}_0(\boldsymbol{k})$. The criterion (5.26) for this representation of $\mathscr{G}$ then reduces to the readily applicable form

$$\sum \chi^p(\{\boldsymbol{Q}\,|\,\boldsymbol{0}\}^2) = \begin{cases} g_0(\boldsymbol{k})\,, & \text{case (a)} \\ 0\,, & \text{case (b)} \\ -g_0(\boldsymbol{k})\,, & \text{case (c)}\,, \end{cases} \tag{5.27}$$

where the summation is over all the elements $\{\boldsymbol{Q}\,|\,\boldsymbol{0}\}$ of the point group $\mathscr{G}_0$ that transform $\boldsymbol{k}$ into a vector equivalent to $-\boldsymbol{k}$, $g_0(\boldsymbol{k})$ is the order of $\mathscr{G}_0(\boldsymbol{k})$, and $\chi^p(\{\boldsymbol{Q}\,|\,\boldsymbol{0}\}^2)$ denotes the character of $\{\boldsymbol{Q}\,|\,\boldsymbol{0}\}^2$ in the irreducible representation $\boldsymbol{\Gamma}^p$ of $\mathscr{G}_0(\boldsymbol{k})$. It may be noted that, although $\{\boldsymbol{Q}\,|\,\boldsymbol{0}\}$ may not be a member of $\mathscr{G}_0(\boldsymbol{k})$, $\{\boldsymbol{Q}\,|\,\boldsymbol{0}\}^2$ necessarily is a member of $\mathscr{G}_0(\boldsymbol{k})$.

Before proving (5.27), it is worthwhile examining what happens if cases (b) or (c) occur in terms of the degeneracy of $E(\boldsymbol{k})$, this degeneracy being considered in the sense of the penultimate paragraph of § 2.1. There are three cases to be considered. They are as follows.

(1) $-\boldsymbol{k}$ is neither in the star of $\boldsymbol{k}$ nor equivalent to $\boldsymbol{k}$. In this case there is no extra degeneracy of $E(\boldsymbol{k})$. That is, there are no more eigenfunctions with this energy $E(\boldsymbol{k})$ *and* this wave vector $\boldsymbol{k}$. The extra overall degeneracy corresponds to $E(\boldsymbol{k})=E(-\boldsymbol{k})$, that is, to eigenvalues corresponding to different wave vectors. This case can only occur if $\mathscr{G}_0$ does not contain I. The reason for this

behaviour is that if $\phi^p_{ks}(r)$, $s=1, 2, ..., l$, are the basis functions of the irreducible representation Γ^p of $\mathscr{G}_o(k)$, then the basis functions of the corresponding irreducible representation Γ^{kp} of the space group $\mathscr{G}$ are $P(\{R_i \mid 0\})\, \phi^p_{ks}(r)$, the rotations $\{R_i \mid 0\}$ generating the star of k. The basis functions of $(\Gamma^{kp})^*$ are then $P(\{R_i \mid 0\})\, \phi^p_{ks}(r)^*$, and none of these have k as their wave vector, if $-k$ is not in the star of k nor is equivalent to k.

(2) $-k$ is in the star of k. In this case the argument given in (1) shows that there is an extra degeneracy of $E(k)$ because of time-reversal symmetry. As $(\Gamma^{kp})^*$ corresponds in this case to wave vector k, as well as to $-k$, $(\Gamma^{kp})^*$ must be equivalent to some representation Γ^{kq}. If $p \neq q$, this corresponds to case (b) above, while if $p=q$, this corresponds to case (c). The extra degeneracy at k is then a degeneracy between the levels $E(k)$ corresponding to the irreducible representations Γ^p and Γ^q of $\mathscr{G}_o(k)$. This is sometimes described in the literature by saying that Γ^p and Γ^q 'stick together' as a result of time-reversal symmetry.

(3) $-k$ is equivalent to k. In this case there is again an extra degeneracy of $E(k)$ due to time-reversal symmetry, and if Γ^{kq} is the representation that is equivalent to $(\Gamma^{kp})^*$, Γ^p and Γ^q stick together.

It often happens in cases (2) and (3) that there are only two irreducible representations of $\mathscr{G}_o(k)$ of type (b), and obviously they are the two representations that stick together. However, it may happen that there are more than two irreducible representations of $\mathscr{G}_o(k)$ of type (b), in which case the analysis has to be carried a stage further in order to find how the representations pair off. This may be done as follows.

In case (2), as $-k$ is in the star of k, there exists an $\{R_i \mid 0\}$ of $\mathscr{G}_o$ such that $R_i k = -k$. There is, by construction, only one such $\{R_i \mid 0\}$ of the set (5.1) with this property. Denote this R_i by R_-. (Of course, any member of the set $R_- \mathscr{G}_o(k)$ has this property, so that the choice of R_- is arbitrary to this extent.) Then if Γ^p and Γ^q are two irreducible representations of $\mathscr{G}_o(k)$ of type (b) that stick together, and χ^p and χ^q denote their characters, then

$$\chi^p(\{\boldsymbol{R} \mid \boldsymbol{0}\})^* = \chi^q(\{\boldsymbol{R}_-^{-1}\boldsymbol{R}\boldsymbol{R}_- \mid \boldsymbol{0}\}) \tag{5.28}$$

for every $\{\boldsymbol{R} \mid \boldsymbol{0}\}$ of $\mathscr{G}_o(\boldsymbol{k})$. Thus, an examination of the character table of $\mathscr{G}_o(\boldsymbol{k})$ is sufficient to determine which $\boldsymbol{\Gamma}^q$ of type (b) pairs off with $\boldsymbol{\Gamma}^p$. The condition (5.28) may be proved in the following way. If $(\boldsymbol{\Gamma}^{kp})^*$ is equivalent to $\boldsymbol{\Gamma}^{kq}$ then the functions $\phi^p_{\boldsymbol{k}s}(\boldsymbol{r})^*$ must be linear combinations of the functions $P(\{\boldsymbol{R}_- \mid \boldsymbol{0}\})\, \phi^q_{\boldsymbol{k}t}(\boldsymbol{r})$. By an appropriate similarity transformation, the $\phi^q_{\boldsymbol{k}t}(\boldsymbol{r})$ may then be arranged so that

$$\phi^p_{\boldsymbol{k}s}(\boldsymbol{r})^* = P(\{\boldsymbol{R}_- \mid \boldsymbol{0}\})\, \phi^q_{\boldsymbol{k}s}(\boldsymbol{r}), \tag{5.29}$$

for every row s. However, for any $\{\boldsymbol{R} \mid \boldsymbol{0}\}$ of $\mathscr{G}_o(\boldsymbol{k})$,

$$P(\{\boldsymbol{R} \mid \boldsymbol{0}\})\, \phi^p_{\boldsymbol{k}s}(\boldsymbol{r})^* = \sum_t \Gamma^p(\{\boldsymbol{R} \mid \boldsymbol{0}\})^*_{ts}\, \phi^p_{\boldsymbol{k}t}(\boldsymbol{r})^*,$$

while

$$\begin{aligned} P(\{\boldsymbol{R} \mid \boldsymbol{0}\})\,[P(\{\boldsymbol{R}_- \mid \boldsymbol{0}\})\, \phi^q_{\boldsymbol{k}s}(\boldsymbol{r})] &\\ = P(\{\boldsymbol{R}_- \mid \boldsymbol{0}\})\, P(\{\boldsymbol{R}_-^{-1}\boldsymbol{R}\boldsymbol{R}_- \mid \boldsymbol{0}\})\, \phi^q_{\boldsymbol{k}s}(\boldsymbol{r}) &\\ = \sum_t \Gamma^q(\{\boldsymbol{R}_-^{-1}\boldsymbol{R}\boldsymbol{R}_- \mid \boldsymbol{0}\})_{ts}\, [P(\{\boldsymbol{R}_- \mid \boldsymbol{0}\})\, \phi^q_{\boldsymbol{k}t}(\boldsymbol{r})], \end{aligned}$$

as $\{\boldsymbol{R}_-^{-1}\boldsymbol{R}\boldsymbol{R}_- \mid \boldsymbol{0}\}$ is a member of $\mathscr{G}_o(\boldsymbol{k})$. Thus (5.29) is only satisfied if

$$\Gamma^p(\{\boldsymbol{R} \mid \boldsymbol{0}\})^*_{ts} = \Gamma^q(\{\boldsymbol{R}_-^{-1}\boldsymbol{R}\boldsymbol{R}_- \mid \boldsymbol{0}\})_{ts}$$

for all t and s, that is, if (5.28) holds.

In case (3), where $-\boldsymbol{k}$ is equivalent to $\boldsymbol{k}$, $(\boldsymbol{\Gamma}^{kp})^*$ is equivalent to $\boldsymbol{\Gamma}^{kq}$ if the functions $\phi^p_{\boldsymbol{k}s}(\boldsymbol{r})^*$ are linear combinations of the $\phi^q_{\boldsymbol{k}t}(\boldsymbol{r})$, so that the condition corresponding to (5.28) is that

$$\chi^p(\{\boldsymbol{R} \mid \boldsymbol{0}\})^* = \chi^q(\{\boldsymbol{R} \mid \boldsymbol{0}\}) \tag{5.30}$$

for every $\{\boldsymbol{R} \mid \boldsymbol{0}\}$ of $\mathscr{G}_o(\boldsymbol{k})$. This condition again involves only the character table of $\mathscr{G}_o(\boldsymbol{k})$.

The rest of this subsection is devoted to the derivation of (5.27) from (5.26). Applications of (5.27) follow in §7.2.

Eq. (5.3) shows that, as $\{\boldsymbol{R} \mid \boldsymbol{t}_n\}^2 = \{\boldsymbol{R}^2 \mid \boldsymbol{R}\boldsymbol{t}_n + \boldsymbol{t}_n\}$,

$$\sum \chi^{kp}(\{\boldsymbol{R} \mid \boldsymbol{t}_n\}^2) = \sum_{\boldsymbol{R},\, \boldsymbol{R}_j,\, \boldsymbol{t}_n} \chi^p(\{\boldsymbol{R}_j^{-1}\boldsymbol{R}^2\boldsymbol{R}_j \mid \boldsymbol{0}\}) \exp\{-\mathrm{i}\boldsymbol{R}_j\boldsymbol{k}\cdot(\boldsymbol{R}\boldsymbol{t}_n + \boldsymbol{t}_n)\}. \quad (5.31)$$

Here the summation on the left-hand side is over all transformations $\{\boldsymbol{R} \mid \boldsymbol{t}_n\}$ of the space group, and so corresponds to the left-hand side of (5.26), while the summation on the right-hand side involves only those $\boldsymbol{R}$ and $\boldsymbol{R}_j$ such that $\{\boldsymbol{R}_j^{-1}\boldsymbol{R}^2\boldsymbol{R}_j \mid \boldsymbol{0}\}$ is a member of $\mathscr{G}_o(\boldsymbol{k})$. However,

$$\sum_{\boldsymbol{t}_n} \exp\{-\mathrm{i}\boldsymbol{R}_j\boldsymbol{k}\cdot(\boldsymbol{R}\boldsymbol{t}_n + \boldsymbol{t}_n)\} = \sum_{\boldsymbol{t}_n} \exp\{-\mathrm{i}(\boldsymbol{R}^{-1} + 1)\,\boldsymbol{R}_j\boldsymbol{k}\cdot\boldsymbol{t}_n\},$$

which, by eq. (4.15) is equal to N if $\boldsymbol{R}^{-1}\boldsymbol{R}_j\boldsymbol{k}$ is equivalent to $-\boldsymbol{R}_j\boldsymbol{k}$, but is zero otherwise. If $\boldsymbol{R}^{-1}\boldsymbol{R}_j\boldsymbol{k}$ is equivalent to $-\boldsymbol{R}_j\boldsymbol{k}$, then $\boldsymbol{R}_j^{-1}\boldsymbol{R}^2\boldsymbol{R}_j$ is necessarily a member of $\mathscr{G}_o(\boldsymbol{k})$, so the condition for non-vanishing is more stringent here than in (5.31).

Let $\boldsymbol{Q}$ be a rotation such that $\boldsymbol{Q}\boldsymbol{k}$ is equivalent to $-\boldsymbol{k}$. Then, if $\boldsymbol{R} = \boldsymbol{R}_j\boldsymbol{Q}\boldsymbol{R}_j^{-1}$, $\boldsymbol{R}^{-1}\boldsymbol{R}_j\boldsymbol{k}$ is equivalent to $-\boldsymbol{R}_j\boldsymbol{k}$, and only such rotations are to be retained in (5.31). Then, as $\boldsymbol{R}_j^{-1}\boldsymbol{R}^2\boldsymbol{R}_j = \boldsymbol{Q}^2$, (5.31) becomes simply

$$\sum \chi^{kp}(\{\boldsymbol{R} \mid \boldsymbol{t}_n\}^2) = NM(\boldsymbol{k}) \sum_{\boldsymbol{Q}} \chi^p(\{\boldsymbol{Q} \mid \boldsymbol{0}\}^2).$$

Eq. (5.27) then follows, as the order of $\mathscr{G}$ is $g = Ng_o = NM(\boldsymbol{k})\, g_o(\boldsymbol{k})$, by (5.2).

§ 7.2. *Application to the space groups O_h^9, O_h^5 and T_d^2*

The application of the theory of § 7.1 to the cubic space groups O_h^9 and O_h^5 will be considered first. For a general point $\boldsymbol{k}$ of the Brillouin zone, the only operator of $\mathscr{G}_o$ that transforms $\boldsymbol{k}$ into a vector equivalent to $-\boldsymbol{k}$ is the inversion operator I. As $I^2 = E$, $\chi(E) = 1$ and $g_o(\boldsymbol{k}) = 1$, eq. (5.27) shows that this is an example of case (a). There are, therefore, no additional degeneracies at general points for these space groups. Similarly, for $\boldsymbol{k}$ on a symmetry plane, I and the rotation through π about the two-fold axis perpendicular to the plane each transform $\boldsymbol{k}$ into a vector equivalent to $-\boldsymbol{k}$. As the squares of both of these operators are equal to E, and as $\chi(E)=1$ for both irreducible representations of $\mathscr{G}_o(\boldsymbol{k})$ and as $g_o(\boldsymbol{k})=2$, eq. (5.27) shows that this is another example of case (a). On applying the same argument to the symmetry axes and the symmetry points, it is found that *no* additional degeneracies due to time-reversal symmetry occur anywhere for the space groups O_h^9 and O_h^5.

For the space group T_d^2 of the zinc blende structure, the situation is quite different. This space group was described in ch. 1 §3.2, where it was mentioned that its crystal lattice is the same as for the face-centred cubic structure. Consequently, it has the Brillouin zone that is shown in fig. 4.5, but the point group $\mathscr{G}_o$ is T_d, a proper subgroup of O_h. The point group T_d does not contain the inversion operator I, so that the degeneracy $E(\boldsymbol{k})=E(-\boldsymbol{k})$ is caused in this case by time-reversal symmetry.

The analysis of further degeneracies for the space group T_d^2 is quite straightforward. For each $\boldsymbol{k}$, the group $\mathscr{G}_o(\boldsymbol{k})$ is the *common* subgroup of T_d and the corresponding group $\mathscr{G}_o(\boldsymbol{k})$ of the space group O_h^5. For a general point $\boldsymbol{k}$ there is no operator in $\mathscr{G}_o$ that transforms $\boldsymbol{k}$ into a vector equivalent to $-\boldsymbol{k}$, so that this is an example of case (b). As $-\boldsymbol{k}$ is neither in the star of $\boldsymbol{k}$ nor equivalent to $\boldsymbol{k}$, this is also an example of case (1) of §7.1, so that the resulting degeneracy is just the degeneracy $E(\boldsymbol{k})=E(-\boldsymbol{k})$ mentioned above.

A more interesting situation occurs for $\boldsymbol{k}$ on the Δ axis of T_d^2, for which $\boldsymbol{k}=(\pi/a)(0,0,2\kappa)$, $0<\kappa<1$, and $\mathscr{G}_o(\boldsymbol{k})$ is the point group C_{2v}, with classes $\mathscr{C}_1=E$, $\mathscr{C}_2=C_{2z}$, $\mathscr{C}_3=IC_{2a}$, $\mathscr{C}_4=IC_{2b}$. Four operations of $\mathscr{G}_o$ transform $\boldsymbol{k}$ into a vector equivalent to $-\boldsymbol{k}$, namely C_{2x}, C_{2y}, IC_{4z} and IC_{4z}^{-1}, whose squares are E, E, C_{2z} and C_{2z} respectively. The application of (5.27) to the character table for C_{2v} given in table A.15 shows that the irreducible representations Δ_1 and Δ_2 correspond to case (a), but Δ_3 and Δ_4 correspond to case (b). (The relevant notation for the irreducible representations is that of the third column of table A.15.) As $-\boldsymbol{k}$ is in the star of $\boldsymbol{k}$, this is an example of case (2) of §7.1, so the energy levels at $\boldsymbol{k}$ corresponding to Δ_3 and Δ_4 stick together. Of course, no other possible pairing could occur here, but it is worthwhile demonstrating how this pairing follows from the condition (5.28). Choosing $\boldsymbol{R}_-$ to be $\boldsymbol{R}(C_{2x})$, as

$$C_{2x}^{-1}EC_{2x} = E\,, \qquad C_{2x}^{-1}C_{2z}C_{2x} = C_{2z}\,,$$
$$C_{2x}^{-1}IC_{2a}C_{2x} = IC_{2b}\,, \qquad C_{2x}^{-1}IC_{2b}C_{2x} = IC_{2a}\,,$$

the character table A.15 then shows that

$$\chi^3(\{\boldsymbol{R}\mid \boldsymbol{0}\})^* = \chi^4(\{\boldsymbol{R}_-^{-1}\boldsymbol{R}\boldsymbol{R}_-\mid \boldsymbol{0}\}) \quad \text{for every} \quad \{\boldsymbol{R}\mid \boldsymbol{0}\} \text{ of } \mathscr{G}_o(\boldsymbol{k}).$$

Similar degeneracies caused by time-reversal symmetry occur at other points $\boldsymbol{k}$ for the space group T_d^2. All of them have been listed by PARMENTER [1955].

Chapter 6

FURTHER ABSTRACT GROUP THEORY

This chapter is devoted to a study of the further abstract group theory that is needed in the development of the theory of the representations of non-symmorphic space groups, which will be given in ch. 7. As the ideas encountered are somewhat more abstract than those met hitherto, all the proofs will be given in detail.

§ 1. REARRANGEMENT THEOREMS

Theorem. Let $\mathscr{G}$ be a group of order g with elements $A_1(=E), A_2, \ldots, A_g$. Then, if A_n is *any* member of $\mathscr{G}$, the set of products $A_nA_1, A_nA_2, \ldots, A_nA_g$ contains every element of $\mathscr{G}$ once and only once.

[*Proof.* It is easily shown that every element of $\mathscr{G}$ is contained in this set, for let A_m be any element of $\mathscr{G}$. Then there exists an element A_r of $\mathscr{G}$, namely $A_r = A_n^{-1}A_m$, such that $A_nA_r = A_m$.

To see that a given element appears only once in the set, consider what happens if A_nA_r and A_nA_s are equal. On multiplying both by A_n^{-1}, one finds that $A_r = A_s$, that is, A_r and A_s are not distinct.]

There is an identical theorem for the set of products evaluated in the opposite order, namely $A_1A_n, A_2A_n, \ldots, A_gA_n$. One implication of these theorems is that, in the group multiplication table, every element of the group appears once, and only once, in every row, and in every column. It will be seen that the group multiplication tables given in tables 1.1 and 1.2 display this necessary property.

§ 2. COSETS

Definitions of cosets. Let $\mathscr{S}$ be a subgroup of order s of a group $\mathscr{G}$ of order g, and suppose that the elements of $\mathscr{S}$ are $S_1(=E)$, $S_2, \ldots, S_s$. If A is an element of $\mathscr{G}$, which may or may not be an element of $\mathscr{S}$, then the set of s elements $S_1A, S_2A, \ldots, S_sA$ is called the *right coset* of $\mathscr{S}$ with respect to A, and is denoted by $\mathscr{S}A$. Similarly, the set of s elements $AS_1, AS_2, \ldots, AS_s$ is called the *left coset* of $\mathscr{S}$ with respect to A, and is correspondingly denoted by $A\mathscr{S}$.

In the discussion of sets, and in particular of cosets, that follows, it should be borne in mind that two sets are said to be *identical* merely if they contain the same members. These members need not be arranged in the same order in both sets.

As an example, suppose that $\mathscr{G}$ is the point group C_{3v} of ch. 1 §2.2, and $\mathscr{S}$ is the subgroup consisting of E and IC_{2y}. Then, from table 1.2, it follows that the right cosets are

$$\left.\begin{aligned} \mathscr{S}E &= \mathscr{S}(IC_{2y}) = \{E, IC_{2y}\} = \mathscr{S}\,, \\ \mathscr{S}C_{3z} &= \mathscr{S}(IC_{2C}) = \{C_{3z}, IC_{2C}\}\,, \\ \mathscr{S}C_{3z}^{-1} &= \mathscr{S}(IC_{2D}) = \{C_{3z}^{-1}, IC_{2D}\}\,, \end{aligned}\right|$$

whereas the left cosets are

$$\left.\begin{aligned} E\mathscr{S} &= (IC_{2y})\,\mathscr{S} = \{E, IC_{2y}\} = \mathscr{S}\,, \\ C_{3z}\mathscr{S} &= (IC_{2D})\,\mathscr{S} = \{C_{3z}, IC_{2D}\}\,, \\ C_{3z}^{-1}\mathscr{S} &= (IC_{2C})\,\mathscr{S} = \{C_{3z}^{-1}, IC_{2C}\}\,. \end{aligned}\right|$$

One important point to be noted is that $(IC_{2C})\,\mathscr{S} \neq \mathscr{S}(IC_{2C})$, which demonstrates that, in general, $A\mathscr{S} \neq \mathscr{S}A$, that is, right and left cosets formed from the same element A are not necessarily identical.

The properties of cosets will now be summarized. The results and proofs are given for right cosets, but exactly the same results apply for left cosets. The above example of the group C_{3v} gives an illustration of these properties.

(a) If A is a member of $\mathscr{S}$, then $\mathscr{S}A$ is merely $\mathscr{S}$.

[*Proof.* This result follows immediately on applying the rearrangement theorem of §1 to $\mathscr{S}$, considering $\mathscr{S}$ as a group.]

(b) If A is not a member of $\mathscr{S}$, then $\mathscr{S}A$ is not a group.

[*Proof.* If $\mathscr{S}A$ were a group, it would contain the identity E. There would then exist an element S_m of $\mathscr{S}$ such that $S_mA=E$. This would imply that $A=S_m^{-1}$, and hence that A is a member of $\mathscr{S}$, which is contrary to the initial assumption about A.]

(c) Every element of $\mathscr{G}$ lies in some right coset.

[*Proof.* Let A be any element of $\mathscr{G}$. Then A is certainly a member of $\mathscr{S}A$, as $A=S_1A=EA$.]

(d) Each right coset contains s distinct elements.

[*Proof.* Suppose that $\mathscr{S}A$ is a coset in which $S_mA=S_nA$, but $S_m \neq S_n$. Post-multiplying by A^{-1} gives $S_m=S_n$, which is a contradiction.]

(e) Two right cosets are either identical or have no elements in common.

[*Proof.* Let $\mathscr{S}A$ and $\mathscr{S}B$ be two right cosets of $\mathscr{S}$, and suppose that they possess a common element. Then it will be shown that $\mathscr{S}A$ and $\mathscr{S}B$ are identical.

Let $S_mA=S_nB$ be the common element of $\mathscr{S}A$ and $\mathscr{S}B$. Then $AB^{-1}=S_m^{-1}S_n$. As $S_m^{-1}S_n$ is a member of $\mathscr{S}$, AB^{-1} must be a member of $\mathscr{S}$. Hence the coset $\mathscr{S}(AB^{-1})$ is identical to $\mathscr{S}$, by property (a) above. Thus the set $\mathscr{S}(AB^{-1})\,B$ is identical to the set $\mathscr{S}B$. However, the set $\mathscr{S}(AB^{-1})B$ consists of the elements $S_1(AB^{-1})\,B=S_1A$, $S_2(AB^{-1})\,B=S_2A, \ldots$, so that $\mathscr{S}(AB^{-1})\,B=\mathscr{S}A$. Hence $\mathscr{S}A$ is identical to $\mathscr{S}B$.]

(f) If B is a member of $\mathscr{S}A$, then $\mathscr{S}B$ is identical to $\mathscr{S}A$. This result shows that the same coset is formed from any member of the coset, so that *any* member of the coset may be taken as the '*coset representative*'. This concept is used in §4 in constructing factor groups.

[*Proof.* As $B=EB=S_1B$, B is certainly a member of $\mathscr{S}B$. Thus, by property (e), if B is a member of $\mathscr{S}A$, then $\mathscr{S}B$ and $\mathscr{S}A$ are identical.]

(g) The number of different cosets i, which is necessarily an integer, satisfies the equation $g = si$.

[*Proof.* By property (d), each right coset contains s distinct elements, so if there are i distinct cosets, then the set of cosets contains si distinct elements. As every one of these elements is a member of $\mathscr{G}$, and by property (c) every member of $\mathscr{G}$ is in this set of cosets, then $g = si$.]

It may now be seen that the set of rotations $\{\boldsymbol{R}_1 \mid \boldsymbol{0}\}$, $\{\boldsymbol{R}_2 \mid \boldsymbol{0}\}, \ldots$ that were introduced in connection with eq. (5.1) are merely a set of coset representations in the decomposition of the point group $\mathscr{G}_o$ into left cosets with respect to its subgroup $\mathscr{G}_o(\boldsymbol{k})$. Eq. (5.2) is just the property (g) above.

§ 3. INVARIANT SUBGROUPS

As in §2, let $\mathscr{S}$ be a subgroup of order s of a group $\mathscr{G}$, and suppose that the elements of $\mathscr{S}$ are $S_1\,(=E), S_2, \ldots, S_s$. Let X be a fixed element of $\mathscr{G}$. Then the set of elements $XS_1X^{-1}, XS_2X^{-1}, \ldots, XS_sX^{-1}$ will be denoted by $X\mathscr{S}X^{-1}$.

Theorem. $X\mathscr{S}X^{-1}$ is a subgroup of $\mathscr{G}$.

[*Proof.* It is obvious that every element of $X\mathscr{S}X^{-1}$ is a member of $\mathscr{G}$, so all that is required is to prove that $X\mathscr{S}X^{-1}$ is a group. This involves showing as follows that the four group postulates of ch. 1 §2.1 hold.

(a) If XS_mX^{-1} and XS_nX^{-1} are any two elements of $X\mathscr{S}X^{-1}$, then $(XS_mX^{-1})\,(XS_nX^{-1}) = X(S_mS_n)\,X^{-1}$, which is also a member of $X\mathscr{S}X^{-1}$, as S_mS_n is a member of $\mathscr{S}$.

(b) If XS_mX^{-1}, XS_nX^{-1} and XS_pX^{-1} are three elements of $X\mathscr{S}X^{-1}$, then

$$(XS_mX^{-1}\cdot XS_nX^{-1})\,XS_pX^{-1} = X(S_mS_nS_p)\,X^{-1} = XS_mX^{-1}\,(XS_nX^{-1}\cdot XS_pX^{-1}).$$

(c) The identity E is a member of $X\mathscr{S}X^{-1}$, as $E=XS_1X^{-1}$.
(d) The inverse of XS_mX^{-1} is $XS_m^{-1}X^{-1}$, which is also a member of $X\mathscr{S}X_m^{-1}$, as S_m^{-1} is a member of $\mathscr{S}$.]

Definition of an invariant subgroup. If the set of elements of the subgroup $X\mathscr{S}X^{-1}$ is the same as the set of elements of $\mathscr{S}$ for *every* element X of $\mathscr{G}$, then $\mathscr{S}$ is said to be an *invariant* subgroup of $\mathscr{G}$.

Invariant subgroups are also sometimes known as normal subgroups or normal divisors. It should be noted that the definition does not imply that $XS_mX^{-1}=S_m$ for every S_m of $\mathscr{S}$ and every X of $\mathscr{G}$, but only that the two sets $X\mathscr{S}X^{-1}$ and $\mathscr{S}$ contain the same elements, but possibly in a different order.

The real importance of invariant subgroups lies in the fact that they are used in the construction of factor groups, as will be shown in §4. However, the following properties are also worth recording.
(a) An invariant subgroup $\mathscr{S}$ of $\mathscr{G}$ consists entirely of *complete* classes of $\mathscr{G}$. The converse is also true.

[*Proof.* Let S_m be any member of $\mathscr{S}$. Then, as $\mathscr{S}$ is an invariant subgroup, XS_mX^{-1} is a member of $\mathscr{S}$ for every X of $\mathscr{G}$. That is, the class containing S_m is entirely contained in $\mathscr{S}$. The proof of the converse is trivial.]

This converse property provides the easiest way of recognizing invariant subgroups of a group $\mathscr{G}$, if the subgroups and classes of $\mathscr{G}$ are known. It shows that the group $\mathscr{G}$ itself and the subgroup consisting only of the identity E are both invariant subgroups of $\mathscr{G}$, although they are rather trivial. It also shows that the only non-trivial invariant subgroup of the point group C_{3v}, whose subgroups and classes were described in ch. 1 §2.3 and ch. 1 §2.4 respectively, is that consisting of E, C_{3z} and C_{3z}^{-1}.
(b) The left and right cosets $A\mathscr{S}$ and $\mathscr{S}A$ of an invariant subgroup are identical for every A of $\mathscr{G}$.

[*Proof.* It will be shown that if B is a member of $\mathscr{S}A$, then B is a member of $A\mathscr{S}$. The above assertion follows on combining this result with the converse result, which may be proved by a similar argument.

If B is a member of $\mathscr{S}A$, there exists an element S_m of $\mathscr{S}$ such that $B=S_mA$. Then $A^{-1}B=A^{-1}S_mA$, which is a member of $A^{-1}\mathscr{S}A$, and so is also a member of $\mathscr{S}$, as $\mathscr{S}$ is an invariant subgroup. As $A^{-1}B$ is therefore a member of $\mathscr{S}$, then $A(A^{-1}B)=B$ is a member of $A\mathscr{S}$.]

§ 4. FACTOR GROUPS

Let $\mathscr{S}$ be an *invariant* subgroup of order s of a group $\mathscr{G}$ of order g, and suppose that the set of distinct right cosets of $\mathscr{S}$ are $\mathscr{S}A_1$ $(=\mathscr{S})$, $\mathscr{S}A_2, \ldots, \mathscr{S}A_i$, where $i=g/s$. Here $A_1, A_2, \ldots, A_i$ are a set of coset representatives, as defined in connection with property (f) of §2. That is, A_j is a member of $\mathscr{S}A_j$ for $1 \leqslant j \leqslant i$.

Now consider the i cosets $\mathscr{S}A_1, \mathscr{S}A_2, \ldots, \mathscr{S}A_i$ as being i distinct entities. That is, each coset is now considered as a *single* entity or element, the contents of the coset being disregarded for the moment. The product of two such entities may be *defined* to be the coset corresponding to the product of the representative elements. That is, *by definition*,

$$\mathscr{S}A_m \cdot \mathscr{S}A_n = \mathscr{S}(A_mA_n). \tag{6.1}$$

First it has to be proved that this definition is meaningful, in the sense that if different coset representatives had been chosen for the cosets on the left-hand side of (6.1), the coset on the right-hand side would be unchanged. This means proving that if A'_m is a member of $\mathscr{S}A_m$ (and so is an alternative coset representative of $\mathscr{S}A_m$), and A'_n is a member of $\mathscr{S}A_n$, then $\mathscr{S}(A_mA_n)=\mathscr{S}(A'_mA'_n)$.

[*Proof.* It will be shown that $A'_mA'_n$ is a member of $\mathscr{S}(A_mA_n)$, from which the required result follows by property (f) of §2. As A'_m is a member of $\mathscr{S}A_m$, and $\mathscr{S}$ is a invariant subgroup, then, by property (b) of §3, A'_m is a member of $A_m\mathscr{S}$. Therefore there exists an element S_j of $\mathscr{S}$ such that $A'_m=A_mS_j$. As A'_n is a member of

$\mathscr{S}A_n$, there also exists an element S_k of $\mathscr{S}$ such that $A'_n = S_k A_n$. Then $A'_m A'_n = A_m S_j S_k A_n = A_m S_l A_n$, where $S_l = S_j S_k$ is also a member of $\mathscr{S}$. However, $A_m S_l$ is a member of $A_m \mathscr{S}$, and hence of $\mathscr{S}A_m$, so that there exists an element S_p of $\mathscr{S}$ such that $A_m S_l = S_p A_m$. Then $A'_m A'_n = S_p A_m A_n$, so that $A'_m A'_n$ is a member of $\mathscr{S}(A_m A_n)$.]

Definition of the factor group $\mathscr{G}/\mathscr{S}$. The factor group $\mathscr{G}/\mathscr{S}$ is defined to consist of the $i = g/s$ elements $\mathscr{S}A_1 (= \mathscr{S})$, $\mathscr{S}A_2, \ldots, \mathscr{S}A_i$, with eq. (6.1) defining the group multiplication operation.

It will be seen that the elements of the factor group are each cosets, which are here regarded as being single entities. Thus one is involved here in a higher level of abstraction. As was shown in property (b) of §3, there is no difference between left and right cosets formed from an invariant subgroup $\mathscr{S}$, so that the elements of the factor group $\mathscr{G}/\mathscr{S}$ could equally well be regarded as being *left* cosets. The proof that the above definition does conform with the four group postulates of ch. 1 §2.1 will now be given.

[*Proof.* (a) By the definition (6.1), the product of two elements of the factor group is itself a coset of $\mathscr{S}$, and so it is also an element of the factor group.
(b) $(\mathscr{S}A_m \cdot \mathscr{S}A_n)\, \mathscr{S}A_p = \mathscr{S}(A_m A_n A_p) = \mathscr{S}A_m (\mathscr{S}A_n \cdot \mathscr{S}A_p)$, so that the associative law is valid.
(c) The identity element of the factor group is $\mathscr{S}$, for the coset representative of $\mathscr{S} = \mathscr{S}A_1$ may be taken to be E, and $\mathscr{S}E \cdot \mathscr{S}A_m = \mathscr{S}(EA_m) = \mathscr{S}A_m$, for all $\mathscr{S}A_m$, $1 \leqslant m \leqslant i$.
(d) The inverse of $\mathscr{S}A_m$ is $\mathscr{S}A_m^{-1}$, which is also a member of $\mathscr{G}/\mathscr{S}$. That this is the inverse follows from the fact that $\mathscr{S}A_m \cdot \mathscr{S}A_m^{-1} = \mathscr{S}(A_m A_m^{-1}) = \mathscr{S}$, and $\mathscr{S}$ is the identity element of $\mathscr{G}/\mathscr{S}$, as noted in (c) above.]

As an example of a factor group, take $\mathscr{G}$ to be the point group C_{3v} and $\mathscr{S}$ to be its invariant subgroup $\{E, C_{3z}, C_{3z}^{-1}\}$. Then $\mathscr{G}/\mathscr{S}$ consists of two elements, namely $\mathscr{S}$ and $\mathscr{S}(IC_{2y}) = \mathscr{S}(IC_{2C}) = \mathscr{S}(IC_{2D}) = \{IC_{2y}, IC_{2C}, IC_{2D}\}$. The group multiplication table is given in table 6.1.

TABLE 6.1

	$\mathscr{S}$	$\mathscr{S}(IC_{2y})$
$\mathscr{S}$	$\mathscr{S}$	$\mathscr{S}(IC_{2y})$
$\mathscr{S}(IC_{2y})$	$\mathscr{S}(IC_{2y})$	$\mathscr{S}$

One important example of a factor group occurs in the theory of space groups. This is, that if $\mathscr{G}$ is a space group, $\mathscr{T}$ its subgroup of pure translations, and $\mathscr{G}_o$ its point group, then $\mathscr{T}$ is an invariant subgroup of $\mathscr{G}$, and $\mathscr{G}/\mathscr{T}$ is isomorphic to $\mathscr{G}_o$.

[*Proof.* $\mathscr{T}$ is an invariant subgroup of $\mathscr{G}$, as if $\{\boldsymbol{R}' \mid \boldsymbol{t}'\}$ is any member of $\mathscr{G}$ and $\{\boldsymbol{E} \mid \boldsymbol{t}_n\}$ is any member of $\mathscr{T}$, then $\{\boldsymbol{R}' \mid \boldsymbol{t}'\} \times \{\boldsymbol{E} \mid \boldsymbol{t}_n\} \{\boldsymbol{R}' \mid \boldsymbol{t}'\}^{-1}$ is a pure translation, and so is a member of $\mathscr{T}$.

A typical element of the factor group $\mathscr{G}/\mathscr{T}$ is the coset $\{\boldsymbol{R} \mid \boldsymbol{t}\} \mathscr{T}$, where $\{\boldsymbol{R} \mid \boldsymbol{t}\}$ is a typical element of $\mathscr{G}$. It is clear that all the transformations in the coset $\{\boldsymbol{R} \mid \boldsymbol{t}\} \mathscr{T}$ have $\boldsymbol{R}$ as their rotational part. Conversely, it is also true that all the elements of $\mathscr{G}$ having $\boldsymbol{R}$ as their rotational part are contained in $\{\boldsymbol{R} \mid \boldsymbol{t}\} \mathscr{T}$. The reason for this is that if $\{\boldsymbol{R} \mid \boldsymbol{t}'\}$ is any such element, then by eqs. (1.6) and (1.7), $\{\boldsymbol{R} \mid \boldsymbol{t}\} \{\boldsymbol{R} \mid \boldsymbol{t}'\}^{-1} = \{\boldsymbol{E} \mid -\boldsymbol{t}' + \boldsymbol{t}\}$, which is a member of $\mathscr{G}$ and is a pure translation, so that it must be a pure primitive translation. Thus $\boldsymbol{t}'$ differs from $\boldsymbol{t}$ by a lattice vector, and so $\{\boldsymbol{R} \mid \boldsymbol{t}'\}$ can be written as a product of $\{\boldsymbol{R} \mid \boldsymbol{t}\}$ and a primitive translation of $\mathscr{T}$. There is, therefore, a one-to-one correspondence between distinct cosets $\{\boldsymbol{R} \mid \boldsymbol{t}\} \mathscr{T}$ and the elements of $\mathscr{G}_o$.

This correspondence is an isomorphism, because if $\boldsymbol{R}$ and $\boldsymbol{R}'$ are any two rotations, their product is $\boldsymbol{R}\boldsymbol{R}'$, and the corresponding product of cosets is $\{\boldsymbol{R} \mid \boldsymbol{t}\} \mathscr{T} \cdot \{\boldsymbol{R}' \mid \boldsymbol{t}'\} \mathscr{T}$. However, by the definition (6.1), this is the coset $\{\boldsymbol{R}\boldsymbol{R}' \mid \boldsymbol{R}\boldsymbol{t}' + \boldsymbol{t}\} \mathscr{T}$, that is, it is the coset corresponding to $\boldsymbol{R}\boldsymbol{R}'$.]

CHAPTER 7

NON-SYMMORPHIC SPACE GROUPS

§ 1. IRREDUCIBLE REPRESENTATIONS OF A NON-SYMMORPHIC SPACE GROUP

The theory for non-symmorphic space groups follows a similar line to that for symmorphic space groups, but with certain concepts and quantities appearing in a more general form. As a non-symmorphic space group $\mathscr{G}$ still contains the group $\mathscr{T}$ of primitive translations as a subgroup, every energy eigenfunction can still be written as a Bloch function. However, the neat division of the space group into translational and rotational parts is no longer possible.

This section deals with certain preliminary ideas leading up to the statement of the fundamental theorem on the irreducible representation of non-symmorphic space groups. The application of this theorem in practice is made much easier by invoking the result given in §2. The consequences of the theorem will then be described in §§3–8, the space group D_{6h}^4 of the hexagonal close-packed structure being used as an example. A comprehensive list of papers describing the irreducible representations of the non-symmorphic space groups is given in appendix 3.

The first quantity needed is the group $\mathscr{G}(\boldsymbol{k})$, which is a generalization of the group $\mathscr{G}_o(\boldsymbol{k})$ that was met in ch. 5.

Definition of $\mathscr{G}(\boldsymbol{k})$, the group of the wave vector $\boldsymbol{k}$. The group $\mathscr{G}(\boldsymbol{k})$ is the *subgroup* of the space group $\mathscr{G}$ that consists of all transformations $\{\boldsymbol{R} \mid \boldsymbol{t}\}$ having the property that

$$\boldsymbol{Rk} = \boldsymbol{k} + \boldsymbol{K}_m ,$$

where $\boldsymbol{K}_m$ is some lattice vector of the reciprocal lattice, as defined in eq. (4.13), which may be zero.

$\mathscr{G}(\boldsymbol{k})$ contains translations as well as rotations. Indeed, $\mathscr{G}(\boldsymbol{k})$ contains the group $\mathscr{T}$ of primitive translations as a subgroup.

Definitions of general points and symmetry points, symmetry axes and symmetry planes of the Brillouin zone. If $\mathscr{G}(\boldsymbol{k})$ is merely equal to $\mathscr{T}$ then $\boldsymbol{k}$ is said to be a general point of the Brillouin zone. If $\boldsymbol{k}$ is such that $\mathscr{G}(\boldsymbol{k})$ is a larger group than those corresponding to all neighbouring points, the $\boldsymbol{k}$ is called a symmetry point of the Brillouin zone. If all the points on a line or plane have the same $\mathscr{G}(\boldsymbol{k})$ that is larger than $\mathscr{T}$, then this line or plane is known as a symmetry axis or plane.

The Brillouin zone for the simple hexagonal lattice is given in fig. 4.6. For the space group D_{6h}^4 the symmetry points are Γ, A, H, K, L and M, the symmetry axes are Δ, Σ, R, P, S, S', T and T', and the symmetry planes are those containing three symmetry axes.

Definition of the star of $\boldsymbol{k}$. Let $\{\boldsymbol{R}_1 \mid \boldsymbol{t}_1\}$ $(=\{\boldsymbol{E} \mid \boldsymbol{0}\})$, $\{\boldsymbol{R}_2 \mid \boldsymbol{t}_2\}, \ldots$ be a set of coset representatives for the decomposition of the space group $\mathscr{G}$ into left cosets with respect to $\mathscr{G}(\boldsymbol{k})$. Then the set of wave vectors $\boldsymbol{k}_1$ $(=\boldsymbol{k})$, $\boldsymbol{k}_2=\boldsymbol{R}_2\boldsymbol{k}, \ldots$ is called the star of $\boldsymbol{k}$.

It follows from property (g) of ch. 6 §2 that the number of coset representatives $M(\boldsymbol{k})$ is given by

$$g = g(\boldsymbol{k})\, M(\boldsymbol{k}), \tag{7.1}$$

where g and $g(\boldsymbol{k})$ are the orders of $\mathscr{G}$ and $\mathscr{G}(\boldsymbol{k})$ respectively.

Fundamental theorem on the basis functions of the irreducible representations of a non-symmorphic space group $\mathscr{G}$. Suppose that $\boldsymbol{k}$ is an allowed $\boldsymbol{k}$-vector and $\phi_{\boldsymbol{k}s}^p(\boldsymbol{r})$ is a Bloch function transforming according to the sth row of the irreducible unitary representation $\boldsymbol{\Gamma}^p$ of $\mathscr{G}(\boldsymbol{k})$ that satisfies the equation

$$\boldsymbol{\Gamma}^p(\{\boldsymbol{E} \mid \boldsymbol{t}_n\}) = \exp(-\mathrm{i}\boldsymbol{k}\cdot\boldsymbol{t}_n)\, \boldsymbol{\Gamma}^p(\{\boldsymbol{E} \mid \boldsymbol{0}\}) \tag{7.2}$$

for every primitive translation $\{\boldsymbol{E} \mid \boldsymbol{t}_n\}$, this representation being of dimension l. Then the set of $lM(\boldsymbol{k})$ functions

$$P(\{\boldsymbol{R}_i \mid \boldsymbol{t}_i\})\, \phi^p_{\boldsymbol{k}s}(\boldsymbol{r}), \qquad i = 1, 2, \ldots, M(\boldsymbol{k}), \qquad s = 1, 2, \ldots, l. \tag{7.3}$$

form a basis for a $lM(\boldsymbol{k})$-dimensional irreducible unitary representation of the space group $\mathscr{G}$. Moreover, *all* the irreducible representations of $\mathscr{G}$ may be obtained in this way by working through all the irreducible representations of $\mathscr{G}(\boldsymbol{k})$ *that satisfy* (7.2) for all the allowed values of $\boldsymbol{k}$ that are in different stars.

It should be noted that not every irreducible representation of $\mathscr{G}(\boldsymbol{k})$ satisfies eq. (7.2). In particular, the identity representation in which every element of $\mathscr{G}(\boldsymbol{k})$ is represented by the one-dimensional unit matrix does not satisfy this equation, unless $\boldsymbol{k}=\boldsymbol{0}$. However, it is clear that Bloch functions with wave vector $\boldsymbol{k}$ can only form basis functions for representations of $\mathscr{G}(\boldsymbol{k})$ that do have this property.

The theorem reduces the problem of finding irreducible representations of $\mathscr{G}$ to that of finding irreducible representations of $\mathscr{G}(\boldsymbol{k})$. However, $\mathscr{G}(\boldsymbol{k})$ is itself a very large group, containing as it does the group $\mathscr{T}$ of primitive translations, and has a complicated structure when $\boldsymbol{k}$ is a non-general point. The result given in the next section reduces the problem still further to that of finding the irreducible representations of a much smaller and more manageable group.

The theorem shows that the irreducible representations of $\mathscr{G}$ can be specified by an allowed $\boldsymbol{k}$-vector and a label, p, specifying an irreducible representation of $\mathscr{G}(\boldsymbol{k})$, and so may be conveniently denoted by $\boldsymbol{\Gamma}^{\boldsymbol{k}p}$. By an argument similar to that leading to eq. (5.3) it follows that the matrix elements in this representation have rows and columns that may each be conveniently distinguished by a pair of indices and that here

$$\Gamma^{\boldsymbol{k}p}(\{\boldsymbol{R} \mid \boldsymbol{t}\})_{j,t;\,i,s} = \begin{cases} 0, & \text{if } \boldsymbol{R}_j^{-1}\boldsymbol{R}\boldsymbol{R}_i\boldsymbol{k} \text{ is not equivalent to } \boldsymbol{k}, \\ \Gamma^p(\{\boldsymbol{R}_j \mid \boldsymbol{t}_j\}^{-1}\,\{\boldsymbol{R} \mid \boldsymbol{t}\}\,\{\boldsymbol{R}_i \mid \boldsymbol{t}_i\})_{ts}, & \text{otherwise}. \end{cases} \tag{7.4}$$

A translational part t_R may be assigned to every rotation of the point group $\mathscr{G}_o$ of the space group $\mathscr{G}$. Every transformation having R as its rotational part is then of the form $\{R \mid t_R + t_n\}$, where t_n is a lattice vector. (For a symmorphic space group t_R can be taken to be 0 for every R. For the hexagonal close-packed space group D_{6h}^4, t_R can be taken to be either 0 or τ, the precise assignment being specified in ch. 1 § 3.3.) Then as

$$\{R_j \mid t_j\}^{-1} \{R \mid t_R + t_n\} \{R_i \mid t_i\} = \{R_j \mid t_j\}^{-1} \{R \mid t_R\} \{R_i \mid t_i\} \{E \mid (RR_i)^{-1} t_n\}$$

it follows from eq. (7.2) that eq. (7.4) can be rewritten as

$$\Gamma^{kp}(\{R \mid t_R + t_n\})_{j,t;i,s} = \begin{cases} 0, & \text{if } R_j^{-1}RR_i k \text{ is not equivalent to } k, \\ \Gamma^p(\{R_j \mid t_j\}^{-1} \{R \mid t_R\} \{R_i \mid t_i\})_{ts} \exp(-\mathrm{i}R_j k \cdot t_n), & \text{otherwise}. \end{cases} \tag{7.5}$$

A proof of the fundamental theorem may be formulated almost exactly on the same lines as the proof given in ch. 5 § 1 for the corresponding theorem on symmorphic space groups. The only difficult point lies in the proof that *all* the irreducible representations of $\mathscr{G}$ may be obtained by the procedure of the theorem. The line of argument given for the corresponding part of the theorem for symmorphic space groups may again be followed, but it requires showing that the sum of the squares of the dimensionalities of the *relevant* irreducible representations of $\mathscr{G}(k)$ is equal to $g_o/M(k)$. An explicit proof of this latter result may be found in the appendix of a paper by SLATER *et al.* [1962]. The original proof of the theorem was given by SEITZ [1936].

§ 2. DETERMINATION OF THE IRREDUCIBLE REPRESENTATIONS OF $\mathscr{G}(k)$ FROM THOSE OF THE FACTOR GROUP $\mathscr{G}(k)/\mathscr{T}(k)$

Definition of the group $\mathscr{T}(k)$. The group $\mathscr{T}(k)$ consists of all primitive translations $\{E \mid t_n\}$ that satisfy the equation

$$\exp(-\mathrm{i}k \cdot t_n) = 1 .$$

Obviously $\mathscr{T}(\boldsymbol{k})$ is a subgroup of the group $\mathscr{T}$ of all primitive translations and therefore it is also a subgroup of $\mathscr{G}(\boldsymbol{k})$. It is, moreover, an *invariant* subgroup of $\mathscr{G}(\boldsymbol{k})$. The reason for this is that if $\{\boldsymbol{R}' \mid \boldsymbol{t}'\}$ is a member of $\mathscr{G}(\boldsymbol{k})$ and $\{\boldsymbol{E} \mid \boldsymbol{t}_n\}$ is a member of $\mathscr{T}(\boldsymbol{k})$, then by eqs. (1.6) and (1.7)

$$\{\boldsymbol{R}' \mid \boldsymbol{t}'\}\{\boldsymbol{E} \mid \boldsymbol{t}_n\}\{\boldsymbol{R}' \mid \boldsymbol{t}'\}^{-1} = \{\boldsymbol{E} \mid \boldsymbol{R}'\boldsymbol{t}_n\}.$$

However $\{\boldsymbol{E} \mid \boldsymbol{R}'\boldsymbol{t}_n\}$ is also a member of $\mathscr{T}(\boldsymbol{k})$ as

$$\begin{aligned}\exp(-\mathrm{i}\boldsymbol{k}\cdot\boldsymbol{R}'\boldsymbol{t}_n) &= \exp(-\mathrm{i}\boldsymbol{R}'^{-1}\boldsymbol{k}\cdot\boldsymbol{t}_n)\\ &= \exp\{-\mathrm{i}(\boldsymbol{k}+\boldsymbol{K}_m)\cdot\boldsymbol{t}_n\},\end{aligned}$$

where $\boldsymbol{K}_m$ is some reciprocal lattice vector as $\{\boldsymbol{R}' \mid \boldsymbol{t}'\}^{-1}$ is also a member of $\mathscr{G}(\boldsymbol{k})$, so that

$$\begin{aligned}\exp(-\mathrm{i}\boldsymbol{k}\cdot\boldsymbol{R}'\boldsymbol{t}_n) &= \exp(-\mathrm{i}\boldsymbol{k}\cdot\boldsymbol{t}_n), \quad \text{by eq. (4.14)},\\ &= 1, \quad \text{as } \{\boldsymbol{E} \mid \boldsymbol{t}_n\} \text{ is a member of } \mathscr{T}(\boldsymbol{k}).\end{aligned}$$

It is possible therefore to form the factor group $\mathscr{G}(\boldsymbol{k})/\mathscr{T}(\boldsymbol{k})$, which has as its elements cosets of the form $\{\boldsymbol{R} \mid \boldsymbol{t}\}\mathscr{T}(\boldsymbol{k})$, the identity element being $\{\boldsymbol{E} \mid \boldsymbol{0}\}\mathscr{T}(\boldsymbol{k})$. The following theorem then gives the required result.

Theorem. Let $\boldsymbol{\Gamma}'$ be an irreducible representation of $\mathscr{G}(\boldsymbol{k})/\mathscr{T}(\boldsymbol{k})$ that satisfies

$$\boldsymbol{\Gamma}'(\{\boldsymbol{E} \mid \boldsymbol{t}_n\}\mathscr{T}(\boldsymbol{k})) = \exp(-\mathrm{i}\boldsymbol{k}\cdot\boldsymbol{t}_n)\,\boldsymbol{\Gamma}'(\{\boldsymbol{E} \mid \boldsymbol{0}\}\mathscr{T}(\boldsymbol{k})) \tag{7.6}$$

for every coset $\{\boldsymbol{E} \mid \boldsymbol{t}_n\}\mathscr{T}(\boldsymbol{k})$ formed from primitive translations. Then the set of matrices $\boldsymbol{\Gamma}(\{\boldsymbol{R} \mid \boldsymbol{t}\})$ defined by

$$\boldsymbol{\Gamma}(\{\boldsymbol{R} \mid \boldsymbol{t}\}) = \boldsymbol{\Gamma}'(\{\boldsymbol{R} \mid \boldsymbol{t}\}\mathscr{T}(\boldsymbol{k})) \tag{7.7}$$

for every $\{\boldsymbol{R} \mid \boldsymbol{t}\}$ of $\mathscr{G}(\boldsymbol{k})$ form an irreducible representation of $\mathscr{G}(\boldsymbol{k})$ which satisfies eq. (7.2). Moreover, all such irreducible representations of $\mathscr{G}(\boldsymbol{k})$ can be constructed in this way.

Not all the irreducible representations of $\mathscr{G}(\boldsymbol{k})/\mathscr{T}(\boldsymbol{k})$ satisfy eq. (7.6). In particular for $\boldsymbol{k}\neq\boldsymbol{0}$ the identity representation does not.

Those irreducible representations of $\mathscr{G}(\boldsymbol{k})/\mathscr{T}(\boldsymbol{k})$ which do satisfy eq. (7.6) will be called the 'relevant' representations. Thus for $\boldsymbol{k} \neq \boldsymbol{0}$ the character table of relevant irreducible representations of $\mathscr{G}(\boldsymbol{k})/\mathscr{T}(\boldsymbol{k})$ is *not* square.

The factor group $\mathscr{G}(\boldsymbol{k})/\mathscr{T}(\boldsymbol{k})$ is not in general isomorphic to a point group, so that its irreducible representations are not so easily found. (The case $\boldsymbol{k}=\boldsymbol{0}$ is exceptional, for $\mathscr{G}(\boldsymbol{0})/\mathscr{T}(\boldsymbol{0})$ is isomorphic to the point group $\mathscr{G}_o$, as will be shown in §3.1.) The standard method for the construction of character tables for these groups may be found in the paper by HERRING [1942]. Alternative methods have more recently been described by ZAK [1960], RAGHAVACHARYULU [1961], RAGHAVACHARYULU and BHAVANACHARYULU [1962], MCINTOSH [1963], OLBRYCHSKI [1963a], KITZ [1965], RUDRA [1965], SLECHTA [1966] and GLÜCK *et al.* [1967]. A comprehensive list of the tabulations of characters for the groups $\mathscr{G}(\boldsymbol{k})/\mathscr{T}(\boldsymbol{k})$ belonging to the non-symmorphic space groups is contained in appendix 3.

The remainder of this section is devoted to the proof of the theorem. First it has to be established that (7.7) defines a representation of $\mathscr{G}(\boldsymbol{k})$. This is so as if $\{\boldsymbol{R} \mid \boldsymbol{t}\}$ and $\{\boldsymbol{R}' \mid \boldsymbol{t}'\}$ are any two members of $\mathscr{G}(\boldsymbol{k})$, then by (7.7)

$$\begin{aligned}
&\boldsymbol{\Gamma}(\{\boldsymbol{R} \mid \boldsymbol{t}\})\,\boldsymbol{\Gamma}(\{\boldsymbol{R}' \mid \boldsymbol{t}'\}) \\
&\quad = \boldsymbol{\Gamma}'(\{\boldsymbol{R} \mid \boldsymbol{t}\}\,\mathscr{T}(\boldsymbol{k}))\,\boldsymbol{\Gamma}'(\{\boldsymbol{R}' \mid \boldsymbol{t}'\}\,\mathscr{T}(\boldsymbol{k})) \\
&\quad = \boldsymbol{\Gamma}'(\{\boldsymbol{R} \mid \boldsymbol{t}\}\,\mathscr{T}(\boldsymbol{k})\cdot\{\boldsymbol{R}' \mid \boldsymbol{t}'\}\,\mathscr{T}(\boldsymbol{k})) \\
&\qquad\qquad\qquad \text{as } \boldsymbol{\Gamma}' \text{ is a representation of } \mathscr{G}(\boldsymbol{k})/\mathscr{T}(\boldsymbol{k}), \\
&\quad = \boldsymbol{\Gamma}'(\{\boldsymbol{R} \mid \boldsymbol{t}\}\,\{\boldsymbol{R}' \mid \boldsymbol{t}'\}\,\mathscr{T}(\boldsymbol{k})), \quad \text{by eq. (6.1)}, \\
&\quad = \boldsymbol{\Gamma}(\{\boldsymbol{R} \mid \boldsymbol{t}\}\,\{\boldsymbol{R}' \mid \boldsymbol{t}'\}), \quad \text{by eq. (7.7)}.
\end{aligned}$$

That the representation is irreducible follows from applying the second theorem of ch. 2 §6, for

$$\sum |\chi(\{\boldsymbol{R} \mid \boldsymbol{t}\})|^2 = t(\boldsymbol{k}) \sum |\chi'(\{\boldsymbol{R} \mid \boldsymbol{t}\}\,\mathscr{T}(\boldsymbol{k}))|^2,$$

where χ and χ' denote the characters of $\boldsymbol{\Gamma}$ and $\boldsymbol{\Gamma}'$ respectively, the summations on the left and right-hand sides are over all the elements of $\mathscr{G}(\boldsymbol{k})$ and $\mathscr{G}(\boldsymbol{k})/\mathscr{T}(\boldsymbol{k})$ respectively, and $t(\boldsymbol{k})$ is the order of $\mathscr{T}(\boldsymbol{k})$. As $\boldsymbol{\Gamma}'$ is an irreducible representation of $\mathscr{G}(\boldsymbol{k})/\mathscr{T}(\boldsymbol{k})$

$$\sum |\chi'(\{\boldsymbol{R} \mid \boldsymbol{t}\}\, \mathscr{T}(\boldsymbol{k}))|^2 = \text{order of } \mathscr{G}(\boldsymbol{k})/\mathscr{T}(\boldsymbol{k}) = g(\boldsymbol{k})/t(\boldsymbol{k}),$$

so that

$$\sum |\chi(\{\boldsymbol{R} \mid \boldsymbol{t}\})|^2 = g(\boldsymbol{k}).$$

The condition (7.2) is satisfied because for any primitive translation $\{E \mid \boldsymbol{t}_n\}$ by eqs. (7.6) and (7.7),

$$\begin{aligned}\boldsymbol{\Gamma}(\{E \mid \boldsymbol{t}_n\}) &= \exp(-\,i\boldsymbol{k}\cdot\boldsymbol{t}_n)\, \boldsymbol{\Gamma}'(\{E \mid \boldsymbol{0}\}\, \mathscr{T}(\boldsymbol{k})) \\ &= \exp(-\,i\boldsymbol{k}\cdot\boldsymbol{t}_n)\, \boldsymbol{\Gamma}(\{E \mid \boldsymbol{0}\}).\end{aligned}$$

The fact that every irreducible representation of $\mathscr{G}(\boldsymbol{k})$ satisfying eq. (7.2) can be constructed using (7.7) is an immediate consequence of the fact that in such an irreducible representation of $\mathscr{G}(\boldsymbol{k})$ the members of $\mathscr{T}(\boldsymbol{k})$ are necessarily represented by a unit matrix, so that all the members of a coset of $\mathscr{T}(\boldsymbol{k})$ are represented by the same matrix.

§ 3. SOME IMMEDIATE CONSEQUENCES OF THE FUNDAMENTAL THEOREM ON THE IRREDUCIBLE REPRESENTATIONS OF A NON-SYMMORPHIC SPACE GROUP

§ 3.1. *Degeneracies of energy levels and the symmetry of* $E(\boldsymbol{k})$

When $\boldsymbol{k}$ is a general point of the Brillouin zone, $\mathscr{G}(\boldsymbol{k})$ is equal to $\mathscr{T}$, whose irreducible representations were found in ch. 4 § 3. In fact, eq. (4.9) shows that there is only one irreducible representation of $\mathscr{T}$ satisfying eq. (7.2), namely $\boldsymbol{\Gamma}^{\boldsymbol{k}}$ itself, and this is one-dimensional. Moreover, as $\mathscr{G}/\mathscr{T}$ is isomorphic to the point group $\mathscr{G}_o$, as demonstrated in ch. 6 § 4, the set of rotational parts $\boldsymbol{R}_i$ of the coset representatives $\{\boldsymbol{R}_i \mid \boldsymbol{t}_i\}$ of the decomposition of $\mathscr{G}$ into left cosets with respect of $\mathscr{G}(\boldsymbol{k})$ constitute the whole of the point group $\mathscr{G}_o$. Thus, in the fundamental theorem of § 1, $l = 1$ and $M(\boldsymbol{k}) = g_o$, the order of $\mathscr{G}_o$, so that the dimension of the corresponding irreducible representation of $\mathscr{G}$ and the degeneracy of the corresponding energy level are both g_o.

As $P(\{\boldsymbol{R}_i \mid \boldsymbol{t}_i\})\, \phi^p_{\boldsymbol{k}s}(\boldsymbol{r})$ corresponds to the wave vector $\boldsymbol{k}_i = \boldsymbol{R}_i\boldsymbol{k}$ of the star of $\boldsymbol{k}$, and as by the definition (4.12) this is an eigenfunction corresponding to the energy eigenvalue $E(\boldsymbol{k}_i)$, it follows that

$$E(\boldsymbol{R}_i\boldsymbol{k}) \equiv E(\boldsymbol{k}_i) = E(\boldsymbol{k})$$

for $i = 1, 2, \ldots, g_o$. Thus $E(\boldsymbol{k})$ has the symmetry of the point group $\mathscr{G}_o$ of the space group $\mathscr{G}$, as was the case for symmorphic space groups, so the same consequences follow.

When $\boldsymbol{k} = \boldsymbol{0}$ the theory also simplifies, for in this case $\mathscr{G}(\boldsymbol{k}) = \mathscr{G}$, $\mathscr{T}(\boldsymbol{k}) = \mathscr{T}$ and hence $\mathscr{G}(\boldsymbol{k})/\mathscr{T}(\boldsymbol{k}) = \mathscr{G}/\mathscr{T}$ which is isomorphic to $\mathscr{G}_o$. As the character tables for all the crystallographic point groups are known and are given in appendix 1, all the irreducible representations are known for $\mathscr{G}(\boldsymbol{k})/\mathscr{T}(\boldsymbol{k})$ for this case. Moreover, they all satisfy eq. (7.6). The star of $\boldsymbol{k} = \boldsymbol{0}$ consists only of $\boldsymbol{k}$, so in the application of the fundamental theorem of § 1, $M(\boldsymbol{0}) = 1$ and the degeneracy of each energy eigenvalue is equal to the dimension of the corresponding irreducible representation of $\mathscr{G}_o$. This is again exactly the same as for symmorphic space groups.

The last case to be considered is that of non-general points where $\mathscr{G}(\boldsymbol{k})$ is smaller than $\mathscr{G}$ but larger than $\mathscr{T}$. The factor group $\mathscr{G}(\boldsymbol{k})/\mathscr{T}(\boldsymbol{k})$ is not isomorphic to a point group. However, as far as the degeneracy and symmetry of energy levels is concerned, the results are the same as for symmorphic space groups. If the star of $\boldsymbol{k}$ contains $M(\boldsymbol{k})$ members and the dimension of an irreducible representation of $\mathscr{G}(\boldsymbol{k})/\mathscr{T}(\boldsymbol{k})$ is l, then by the fundamental theorem the corresponding energy level is $lM(\boldsymbol{k})$-fold degenerate, this degeneracy being made up as follows:

(a) Using a similar argument to that given above for a general point it follows that

$$E(\boldsymbol{k}_i) = E(\boldsymbol{k}), \qquad i = 1, 2, \ldots, M(\boldsymbol{k}),$$

so that $E(\boldsymbol{k})$ has yet again the symmetry of the point group $\mathscr{G}_o$.

(b) Each $E(\boldsymbol{k}_i)$ is in addition 'l-fold degenerate', in the sense that there are l linearly independent Bloch eigenfunctions of $H(\boldsymbol{r})$ corresponding to this eigenvalue *and* to this particular wave vector $\boldsymbol{k}_i$.

The exact correspondence of these results with those on symmorphic space groups allows the description of degeneracies given in the last part of ch. 5 § 2.1 to be immediately applied to non-symmorphic space groups as well. The symmetry $E(\boldsymbol{k})=E(-\boldsymbol{k})$ also persists even in the case when $\mathscr{G}_o$ does not contain the inversion operator I because of time-reversal symmetry.

§ 3.2. *A matrix element theorem for Bloch basis functions*

The matrix element theorem for Bloch basis functions and its consequences that were described in ch. 5 § 2.2 apply equally well for non-symmorphic space groups. The only small alteration that must be made is that the Bloch functions $\phi_{\boldsymbol{k}i}^p(\boldsymbol{r})$ now transform as the ith row of the pth irreducible unitary representation of $\mathscr{G}(\boldsymbol{k})$ that satisfies eq. (7.2).

§ 4. THE IRREDUCIBLE REPRESENTATIONS OF $\mathscr{G}(\boldsymbol{k})/\mathscr{T}(\boldsymbol{k})$ FOR THE HEXAGONAL CLOSE-PACKED SPACE GROUP D_{6h}^4

§ 4.1. *Introduction*

The hexagonal close-packed space group D_{6h}^4 will be considered as an example of a non-symmorphic space group. The space group was described in ch. 1 § 3.3 and its Brillouin zone was given in fig. 4.6. The results that will be described are based on the work of HERRING [1942]. The correspondence between Herring's notation and the more common notation that is used in this book is given in table 1.6.

The theorems of § 1 and § 2 show that an irreducible representation of the space group $\mathscr{G}$ is specified by an allowed $\boldsymbol{k}$-vector and a label, p, of a relevant irreducible representation of $\mathscr{G}(\boldsymbol{k})/\mathscr{T}(\boldsymbol{k})$. As with symmorphic space groups, the symmetry points and symmetry axes are denoted by capital letters, and the relevant irreducible representations of $\mathscr{G}(\boldsymbol{k})/\mathscr{T}(\boldsymbol{k})$ denoted by the appropriate capital letter with a subscript attached.

§ 4.2. *Symmetry points*

For the point Γ for which $\boldsymbol{k}=\boldsymbol{0}$, the factor group $\mathscr{G}(\boldsymbol{k})/\mathscr{T}(\boldsymbol{k})$ is equal to $\mathscr{G}/\mathscr{T}$ which is isomorphic to the point group $\mathscr{G}_o$, which is here D_{6h}. The character table for this point group is given in table A.4, which contains Herring's notation for the irreducible representations. All that remains to make the description really complete for this point Γ is to make explicit the contents of the classes $\mathscr{C}_1$ to $\mathscr{C}_{12}$ of table A.4 in terms of the elements of the factor group $\mathscr{G}/\mathscr{T}$. These elements are cosets, but for brevity it is a set of coset representatives that are listed in table 7.1. An expression like $\{C_{6z}, C_{6z}^{-1} \mid \boldsymbol{\tau}\}$ stands for $\{C_{6z} \mid \boldsymbol{\tau}\}$ and $\{C_{6z}^{-1} \mid \boldsymbol{\tau}\}$. The vector $\boldsymbol{\tau}$ is as defined in eq. (1.13).

TABLE 7.1

Classes of the factor group $\mathscr{G}(\boldsymbol{k})/\mathscr{T}(\boldsymbol{k})$ for the point Γ of the hexagonal close-packed space group D_{6h}^4

Class	Coset representatives	Class	Coset representatives
$\mathscr{C}_1$	$\{E \mid \boldsymbol{0}\}$	$\mathscr{C}_7$	$\{I \mid \boldsymbol{\tau}\}$
$\mathscr{C}_2$	$\{C_{6z}, C_{6z}^{-1} \mid \boldsymbol{\tau}\}$	$\mathscr{C}_8$	$\{IC_{6z}, IC_{6z}^{-1} \mid \boldsymbol{0}\}$
$\mathscr{C}_3$	$\{C_{3z}, C_{3z}^{-1} \mid \boldsymbol{0}\}$	$\mathscr{C}_9$	$\{IC_{3z}, IC_{3z}^{-1} \mid \boldsymbol{\tau}\}$
$\mathscr{C}_4$	$\{C_{2z} \mid \boldsymbol{\tau}\}$	$\mathscr{C}_{10}$	$\{IC_{2z} \mid \boldsymbol{0}\}$
$\mathscr{C}_5$	$\{C_{2x}, C_{2A}, C_{2B} \mid \boldsymbol{\tau}\}$	$\mathscr{C}_{11}$	$\{IC_{2x}, IC_{2A}, IC_{2B} \mid \boldsymbol{0}\}$
$\mathscr{C}_6$	$\{C_{2y}, C_{2C}, C_{2D} \mid \boldsymbol{0}\}$	$\mathscr{C}_{12}$	$\{IC_{2y}, IC_{2C}, IC_{2D} \mid \boldsymbol{\tau}\}$

For the point A for which $\boldsymbol{k}=(0, 0, \pi/c)$, $\mathcal{T}(\boldsymbol{k})$ contains all the primitive translations $\{\boldsymbol{E} \mid n_1\boldsymbol{a}_1+n_2\boldsymbol{a}_2+n_3\boldsymbol{a}_3\}$, except those for which n_1 is an odd integer. The factor group $\mathcal{T}/\mathcal{T}(\boldsymbol{k})$ has, therefore, two elements, namely $\{\boldsymbol{E} \mid \boldsymbol{0}\}\, \mathcal{T}(\boldsymbol{k})$ and $\{\boldsymbol{E} \mid \boldsymbol{a}_1\}\, \mathcal{T}(\boldsymbol{k})$. The group $\mathcal{G}(\boldsymbol{k})$ is again equal to $\mathcal{G}$, so that the factor group $\mathcal{G}(\boldsymbol{k})/\mathcal{T}(\boldsymbol{k})$ for A contains twice as many elements as for Γ. Table 7.2 gives the character table of $\mathcal{G}(\boldsymbol{k})/\mathcal{T}(\boldsymbol{k})$ for A, only those irreducible representations which satisfy eq. (7.6) being included. The column on the left-hand side of this table lists the classes of $\mathcal{G}(\boldsymbol{k})/\mathcal{T}(\boldsymbol{k})$ in

TABLE 7.2

Character table of the factor group $\mathcal{G}(\boldsymbol{k})/\mathcal{T}(\boldsymbol{k})$ for the point A of the hexagonal close-packed space group D_{6h}^4

	A_1	A_2	A_3
$\{E \mid \boldsymbol{0}\}$	2	2	4
$\{E \mid \boldsymbol{a}_1\}$	-2	-2	-4
$\{C_{6z}, C_{6z}^{-1} \mid \boldsymbol{\tau}, \boldsymbol{\tau}+\boldsymbol{a}_1\}$	0	0	0
$\{C_{3z}, C_{3z}^{-1} \mid \boldsymbol{0}\}$	2	2	-2
$\{C_{3z}, C_{3z}^{-1} \mid \boldsymbol{a}_1\}$	-2	-2	2
$\{C_{2z} \mid \boldsymbol{\tau}, \boldsymbol{\tau}+\boldsymbol{a}_1\}$	0	0	0
$\{C_{2x}, C_{2A}, C_{2B} \mid \boldsymbol{\tau}, \boldsymbol{\tau}+\boldsymbol{a}_1\}$	0	0	0
$\{C_{2y}, C_{2C}, C_{2D} \mid \boldsymbol{0}, \boldsymbol{a}_1\}$	0	0	0
$\{I \mid \boldsymbol{\tau}, \boldsymbol{\tau}+\boldsymbol{a}_1\}$	0	0	0
$\{IC_{6z}, IC_{6z}^{-1} \mid \boldsymbol{0}, \boldsymbol{a}_1\}$	0	0	0
$\{IC_{3z}, IC_{3z}^{-1} \mid \boldsymbol{\tau}, \boldsymbol{\tau}+\boldsymbol{a}_1\}$	0	0	0
$\{IC_{2z} \mid \boldsymbol{0}, \boldsymbol{a}_1\}$	0	0	0
$\{IC_{2x}, IC_{2A}, IC_{2B} \mid \boldsymbol{0}\}$	2	-2	0
$\{IC_{2x}, IC_{2A}, IC_{2B} \mid \boldsymbol{a}_1\}$	-2	2	0
$\{IC_{2y}, IC_{2C}, IC_{2D} \mid \boldsymbol{\tau}, \boldsymbol{\tau}+\boldsymbol{a}_1\}$	0	0	0

terms of coset representatives, an expression like $\{C_{6z}, C_{6z}^{-1} \mid \tau, \tau+a_1\}$ standing for the four coset representatives $\{C_{6z} \mid \tau\}$, $\{C_{6z}^{-1} \mid \tau\}$, $\{C_{6z} \mid \tau+a_1\}$ and $\{C_{6z}^{-1} \mid \tau+a_1\}$.

Character tables of $\mathscr{G}(\boldsymbol{k})/\mathscr{T}(\boldsymbol{k})$ for all the symmetry points of the Brillouin zone are given in the paper of HERRING [1942].

§ 4.3. *Symmetry axes*

For the symmetry axes HERRING [1942] does not give the complete character tables of $\mathscr{G}(\boldsymbol{k})/\mathscr{T}(\boldsymbol{k})$ explicitly, but he does give enough information for them to be easily obtained. To illustrate the method, the axis Δ, on which $\boldsymbol{k}=(\pi/c)\,(0, 0, \kappa)$, $0<\kappa<1$, will be considered. For a general value of κ in this interval, $\mathscr{T}(\boldsymbol{k})$ contains all translations of the form $\{E \mid n_2 a_2+n_3 a_3\}$, but none of the form $\{E \mid n_1 a_1\}$ for $n_1 \neq 0$. Thus $\{E \mid n_1 a_1\}$ for $n_1=1, 2, \ldots, N_1-1$, N_1 being as defined in eq. (4.1), form a sub-set of the coset representatives of $\mathscr{G}(\boldsymbol{k})/\mathscr{T}(\boldsymbol{k})$. However, their representations satisfying eq. (7.6) can be obtained immediately once the dimension of the representation is known, for eq. (7.6) itself requires that

$$\Gamma'(\{E \mid n_1 a_1\}\mathscr{T}(\boldsymbol{k})) = \exp(-\,i\boldsymbol{k}\cdot n_1 a_1)\,\Gamma'(\{E \mid 0\}\mathscr{T}(\boldsymbol{k}))\,.$$

In addition to this set of N_1-1 coset representatives, $\mathscr{G}(\boldsymbol{k})/\mathscr{T}(\boldsymbol{k})$ also possesses the following coset representatives:

$$\left.\begin{array}{l}\{C_{6z}, C_{6z}^{-1} \mid \tau\}, \{C_{3z}, C_{3z}^{-1} \mid 0\}, \{C_{2z} \mid \tau\}, \\ \{IC_{2x}, IC_{2A}, IC_{2B} \mid 0\}, \quad \{IC_{2y}, IC_{2C}, IC_{2D} \mid \tau\}.\end{array}\right| \qquad (7.8)$$

All the remaining coset representatives of distinct cosets are products of the members of this set (7.8) with the members of the set $\{E \mid n_1 a_1\}$ for $n_1=1, 2, \ldots, N_1-1$. Thus a representation is completely specified once the matrices representing the members of the set (7.8) are known.

In order to specify the characters of these matrices, HERRING [1942] lists in tables IV to IX the character tables for the factor groups having the form $\mathscr{G}(\boldsymbol{k})/\mathscr{T}(\boldsymbol{k}_s)$ in the limit in which $\boldsymbol{k}\to\boldsymbol{k}_s$.

Here $\boldsymbol{k}_s$ is the wave vector corresponding to a symmetry point at the end of the symmetry axis under consideration. The group $\mathscr{T}(\boldsymbol{k}_s)$ is constructed for the symmetry point $\boldsymbol{k}_s$, and so in general contains as a proper subgroup the group $\mathscr{T}(\boldsymbol{k})$ for wave vectors $\boldsymbol{k}$ on the symmetry axis itself. However, the group $\mathscr{G}(\boldsymbol{k})$ is constructed for $\boldsymbol{k}$ on the symmetry axis, and it is smaller in general than the group $\mathscr{G}(\boldsymbol{k}_s)$.

As $\mathscr{G}(\boldsymbol{k})$ is the *same* group for all $\boldsymbol{k}$ on a symmetry axis, by the definition of a symmetry axis, the factor groups $\mathscr{G}(\boldsymbol{k})/\mathscr{T}(\boldsymbol{k})$ and $\mathscr{G}(\boldsymbol{k})/\mathscr{T}(\boldsymbol{k}_s)$ remain the same for all $\boldsymbol{k}$ on the axis. However, only representations of $\mathscr{G}(\boldsymbol{k})/\mathscr{T}(\boldsymbol{k})$ satisfying eq. (7.6) are relevant, and this condition is $\boldsymbol{k}$-dependent. Thus, from the *whole* set of irreducible representations of $\mathscr{G}(\boldsymbol{k})/\mathscr{T}(\boldsymbol{k})$, which is the same set for all points along the axis, a *varying* set of irreducible representations has to be taken as $\boldsymbol{k}$ varies. However, one may think of these irreducible representations being *labelled* differently at different points on the axis in such a way that each irreducible representation with a particular index is a *continuous* function of $\boldsymbol{k}$ for $\boldsymbol{k}$ on the axis. This is the labelling convention that is universally employed.

As $\mathscr{G}(\boldsymbol{k})/\mathscr{T}(\boldsymbol{k}_s)$ is isomorphic to a subgroup of $\mathscr{G}(\boldsymbol{k})/\mathscr{T}(\boldsymbol{k})$, every representation of $\mathscr{G}(\boldsymbol{k})/\mathscr{T}(\boldsymbol{k})$ gives a representation of $\mathscr{G}(\boldsymbol{k})/\mathscr{T}(\boldsymbol{k}_s)$. Moreover, every irreducible representation of $\mathscr{G}(\boldsymbol{k})/\mathscr{T}(\boldsymbol{k})$ satisfying (7.6) gives an irreducible representation of $\mathscr{G}(\boldsymbol{k})/\mathscr{T}(\boldsymbol{k}_s)$ satisfying (7.6), and there is a one-to-one correspondence between these irreducible representations. Thus, form a knowledge of the relevant characters of $\mathscr{G}(\boldsymbol{k})/\mathscr{T}(\boldsymbol{k}_s)$ in the limit $\boldsymbol{k}\to\boldsymbol{k}_s$, the relevant characters of $\mathscr{G}(\boldsymbol{k})/\mathscr{T}(\boldsymbol{k})$ may be determined if their $\boldsymbol{k}$-dependences are known.

All that remains then is to find the $\boldsymbol{k}$-dependences of the matrices of a representation, which may be done as follows. Suppose that $\{\boldsymbol{R} \mid \boldsymbol{t}\}$ is a typical coset representative of $\mathscr{G}(\boldsymbol{k})/\mathscr{T}(\boldsymbol{k})$ of the type appearing in the set (7.8). Then there exists an integer m such that $\boldsymbol{R}^m=\boldsymbol{E}$. Let $\boldsymbol{t}_n$ be the lattice vector defined by $\{\boldsymbol{R} \mid \boldsymbol{t}\}^m=\{\boldsymbol{E} \mid \boldsymbol{t}_n\}$.

Then

$$\begin{aligned}\boldsymbol{\Gamma}'(\{\boldsymbol{R}\mid\boldsymbol{t}\}\,\mathscr{T}(\boldsymbol{k}))^m &= \boldsymbol{\Gamma}'(\{\boldsymbol{E}\mid\boldsymbol{t}_n\}\,\mathscr{T}(\boldsymbol{k}))\\ &= \exp(-\,\mathrm{i}\boldsymbol{k}\cdot\boldsymbol{t}_n)\,\boldsymbol{\Gamma}'(\{\boldsymbol{E}\mid\boldsymbol{0}\}\,\mathscr{T}(\boldsymbol{k})),\quad \text{by (7.6)},\end{aligned}$$

so that

$$\chi'(\{\boldsymbol{R}\mid\boldsymbol{t}\}\,\mathscr{T}(\boldsymbol{k})) = \exp(-\,\mathrm{i}\boldsymbol{k}\cdot\boldsymbol{t}_n/m)\,\mathrm{trace}\,[\boldsymbol{\Gamma}'(\{\boldsymbol{E}\mid\boldsymbol{0}\}\,\mathscr{T}(\boldsymbol{k}))]^{1/m}.$$

The matrix $\boldsymbol{\Gamma}'(\{\boldsymbol{E}\mid\boldsymbol{0}\}\,\mathscr{T}(\boldsymbol{k}))$ will have more than one mth root, but the continuity property mentioned above implies that in the corresponding representation of the limit of $\mathscr{G}(\boldsymbol{k})/\mathscr{T}(\boldsymbol{k}_s)$ as $\boldsymbol{k}\to\boldsymbol{k}_s$,

$$\chi'(\{\boldsymbol{R}\mid\boldsymbol{t}\}\,\mathscr{T}(\boldsymbol{k}_s)) = \exp(-\,\mathrm{i}\boldsymbol{k}_s\cdot\boldsymbol{t}_n/m)\,\mathrm{trace}\,[\boldsymbol{\Gamma}'(\{\boldsymbol{E}\mid\boldsymbol{0}\}\,\mathscr{T}(\boldsymbol{k}_s))]^{1/m},$$

where the same mth root appears, so that

$$\chi'(\{\boldsymbol{R}\mid\boldsymbol{t}\}\,\mathscr{T}(\boldsymbol{k})) = \exp\{-\,\mathrm{i}(\boldsymbol{k}-\boldsymbol{k}_s)\cdot\boldsymbol{t}_n/m\}\,\chi'(\{\boldsymbol{R}\mid\boldsymbol{t}\}\,\mathscr{T}(\boldsymbol{k}_s)). \tag{7.9}$$

This is the required formula for finding the characters of cosets of the type (7.8) of $\mathscr{G}(\boldsymbol{k})/\mathscr{T}(\boldsymbol{k})$ from Herring's tables.

As an example, consider the axis Δ and the coset representative $\{C_{6z}\mid\boldsymbol{\tau}\}$ from (7.8). By repeated application of eq. (1.6), $\{C_{6z}\mid\boldsymbol{\tau}\}^6 = \{\boldsymbol{E}\mid 3\boldsymbol{a}_1\}$. Taking $\boldsymbol{k}_s=\boldsymbol{0}$, eq. (7.9) gives for each relevant representation

$$\chi'(\{C_{6z}\mid\boldsymbol{\tau}\}\,\mathscr{T}(\boldsymbol{k})) = \exp(-\,\tfrac{1}{2}\mathrm{i}\kappa c)\,\chi'(\{C_{6z}\mid\boldsymbol{\tau}\}\,\mathscr{T}(\boldsymbol{0})).$$

The last factor here may be obtained from the table IV of HERRING [1942], which is reproduced in table 7.3. The set of characters found this way for the irreducible representations of $\mathscr{G}(\boldsymbol{k})/\mathscr{T}(\boldsymbol{k})$ satisfying eq. (7.6) for $\{\boldsymbol{E}\mid\boldsymbol{0}\}$ and the coset representatives (7.8) are given in table 7.4.

The axis R on which $\boldsymbol{k}=(0, k_y, \pi/c)$, $0<k_y<2\pi/a\sqrt{3}$, provides an example in which this procedure applies particularly easily, for the coset representatives of $\mathscr{G}(\boldsymbol{k})/\mathscr{T}(\boldsymbol{k})$ are just the set

$$\{E\mid\boldsymbol{0}\},\quad \{C_{2y}\mid\boldsymbol{0}\},\quad \{IC_{2z}\mid\boldsymbol{0}\}\quad\text{and}\quad \{IC_{2x}\mid\boldsymbol{0}\}, \tag{7.10}$$

and products of this set with pure translations. As none of the set (7.10) contain translational parts, eq. (7.9) shows that the characters

TABLE 7.3

Character table of the factor group $\mathscr{G}(\boldsymbol{k})/\mathscr{T}(\boldsymbol{0})$ in the limit $\boldsymbol{k}\to\boldsymbol{0}$ for $\boldsymbol{k}$ on the axis Δ of the hexagonal close-packed space group D^4_{6h}

	Δ_1	Δ_2	Δ_3	Δ_4	Δ_5	Δ_6
$\{E \mid \boldsymbol{0}\}$	1	1	1	1	2	2
$\{C_{6z}, C_{6z}^{-1} \mid \boldsymbol{\tau}\}$	1	-1	1	-1	-1	1
$\{C_{3z}, C_{3z}^{-1} \mid \boldsymbol{0}\}$	1	1	1	1	-1	-1
$\{C_{2z} \mid \boldsymbol{\tau}\}$	1	-1	1	-1	2	-2
$\{IC_{2x}, IC_{2A}, IC_{2B} \mid \boldsymbol{0}\}$	1	1	-1	-1	0	0
$\{IC_{2y}, IC_{2C}, IC_{2D} \mid \boldsymbol{\tau}\}$	1	-1	-1	1	0	0

of this set are $\boldsymbol{k}$-independent. The characters of the set (7.10) are given in table 7.5. The energy levels corresponding to R_1 and R_3 and to R_2 and R_4 are degenerate in pairs by time-reversal symmetry, as will be shown in §8.

TABLE 7.4

Character table of the factor group $\mathscr{G}(\boldsymbol{k})/\mathscr{T}(\boldsymbol{k})$ for $\boldsymbol{k}$ on the axis Δ of the hexagonal close-packed space group D^4_{6h}. Here $\boldsymbol{k}=(\pi/c)\,(0,0,\kappa)$, $0<\kappa<1$, and $\eta=\exp(-\frac{1}{2}i\kappa c)$

	Δ_1	Δ_2	Δ_3	Δ_4	Δ_5	Δ_6
$\{E \mid \boldsymbol{0}\}$	1	1	1	1	2	2
$\{C_{6z}, C_{6z}^{-1} \mid \boldsymbol{\tau}\}$	η	$-\eta$	η	$-\eta$	$-\eta$	η
$\{C_{3z}, C_{3z}^{-1} \mid \boldsymbol{0}\}$	1	1	1	1	-1	-1
$\{C_{2z} \mid \boldsymbol{\tau}\}$	η	$-\eta$	η	$-\eta$	2η	-2η
$\{IC_{2x}, IC_{2A}, IC_{2B} \mid \boldsymbol{0}\}$	1	1	-1	-1	0	0
$\{IC_{2y}, IC_{2C}, IC_{2D} \mid \boldsymbol{\tau}\}$	η	$-\eta$	$-\eta$	η	0	0

TABLE 7.5

Character table of the factor group $\mathscr{G}(\boldsymbol{k})/\mathscr{T}(\boldsymbol{k})$ for $\boldsymbol{k}$ on the axis R of the hexagonal close-packed space group D_{6h}^4. Here $\boldsymbol{k}=(0, k_y, \pi/c)$, $0<k_y<2\pi/a\sqrt{3}$

	R_1	R_2	R_3	R_4
$\{E \mid \boldsymbol{0}\}$	1	1	1	1
$\{C_{2y} \mid \boldsymbol{0}\}$	1	1	-1	-1
$\{IC_{2z} \mid \boldsymbol{0}\}$	1	-1	-1	1
$\{IC_{2x} \mid \boldsymbol{0}\}$	1	-1	1	-1

§ 4.4. *Symmetry planes*

For a point $\boldsymbol{k}$ on a symmetry plane, $\mathscr{G}(\boldsymbol{k})$ consists of a reflection $\{IC_{2i} \mid \boldsymbol{t}'\}$, where $\boldsymbol{t}'$ may be zero, $\mathscr{T}$, and products of $\{IC_{2i} \mid \boldsymbol{t}'\}$ with the elements of $\mathscr{T}$. It is easiest in this case to consider the irreducible representations of $\mathscr{G}(\boldsymbol{k})$ directly, rather than considering those of $\mathscr{G}(\boldsymbol{k})/\mathscr{T}(\boldsymbol{k})$.

It will now be shown that $\mathscr{G}(\boldsymbol{k})$ has two irreducible representations that satisfy eq. (7.2) and both are one-dimensional. $\mathscr{G}(\boldsymbol{k})$ contains $2N$ elements, where N is the number of elements of $\mathscr{T}$, as defined in ch. 4 §2. The sum $\sum |\chi(\{E \mid \boldsymbol{t_n}\})|^2$ for the pure translations of $\mathscr{G}(\boldsymbol{k})$ alone for a representation satisfying eq. (7.2) is Nl^2, l being the dimension of the representation. Thus, by the second theorem of ch. 2 §6, if this representation is irreducible, then $Nl^2 \leqslant 2N$, so that l can only be equal to one. Thus, for each irreducible representation of $\mathscr{G}(\boldsymbol{k})$ satisfying (7.2),

$$\boldsymbol{\Gamma}(\{E \mid \boldsymbol{t_n}\}) = (\exp(-\mathrm{i}\boldsymbol{k}\cdot\boldsymbol{t_n})). \tag{7.11}$$

Moreover, if $\{IC_{2i} \mid \boldsymbol{t}'+\boldsymbol{t_n}\}$ is a typical product involving IC_{2i} then

$$\{IC_{2i} \mid \boldsymbol{t}'+\boldsymbol{t_n}\}^2 = \{\boldsymbol{E} \mid (1+IC_{2i})(\boldsymbol{t}'+\boldsymbol{t_n})\},$$

which is necessarily a pure primitive translation, so that by (7.11)

$$\boldsymbol{\Gamma}(\{IC_{2i} \mid \boldsymbol{t}' + \boldsymbol{t}_n\}) = (\pm \exp\{-\tfrac{1}{2}i\boldsymbol{k}\cdot(1 + IC_{2i})(\boldsymbol{t}' + \boldsymbol{t}_n)\}). \quad (7.12)$$

It is easily verified that the two sets of matrices corresponding to the two possible signs in (7.12) give two representations, both of which are irreducible and satisfy (7.2).

As an example, consider a point on the hexagonal face ALH of the Brillouin zone, as shown in fig. 4.6, for which $\boldsymbol{k}=(k_x, k_y, \pi/c)$. $\mathscr{G}(\boldsymbol{k})$ then contains the reflection $\{IC_{2z} \mid \boldsymbol{0}\}$, and by (7.12), $\boldsymbol{\Gamma}(\{IC_{2z} \mid \boldsymbol{0}\})=(\pm 1)$. The energy levels corresponding to these two irreducible representations always stick together because of time-reversal symmetry, as will be demonstrated in §8, giving a two-fold degeneracy over the whole of the hexagonal face of the Brillouin zone.

§ 5. SELECTION RULES FOR NON-SYMMORPHIC SPACE GROUPS

§ 5.1. *Formulation in terms of the groups $\mathscr{G}(\boldsymbol{k})$*

The theory of selection rules given in ch. 5 §4.1 for symmorphic space groups can be easily extended to non-symmorphic space groups. As in ch. 5 §4.1, if the initial state eigenfunction $\psi_i(\boldsymbol{r})$, the final state eigenfunction $\psi_f(\boldsymbol{r})$ and the perturbation $V'(\boldsymbol{r})$ transform as the irreducible representations $\boldsymbol{\Gamma}^{\boldsymbol{k}p}$, $\boldsymbol{\Gamma}^{\boldsymbol{k}''p''}$ and $\boldsymbol{\Gamma}^{\boldsymbol{k}'p'}$ of the space group $\mathscr{G}$ respectively, then $(\psi_f, V'(\boldsymbol{r})\,\psi_i)=0$ if $n^{\boldsymbol{k}''p''}_{\boldsymbol{k}p,\boldsymbol{k}'p'}=0$, where $n^{\boldsymbol{k}''p''}_{\boldsymbol{k}p,\boldsymbol{k}'p'}$ is given by eq. (5.4). Eqs. (7.5) and (4.15) then imply that the analogue of eq. (5.5) is

$$\begin{aligned} n^{\boldsymbol{k}''p''}_{\boldsymbol{k}p,\boldsymbol{k}'p'} = (1/g_o) \sum_{\boldsymbol{R}} \sum_{\boldsymbol{R}_i} \sum_{\boldsymbol{R}'_j} \sum_{\boldsymbol{R}''_l} & \chi^p(\{\boldsymbol{R}_i \mid \boldsymbol{t}_i\}^{-1} \{\boldsymbol{R} \mid \boldsymbol{t}_{\boldsymbol{R}}\} \{\boldsymbol{R}_i \mid \boldsymbol{t}_i\}) \\ & \times \chi^{p'}(\{\boldsymbol{R}'_j \mid \boldsymbol{t}'_j\}^{-1} \{\boldsymbol{R} \mid \boldsymbol{t}_{\boldsymbol{R}}\} \{\boldsymbol{R}'_j \mid \boldsymbol{t}'_j\}) \\ & \times \chi^{p''}(\{\boldsymbol{R}''_l \mid \boldsymbol{t}''_l\}^{-1} \{\boldsymbol{R} \mid \boldsymbol{t}_{\boldsymbol{R}}\} \{\boldsymbol{R}''_l \mid \boldsymbol{t}''_l\}) \\ & \times J_{\boldsymbol{k}}(\boldsymbol{R}_i^{-1}\boldsymbol{R}\boldsymbol{R}_i)\, J_{\boldsymbol{k}'}(\boldsymbol{R}'^{-1}_j\boldsymbol{R}\boldsymbol{R}'_j)\, J_{\boldsymbol{k}''}(\boldsymbol{R}''^{-1}_l\boldsymbol{R}\boldsymbol{R}''_l) \\ & \times \delta(\boldsymbol{R}_i\boldsymbol{k} + \boldsymbol{R}'_j\boldsymbol{k}' - \boldsymbol{R}''_l\boldsymbol{k}'' - \boldsymbol{K}_m), \end{aligned} \quad (7.13)$$

where J_k is defined in eq. (5.6). Here χ^p, $\chi^{p'}$ and $\chi^{p''}$ are the characters of the irreducible representations of $\mathscr{G}(\boldsymbol{k})$, $\mathscr{G}(\boldsymbol{k}')$ and $\mathscr{G}(\boldsymbol{k}'')$ corresponding to $\boldsymbol{\Gamma}^{kp}$, $\boldsymbol{\Gamma}^{k'p'}$ and $\boldsymbol{\Gamma}^{k''p''}$ respectively, and the rotations $\boldsymbol{R}_i$, $\boldsymbol{R}'_j$ and $\boldsymbol{R}''_l$ generate the stars of $\boldsymbol{k}$, $\boldsymbol{k}'$ and $\boldsymbol{k}''$ respectively. The Kronecker delta factor indicates that the only non-zero contributions to the process come from the terms where $\boldsymbol{R}_i\boldsymbol{k}+\boldsymbol{R}'_j\boldsymbol{k}'-\boldsymbol{R}''_l\boldsymbol{k}''$ is equal to a lattice vector $\boldsymbol{K}_m$ of the reciprocal lattice, which may be zero. If at least one non-zero contribution exists, the labelling of vectors in the stars may be chosen so that

$$\boldsymbol{k}+\boldsymbol{k}'-\boldsymbol{k}''=\boldsymbol{K}_m\,, \tag{7.14}$$

where $\boldsymbol{K}_m$ may be zero. An example of this relabelling was given in ch.5 §4.1.

If $\mathscr{G}(\boldsymbol{k}, \boldsymbol{k}', \boldsymbol{k}'')$ is defined to be the group of elements that are common to $\mathscr{G}(\boldsymbol{k})$, $\mathscr{G}(\boldsymbol{k}')$ and $\mathscr{G}(\boldsymbol{k}'')$, then

$$n^{k''p''}_{kp,\,k'p'}=(1/g')\sum\chi^{p}(\{\boldsymbol{R}\mid\boldsymbol{t_R}\})\,\chi^{p'}(\{\boldsymbol{R}\mid\boldsymbol{t_R}\})\,\chi^{p''}(\{\boldsymbol{R}\mid\boldsymbol{t_R}\})^{*}\,. \tag{7.15}$$

Here the sum is over all the elements $\{\boldsymbol{R}\mid\boldsymbol{t_R}\}$ of $\mathscr{G}(\boldsymbol{k}, \boldsymbol{k}', \boldsymbol{k}'')$, *only one* transformation $\{\boldsymbol{R}\mid\boldsymbol{t_R}\}$ being included for each rotation $\boldsymbol{R}$ of $\mathscr{G}(\boldsymbol{k}, \boldsymbol{k}', \boldsymbol{k}'')$, and g' is the number of such terms in the summation. The selection rules may be expressed compactly in the form of equations like (5.9), but the cautionary remarks made about the literal interpretation of such equations still apply.

The proof of (7.15) given by ZAK [1962] is a straight-forward generalization of the proof of (5.8) that was given in ch. 5 §4.1. Another direct proof has been given by BRADLEY [1966]. The formula (7.15) is equivalent to the result obtained by LAX and HOPFIELD [1961] by a less direct argument. More complicated methods of finding selection rules have been proposed by ELLIOTT and LOUDON [1960] and by BIRMAN [1962]. The method of ELLIOTT and LOUDON [1960] has the disadvantage of requiring a knowledge of *all* the irreducible representations of $\mathscr{G}(\boldsymbol{k})/\mathscr{T}(\boldsymbol{k})$, $\mathscr{G}(\boldsymbol{k}')/\mathscr{T}(\boldsymbol{k}')$ and $\mathscr{G}(\boldsymbol{k}'')/\mathscr{T}(\boldsymbol{k}'')$, and not merely those satisfying eq. (7.6) and the equivalent equations for $\boldsymbol{k}'$ and $\boldsymbol{k}''$. The method of BIRMAN [1962] is, however, of wider applicability. The relationships between these various methods for calculating selection rules have been studied by LAX [1965] and BIRMAN [1966].

The effect of time-reversal symmetry on selection rules has been examined by LAX [1962] and KARAVAEV [1965].

Selection rules for the diamond structure space group O_h^7 have been tabulated by ELLIOTT and LOUDON [1960], LAX and HOPFIELD [1961], BIRMAN [1962, 1963] and BALKANSKI and NUSIMOVICI [1964]. Selection rules for the wurtzite structure space group C_{6v}^4 have been given by NUSIMOVICI [1965].

§ 5.2. *Direct optical transitions*

The theory given in ch. 5 §4.2 shows that the perturbing term in the Hamiltonian corresponding to direct optical transitions is $V'(\boldsymbol{r})=\boldsymbol{a}_0\cdot\boldsymbol{p}$, and that $\partial/\partial x$, $\partial/\partial y$ and $\partial/\partial z$ transform as rows of a three-dimensional representation of $\mathscr{G}$ corresponding to $\boldsymbol{k}'=\boldsymbol{0}$. In the case of the hexagonal close-packed space group D_{6h}^4, $\mathscr{G}(\boldsymbol{0})/\mathscr{T}(\boldsymbol{0})$ is isomorphic to the point group D_{6h}, which has *no* three-dimensional irreducible representations. Closer examination shows that $\partial/\partial x$ and $\partial/\partial y$ transform as the rows of Γ_6^-, while $\partial/\partial z$ transforms as Γ_2^-. The notation is that of table A.4. Thus, $V'(\boldsymbol{r})$ transforms as Γ_6^- if $\boldsymbol{a}_0$ is in the plane Oxy, but as Γ_2^- if $\boldsymbol{a}_0$ is parallel to Oz, so the selection rules depend on the direction of polarization of the light.

Condition (7.14) implies that $\boldsymbol{k}''=\boldsymbol{k}$, so that, on an energy band diagram of $E(\boldsymbol{k})$ plotted against $\boldsymbol{k}$, the transitions are again 'vertical'. As $\mathscr{G}(\boldsymbol{0})$ is $\mathscr{G}$, which contains $\mathscr{G}(\boldsymbol{k})=\mathscr{G}(\boldsymbol{k}'')$ as a subgroup, the group $\mathscr{G}(\boldsymbol{k}, \boldsymbol{k}', \boldsymbol{k}'')$ of eq. (7.15) is in this case just $\mathscr{G}(\boldsymbol{k})$. The direct optical selection rules for the space group D_{6h}^4 have been listed by CORNWELL [1966].

§ 6. PROPERTIES OF ENERGY BANDS

§ 6.1. *Continuity and compatibility of the irreducible representations of* $\mathscr{G}(\boldsymbol{k})/\mathscr{T}(\boldsymbol{k})$

The discussion of ch. 5 § 5.1 on continuity of irreducible representations for a symmorphic space group also applies to non-symmorphic space groups, the only difference being that the group $\mathscr{G}(\boldsymbol{k})/\mathscr{T}(\boldsymbol{k})$ now takes the place of $\mathscr{G}_o(\boldsymbol{k})$. The continuity of representations is a consequence of the labelling convention established in § 4.3, namely that on a symmetry axis the matrices of an irreducible representation with a particular label, p, are continuous functions of $\boldsymbol{k}$.

The compatibility relations at a symmetry point $\boldsymbol{k}_s$ are found simply from the reduction of the irreducible representations of $\mathscr{G}(\boldsymbol{k}_s)/\mathscr{T}(\boldsymbol{k}_s)$ with respect to the relevant irreducible representations in the limit as $\boldsymbol{k}\to\boldsymbol{k}_s$ of $\mathscr{G}(\boldsymbol{k})/\mathscr{T}(\boldsymbol{k}_s)$. As shown in ch. 2 § 6, only the characters of both these groups are needed to determine the reduction, and for the space group D_{6h}^4 all the necessary character

TABLE 7.6

Compatibility relations between the symmetry point Γ and the symmetry axis Δ of the hexagonal close-packed space group D_{6h}^4

Γ_1^+	Γ_2^+	Γ_3^+	Γ_4^+	Γ_5^+	Γ_6^+
Δ_1	Δ_3	Δ_2	Δ_4	Δ_5	Δ_6

Γ_1^-	Γ_2^-	Γ_3^-	Γ_4^-	Γ_5^-	Γ_6^-
Δ_3	Δ_1	Δ_4	Δ_2	Δ_5	Δ_6

tables are given in the paper of HERRING [1942]. As an example, the compatibility relations between the point Γ and the axis Δ are given in table 7.6, which has been obtained using tables 7.1, A.4 and 7.3.

§ 6.2. *Fine detail of the symmetry of energy bands, their approximate construction, and their critical points*

The conclusions drawn in ch. 5 § 5.2 about the fine detail of the symmetry of energy bands are merely consequences of the fact that $E(\boldsymbol{k})$ has the symmetry of the point group $\mathscr{G}_o$. As was demonstrated in § 3.1, this symmetry exists for non-symmorphic as well as symmorphic space groups, so these conclusions apply equally well to the non-symmorphic case.

The procedure described in ch. 5 § 5.3 for locating and classifying critical points and the method described in ch. 5 § 5.4 for obtaining a picture of the whole energy band structure from a knowledge of the energy levels at the symmetry points of the Brillouin zone can also be applied in the same way to non-symmorphic space groups.

§ 7. SYMMETRIZED WAVE FUNCTIONS

There is no difficulty in applying the projection operator technique of ch. 3 § 5.2 to any chosen Bloch function with wave vector $\boldsymbol{k}$ to obtain Bloch basis functions of the irreducible representations of $\mathscr{G}(\boldsymbol{k})$. The fundamental theorem of § 1 then gives the basis functions of the irreducible representations of the space group $\mathscr{G}$ immediately.

Let $\phi_{\boldsymbol{k}}(\boldsymbol{r}) = \exp(\mathrm{i}\boldsymbol{k}\cdot\boldsymbol{r})\, u_{\boldsymbol{k}}(\boldsymbol{r})$ be any Bloch function with wave vector $\boldsymbol{k}$. Then

$$\mathscr{P}^p_{ss}\phi_{\boldsymbol{k}}(\boldsymbol{r}) = \{l/g(\boldsymbol{k})\} \sum \Gamma^p(\{\boldsymbol{R} \mid \boldsymbol{t}\})^*_{ss}\, P(\{\boldsymbol{R} \mid \boldsymbol{t}\})\, \phi_{\boldsymbol{k}}(\boldsymbol{r})$$

transforms as the sth row of the irreducible representation $\boldsymbol{\Gamma}^p$ of $\mathscr{G}(\boldsymbol{k})$, where l is the dimension of $\boldsymbol{\Gamma}^p$, $g(\boldsymbol{k})$ is the order of $\mathscr{G}(\boldsymbol{k})$, and the summation is over all the elements of $\mathscr{G}(\boldsymbol{k})$. However, for any

$\{E \mid t_n\}$ of $\mathscr{T}(k)$, $P(\{E \mid t_n\})\, \phi_k(r) = \phi_k(r)$, which follows from eq. (4.10) and the definition of $\mathscr{T}(k)$ in § 2. Therefore, as all the members of a coset of $\mathscr{G}(k)$ with respect to $\mathscr{T}(k)$ are by (7.7) represented by the same matrix,

$$\mathscr{P}^p_{ss}\phi_k(r) = \{l/g'\} \sum \Gamma'^p(\{R \mid t\}\, \mathscr{T}(k))^*_{ss}\, P(\{R \mid t\})\, \phi_k(r)\,. \tag{7.16}$$

Here the summation is over a set of coset representatives $\{R \mid t\}$ of $\mathscr{G}(k)/\mathscr{T}(k)$, g' is the order of $\mathscr{G}(k)/\mathscr{T}(k)$, and Γ'^p is the corresponding irreducible representation of $\mathscr{G}(k)/\mathscr{T}(k)$.

The projection operator of (7.16) may be applied very easily to plane waves, for if $\phi_{km}(r) = \exp\{i(k+K_m)\cdot r\}$, then

$$\begin{aligned} P(\{R \mid t\})\, \phi_{km}(r) &= \exp\{i(k + K_m)\cdot R^{-1}(r - t)\} \\ &= \exp\{iR(k + K_m)\cdot(r - t)\}\,. \end{aligned}$$

The application of the projection operator (7.16) to linear combinations of atomic orbitals is slightly more complex, for it is clear that the symmetrized L.C.A.O.s will contain atomic orbitals centred on every atomic site, and not merely on those that are also lattice sites. For the hexagonal close-packed structure, with 2 atoms per lattice point, it is convenient to define 2 atomic orbitals $\psi_{mj}(r)$ from each atomic orbital $\psi_m(r)$ centred on $r = 0$ by $\psi_{m1}(r) = \psi_m(r)$ and $\psi_{m2}(r) = \psi_m(r - \tau)$. Then, if the Bloch sums are defined, as in (5.18), by

$$\psi_{kmj}(r) = M^{-\frac{1}{2}} \sum_{t_n} \exp(ik\cdot t_n)\, \psi_{mj}(r - t_n)\,,$$

it follows by essentially the same argument as that given in ch. 5 § 6.4 that

$$P(\{R \mid t\})\, \psi_{km1}(r) = \begin{cases} M^{-\frac{1}{2}} \sum\limits_{t_n} \exp(ik\cdot t_n)\, P(\{R \mid 0\})\, \psi_{m1}(r_n), & \text{if} \quad t = 0 \\ M^{-\frac{1}{2}} \exp\{ik\cdot(t - R^{-1}t)\} \sum\limits_{t_n} & \\ \qquad \times \exp(ik\cdot t_n)\, P(\{R \mid 0\})\, \psi_{m2}(r_n), & \text{if} \quad t = \tau\,, \end{cases} \tag{7.17}$$

where r_n is replaced by $r - t_n$ after the $P(\{R \mid 0\})$ operation has taken place. The expressions on the right-hand side of (7.17) are also Bloch sums. Combining (7.16) with (7.17) allows the symmetrized L.C.A.O.s to be projected out of $\psi_{km1}(\boldsymbol{r})$. All the symmetrized L.C.A.O.s may be obtained in this way. As in ch. 5 § 6.4, the atomic orbitals can be arranged to transform as spherical harmonics, and consequently they may be labelled appropriately. A detailed study of the tight-binding method as applied to the hexagonal close-packed space group D_{6h}^4 has been made by MIASEK [1957a, 1957b, 1958] and by ERDMANN [1960]. A development of the projection operator technique, which allows the symmetrized L.C.A.O.s for a non-symmorphic space group to be constructed by an electronic computer, has been described by FLODMARK [1963].

Symmetrized orthogonalized plane waves may be obtained from the appropriate symmetrized L.C.A.O.s and symmetrized plane waves, as in ch. 5 § 6.5.

The construction of lattice harmonics for use in the cellular method for the hexagonal close-packed space group D_{6h}^4 is described in articles by ALTMANN and BRADLEY [1965a, 1966b]. TREUSCH and SANDROCK [1966] have given a treatment of the method of KOHN and ROSTOKER [1954] that is suitable for non-symmorphic space groups.

§ 8. TIME-REVERSAL SYMMETRY

As in symmorphic space groups, time-reversal symmetry can cause extra degeneracies, the simplest being that $E(\boldsymbol{k}) = E(-\boldsymbol{k})$ for all $\boldsymbol{k}$. In ch. 5 § 4.1 criteria were presented for determining when additional degeneracies occur. The only point requiring modification for a non-symmorphic space group is the criterion (5.27). An irreducible representation of $\mathscr{G}$ is now specified by a wave vector $\boldsymbol{k}$ and an irreducible representation $\boldsymbol{\Gamma}^p$ of $\mathscr{G}(\boldsymbol{k})$. For this representation the criterion corresponding to (5.27) has the form

$$\sum \chi^p(\{\boldsymbol{Q} \mid \boldsymbol{t}\}^2) = \begin{cases} g_o/M(\boldsymbol{k}), & \text{case (a)}, \\ 0, & \text{case (b)}, \\ -g_o/M(\boldsymbol{k}), & \text{case (c)}, \end{cases} \tag{7.18}$$

where the summation is over the set of coset representatives $\{\boldsymbol{Q} \mid \boldsymbol{t}\}$ of $\mathscr{G}/\mathscr{T}$ that have the property that $\boldsymbol{Qk}$ is equivalent to $-\boldsymbol{k}$. In (7.18) g_o is the order of $\mathscr{G}_o$, $M(\boldsymbol{k})$ is the number of vectors in the star of $\boldsymbol{k}$, and $\chi^p(\{\boldsymbol{Q} \mid \boldsymbol{t}\}^2)$ denotes the character of $\{\boldsymbol{Q} \mid \boldsymbol{t}\}^2$ in the irreducible representation $\boldsymbol{\Gamma}^p$ of $\mathscr{G}(\boldsymbol{k})$. Although $\{\boldsymbol{Q} \mid \boldsymbol{t}\}$ may not be a member of $\mathscr{G}(\boldsymbol{k})$, $\{\boldsymbol{Q} \mid \boldsymbol{t}\}^2$ necessarily is a member of $\mathscr{G}(\boldsymbol{k})$. The criterion (7.18) was first derived by HERRING [1937a], the derivation from (5.26) running parallel to that given in ch. 5 §7.1.

The analysis of an extra degeneracy corresponding to the cases (b) and (c) in terms of a possible extra degeneracy of $E(\boldsymbol{k})$ follows *exactly* as in ch. 5 §7.1. The condition in case (2) that replaces (5.28) is that, if $\boldsymbol{\Gamma}^p$ and $\boldsymbol{\Gamma}^q$ are two irreducible representations of $\mathscr{G}(\boldsymbol{k})$ of type (b) that stick together, and χ^p and χ^q denote their characters, then

$$\chi^p(\{\boldsymbol{R} \mid \boldsymbol{t}\})^* = \chi^q(\{\boldsymbol{R}_- \mid \boldsymbol{t}_-\}^{-1}\{\boldsymbol{R} \mid \boldsymbol{t}\}\{\boldsymbol{R}_- \mid \boldsymbol{t}_-\}) \qquad (7.19)$$

for every $\{\boldsymbol{R} \mid \boldsymbol{t}\}$ of $\mathscr{G}(\boldsymbol{k})$. Here $\{\boldsymbol{R}_- \mid \boldsymbol{t}_-\}$ is the coset representative of the decomposition of $\mathscr{G}$ into left cosets with respect to $\mathscr{G}(\boldsymbol{k})$ such that $\boldsymbol{R}_-\boldsymbol{k} = -\boldsymbol{k}$. The condition in case (3) that replaces (5.30) is simply that

$$\chi^p(\{\boldsymbol{R} \mid \boldsymbol{t}\})^* = \chi^q(\{\boldsymbol{R} \mid \boldsymbol{t}\}) \qquad (7.20)$$

for every $\{\boldsymbol{R} \mid \boldsymbol{t}\}$ of $\mathscr{G}(\boldsymbol{k})$. It is sufficient to test (7.19) and (7.20) with only those elements $\{\boldsymbol{R} \mid \boldsymbol{t}\}$ of $\mathscr{G}(\boldsymbol{k})$ that are coset representatives of $\mathscr{G}(\boldsymbol{k})/\mathscr{T}(\boldsymbol{k})$ but are not pure translations.

Examples of extra degeneracies caused by time-reversal symmetry occur in the close-packed hexagonal space group D_{6h}^4. Consider first the symmetry axis R, for which the factor group $\mathscr{G}(\boldsymbol{k})/\mathscr{T}(\boldsymbol{k})$ was described in §4.3. On R, $\boldsymbol{k} = (0, k_y, \pi/c)$, $0 < k_y < 2\pi/a\sqrt{3}$, so that the appropriate set of operations $\{\boldsymbol{Q} \mid \boldsymbol{t}\}$ of (7.18) are $\{I \mid \boldsymbol{\tau}\}$, $\{C_{2z} \mid \boldsymbol{\tau}\}$, $\{IC_{2y} \mid \boldsymbol{\tau}\}$ and $\{C_{2x} \mid \boldsymbol{\tau}\}$, whose squares are $\{\boldsymbol{E} \mid \boldsymbol{0}\}$, $\{\boldsymbol{E} \mid \boldsymbol{a}_1\}$, $\{\boldsymbol{E} \mid \boldsymbol{a}_1 + \boldsymbol{a}_2\}$ and $\{\boldsymbol{E} \mid \boldsymbol{a}_2\}$ respectively. The characters of these latter operations are all given directly by eq. (7.2), from which it follows that case (b) applies for R_1, R_2, R_3 and R_4. The vector

$-\boldsymbol{k}$ is in the star of $\boldsymbol{k}$, so that this is an example of case (2), in which the energy levels corresponding to R_1, R_2, R_3 and R_4 stick together in pairs. The condition (7.19) may now be used to find exactly how they pair off. Taking $\{\boldsymbol{R}_- \mid \boldsymbol{t}_-\}$ to be $\{I \mid \boldsymbol{\tau}\}$, (7.19) requires that

$$\begin{aligned}\chi^p(\{\boldsymbol{R} \mid \boldsymbol{0}\})^* &= \chi^q(\{\boldsymbol{R} \mid -\boldsymbol{R\tau} + \boldsymbol{\tau}\}) \\ &= \chi^q(\{\boldsymbol{R} \mid \boldsymbol{0}\})\, \chi^q(\{\boldsymbol{E} \mid -\boldsymbol{\tau} + \boldsymbol{R}^{-1}\boldsymbol{\tau}\}) \\ &= \theta^q(\boldsymbol{R}),\end{aligned}$$

for every $\{\boldsymbol{R} \mid \boldsymbol{0}\}$ of $\mathscr{G}(\mathrm{k})/\mathscr{T}(\mathrm{k})$, where $\theta^q(\boldsymbol{R}) = \chi^q(\{\boldsymbol{R} \mid \boldsymbol{0}\}) \exp \{-\mathrm{i}\boldsymbol{k}\cdot(\boldsymbol{R}^{-1}\boldsymbol{\tau} - \boldsymbol{\tau})\}$. The quantities $\theta^q(\boldsymbol{R})$ are tabulated in table 7.7. Comparison with table 7.5 shows that the two degenerate pairs are R_1, R_3 and R_2, R_4 respectively.

TABLE 7.7

	θ^1	θ^2	θ^3	θ^4
$\{E \mid \boldsymbol{0}\}$	1	1	1	1
$\{C_{2y} \mid \boldsymbol{0}\}$	-1	-1	1	1
$\{IC_{2z} \mid \boldsymbol{0}\}$	-1	1	1	-1
$\{IC_{2x} \mid \boldsymbol{0}\}$	1	-1	1	-1

These degeneracies are intimately related to the two-fold degeneracy that exists over the whole of the hexagonal face of the Brillouin zone, which is another consequence of time-reversal symmetry. To demonstrate this degeneracy, consider again the point $\boldsymbol{k} = (k_x, k_y, \pi/c)$, whose group $\mathscr{G}(\boldsymbol{k})$ was described in §4.4. The only operations $\{\boldsymbol{Q} \mid \boldsymbol{t}\}$ of eq. (7.18) for this $\boldsymbol{k}$ are $\{I \mid \boldsymbol{\tau}\}$ and $\{C_{2z} \mid \boldsymbol{\tau}\}$. As their squares are $\{\boldsymbol{E} \mid \boldsymbol{0}\}$ and $\{\boldsymbol{E} \mid \boldsymbol{a}_1\}$, whose characters are $+1$ and -1 respectively, by eq. (7.2), for both relevant irreducible representations, this is another example of case (b) and case (2). The energy levels at $\boldsymbol{k}$ of the two non-equivalent one-dimensional irreducible representations therefore stick together in pairs.

This degeneracy over the whole of the hexagonal face of the Brillouin zone of the close-packed hexagonal space group was first noted by HUND [1936]. One consequence of the degeneracy is that

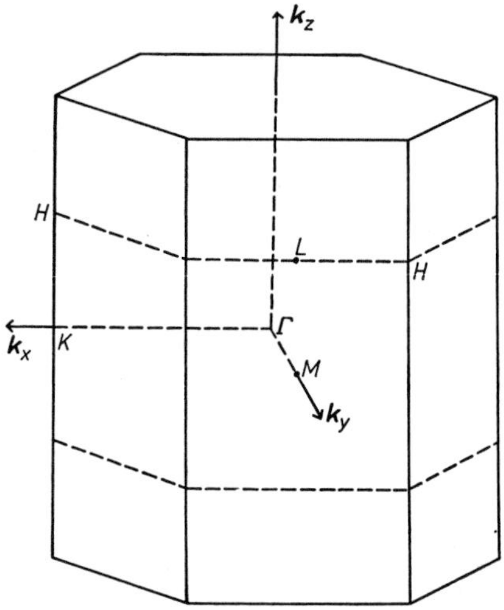

Fig. 7.1. The 'double zone' of the hexagonal close-packed space group D_{6h}^4.

in calculations of the density of states and the Fermi surface it is convenient to work with the 'double zone', which is shown in fig. 7.1, rather than the Brillouin zone of fig. 4.6. This procedure was followed, for example, by CORNWELL [1961] and by ALTMANN and BRADLEY [1964, 1967].

CHAPTER 8

DOUBLE GROUPS AND SPIN–ORBIT COUPLING

§ 1. INTRODUCTION

In chs. 3 to 7 the theory was developed for *scalar* wave functions. This theory is completely satisfactory if the Hamiltonian does not contain any spin dependent terms, for then the wave function for an electron can simply be taken as the product of an 'orbital' wave function with one of two possible spin functions. The only effect of these two spin functions is then to double every 'orbital' degeneracy.

When spin–orbit coupling is taken into account, the Hamiltonian contains spin-dependent terms. The foregoing theory then has to be modified, but, as will be shown in this chapter, the same basic features remain. The only major difference is the use of 'double' groups instead of 'single' groups.

The most important part of this chapter is the development in §3 and §4 of the transformation theory for spinor wave functions along the same lines as that given in ch. 3 §§2, 3 for scalar functions. The very minor modifications required in the theory of ch. 3 §§4–6 are then briefly reviewed in §5. The general properties of double groups are described in §6, and §7 gives the generalization of the theory of the group $\mathscr{T}$ of pure primitive translations that appeared in ch. 4. The double symmorphic and double non-symmorphic space groups are then considered separately in §8 and §9 in what are essentially generalizations of the theories given in chs. 5 and 7 respectively.

The theory developed in this chapter is based on the usual Pauli–Schrödinger formalism. However, JOHNSTON [1958] has shown that the theory may be deduced directly from the Dirac equation, a procedure which has certain advantages, one of which being that the proofs of the time-reversal theorems of WIGNER [1932, 1959] that are mentioned in § 8.3 and § 9.2 may be simplified.

The incorporation of spinor functions into the theory is made possible by the homomorphism between the group of proper rotations in three dimensions and the group SU_2. It is with this homomorphism that the development will begin.

§ 2. THE HOMOMORPHISM BETWEEN THE GROUP OF PROPER ROTATIONS IN THREE DIMENSIONS AND THE GROUP SU_2

The group SU_2 is defined as the group of all 2×2 unitary matrices having determinant unity, with ordinary matrix multiplication as the operation of group multiplication. It will be shown that this group is homomorphic to the group R_3 of proper rotations in three dimensions, there being two elements of SU_2 corresponding to each element R_3. Both of these groups are of infinite order, but the same result is true for all the finite subgroups of R_3, which are the relevant groups for solid state physics.

The homomorphism is most easily established indirectly by considering the 2×2 traceless Hermitian matrix $\boldsymbol{m}(\boldsymbol{r})$, whose components are defined in terms of the position vector $\boldsymbol{r}=(x, y, z)$ by

$$\boldsymbol{m}(\boldsymbol{r}) = \begin{pmatrix} z & x - \mathrm{i}y \\ x + \mathrm{i}y & -z \end{pmatrix}.$$

Then if $\boldsymbol{u}$ is any element of SU_2, $\boldsymbol{u}\boldsymbol{m}(\boldsymbol{r})\,\boldsymbol{u}^{-1}$ is also a traceless Hermitian matrix. (The zero trace is merely a consequence of the fact that the trace of a matrix is unaltered by a similarity transformation, while the Hermitian property follows as $\tilde{\boldsymbol{m}}(\boldsymbol{r})^* = \boldsymbol{m}(\boldsymbol{r})$ and $\tilde{\boldsymbol{u}}^* = \boldsymbol{u}^{-1}$.) Thus one may write

$$\boldsymbol{m}(\boldsymbol{r}') = \begin{pmatrix} z' & x' - \mathrm{i}y' \\ x' + \mathrm{i}y' & -z' \end{pmatrix} = \boldsymbol{u}\boldsymbol{m}(\boldsymbol{r})\,\boldsymbol{u}^{-1}. \tag{8.1}$$

This defines the components of the vector $\boldsymbol{r}'=(x', y', z')$. On evaluating the right-hand side of (8.1) and equating corresponding matrix elements, there is generated a linear transformation from $\boldsymbol{r}$ to $\boldsymbol{r}'$ given by

$$\boldsymbol{r}' = \boldsymbol{R}(\boldsymbol{u})\,\boldsymbol{r}\,, \tag{8.2}$$

$\boldsymbol{R}(\boldsymbol{u})$ being a 3×3 matrix which depends on $\boldsymbol{u}$. The transformation (8.2) represents a *rotation*, because lengths and angles are preserved. This follows because if $\boldsymbol{r}_1$ and $\boldsymbol{r}_2$ are any two vectors and $\boldsymbol{r}_1'=\boldsymbol{R}(\boldsymbol{u})\boldsymbol{r}_1$ and $\boldsymbol{r}_2'=\boldsymbol{R}(\boldsymbol{u})\boldsymbol{r}_2$, then

$$\boldsymbol{r}_1'\cdot\boldsymbol{r}_2' = \tfrac{1}{2}\mathrm{Tr}\,\{\boldsymbol{m}(\boldsymbol{r}_1')\,\boldsymbol{m}(\boldsymbol{r}_2')\} = \tfrac{1}{2}\mathrm{Tr}\,\{\boldsymbol{m}(\boldsymbol{r}_1)\,\boldsymbol{m}(\boldsymbol{r}_2)\} = \boldsymbol{r}_1\cdot\boldsymbol{r}_2\,.$$

Explicit evaluation of $\boldsymbol{R}(\boldsymbol{u})$ shows that it represents a *proper* rotation, and that all proper rotations can appear this way.

The homomorphism is established by showing that if $\boldsymbol{u}_1$ and $\boldsymbol{u}_2$ are members of SU_2 then $\boldsymbol{R}(\boldsymbol{u}_2\boldsymbol{u}_1)=\boldsymbol{R}(\boldsymbol{u}_2)\;\boldsymbol{R}(\boldsymbol{u}_1)$. This result is true because if $\boldsymbol{m}(\boldsymbol{r}')=\boldsymbol{u}_1\boldsymbol{m}(\boldsymbol{r})\,\boldsymbol{u}_1^{-1}$ and $\boldsymbol{m}(\boldsymbol{r}'')=\boldsymbol{u}_2\boldsymbol{m}(\boldsymbol{r}')\,\boldsymbol{u}_2^{-1}$, then $\boldsymbol{m}(\boldsymbol{r}'')=(\boldsymbol{u}_2\boldsymbol{u}_1)\,\boldsymbol{m}(\boldsymbol{r})\,(\boldsymbol{u}_2\boldsymbol{u}_1)^{-1}$. However, the homomorphism is not an isomorphism, for (8.1) shows that $\boldsymbol{u}$ and $-\boldsymbol{u}$ both correspond to the same rotation, that is, $\boldsymbol{R}(\boldsymbol{u})=\boldsymbol{R}(-\boldsymbol{u})$.

It is often convenient to invert the relationship between $\boldsymbol{u}$ and $\boldsymbol{R}(\boldsymbol{u})$, and to regard $\boldsymbol{u}$ as being dependent on $\boldsymbol{R}$, denoting it appropriately by $\boldsymbol{u}(\boldsymbol{R})$. Then $\boldsymbol{u}(\boldsymbol{R})$ and $-\boldsymbol{u}(\boldsymbol{R})$ are said to constitute a 'two-valued' representation of $\boldsymbol{R}$.

An explicit expression for the matrix elements of $\boldsymbol{R}(\boldsymbol{u})$ in terms of the matrix $\boldsymbol{u}$ can be found very easily as follows. Let

$$\boldsymbol{\sigma}_1 = \begin{pmatrix} 0 & 1 \\ 1 & 0 \end{pmatrix}, \qquad \boldsymbol{\sigma}_2 = \begin{pmatrix} 0 & -\mathrm{i} \\ \mathrm{i} & 0 \end{pmatrix}, \qquad \boldsymbol{\sigma}_3 = \begin{pmatrix} 1 & 0 \\ 0 & -1 \end{pmatrix} \tag{8.3}$$

be the three Pauli spin matrices, for which it may be verified that

$$\mathrm{Tr}\,\{\boldsymbol{\sigma}_j\boldsymbol{\sigma}_k\} = 2\delta_{jk}\,. \tag{8.4}$$

Then

$$\boldsymbol{m}(\boldsymbol{r}) = \boldsymbol{\sigma}_1 x + \boldsymbol{\sigma}_2 y + \boldsymbol{\sigma}_3 z$$

and

$$\boldsymbol{m}(\boldsymbol{r}') = \boldsymbol{\sigma}_1 x' + \boldsymbol{\sigma}_2 y' + \boldsymbol{\sigma}_3 z',$$

so that

$$\boldsymbol{\sigma}_1 x' + \boldsymbol{\sigma}_2 y' + \boldsymbol{\sigma}_3 z' = \boldsymbol{u}\boldsymbol{\sigma}_1\boldsymbol{u}^{-1} x + \boldsymbol{u}\boldsymbol{\sigma}_2\boldsymbol{u}^{-1} y + \boldsymbol{u}\boldsymbol{\sigma}_3\boldsymbol{u}^{-1} z. \quad (8.5)$$

Multiplying this equation through by $\boldsymbol{\sigma}_1$, $\boldsymbol{\sigma}_2$ and $\boldsymbol{\sigma}_3$ in turn, taking the trace and using eq. (8.4) gives expressions for x', y' and z' in terms of x, y and z. Comparing with eq. (8.2) then gives

$$\boldsymbol{R}(\boldsymbol{u})_{jk} = \tfrac{1}{2}\,\mathrm{Tr}\,\{\boldsymbol{\sigma}_j \boldsymbol{u} \boldsymbol{\sigma}_k \boldsymbol{u}^{-1}\} \quad (8.6)$$

as the required result.

The formulae for the inverse of (8.6), which are very useful but more complicated, can be obtained as follows. Any 2×2 unitary matrix $\boldsymbol{u}$ can be written in the form

$$\boldsymbol{u} = u_0 \boldsymbol{E} + \mathrm{i} \sum_{j=1}^{3} u_j \boldsymbol{\sigma}_j, \quad (8.7)$$

$\boldsymbol{E}$ being the 2×2 unit matrix. The unitary condition $\tilde{\boldsymbol{u}}^* = \boldsymbol{u}^{-1}$ requires that u_0, u_1, u_2 and u_3 must be all real, and the condition that det $\boldsymbol{u} = +1$ requires that

$$\sum_{j=0}^{3} u_j^2 = 1. \quad (8.8)$$

The coefficients u_j are found by substituting (8.7) into (8.6), which gives on using (8.4) and the similar relations

$$\mathrm{Tr}\,\{\boldsymbol{\sigma}_j \boldsymbol{\sigma}_k \boldsymbol{\sigma}_l\} = 2\mathrm{i}\varepsilon_{jkl},$$
$$\mathrm{Tr}\,\{\boldsymbol{\sigma}_j \boldsymbol{\sigma}_k \boldsymbol{\sigma}_l \boldsymbol{\sigma}_m\} = 2(\delta_{jk}\delta_{lm} - \delta_{jl}\delta_{km} + \delta_{jm}\delta_{kl}),$$

that

$$R_{jk} = \delta_{jk}\left(u_0^2 - \sum_{l=1}^{3} u_l^2\right) + 2u_0 \sum_{l=1}^{3} u_l \varepsilon_{jkl} + 2u_j u_k, \quad (8.9)$$

where

$$\varepsilon_{jkl} = \begin{cases} +1, & \text{if } j, k \text{ and } l \text{ are an even permutation of 1, 2 and 3}, \\ -1, & \text{if } j, k \text{ and } l \text{ are an odd permutation of 1, 2 and 3}, \\ 0, & \text{otherwise}. \end{cases}$$

The first consequence of eq. (8.9) is that

$$\operatorname{Tr} \boldsymbol{R} = 3u_0^2 - \sum_{l=1}^{3} u_l^2 , \tag{8.10}$$

so that by eq. (8.8),

$$u_0^2 = \tfrac{1}{4}(\operatorname{Tr} \boldsymbol{R} + 1), \tag{8.11}$$

which determines u_0^2.

If $u_0 \neq 0$ then eq. (8.9) gives

$$4u_0 u_m = \sum_{j=1}^{3} \sum_{k=1}^{3} \varepsilon_{mjk} R_{jk} ,$$

TABLE 8.1

$\boldsymbol{R}_p$	u_0	u_1	u_2	u_3	$\boldsymbol{R}_p$	u_0	u_1	u_2	u_3
E	1	0	0	0	C_{4y}^{-1}	$2^{-\frac{1}{2}}$	0	$-2^{-\frac{1}{2}}$	0
$C_{3\alpha}$	$\frac{1}{2}$	$-\frac{1}{2}$	$-\frac{1}{2}$	$\frac{1}{2}$	C_{4z}^{-1}	$2^{-\frac{1}{2}}$	0	0	$-2^{-\frac{1}{2}}$
$C_{3\beta}$	$\frac{1}{2}$	$-\frac{1}{2}$	$\frac{1}{2}$	$-\frac{1}{2}$	C_{2a}	0	$2^{-\frac{1}{2}}$	$2^{-\frac{1}{2}}$	0
$C_{3\gamma}$	$\frac{1}{2}$	$\frac{1}{2}$	$-\frac{1}{2}$	$-\frac{1}{2}$	C_{2b}	0	$-2^{-\frac{1}{2}}$	$2^{-\frac{1}{2}}$	0
$C_{3\delta}$	$\frac{1}{2}$	$\frac{1}{2}$	$\frac{1}{2}$	$\frac{1}{2}$	C_{2c}	0	$2^{-\frac{1}{2}}$	0	$2^{-\frac{1}{2}}$
$C_{3\alpha}^{-1}$	$\frac{1}{2}$	$\frac{1}{2}$	$\frac{1}{2}$	$-\frac{1}{2}$	C_{2d}	0	$-2^{-\frac{1}{2}}$	0	$2^{-\frac{1}{2}}$
$C_{3\beta}^{-1}$	$\frac{1}{2}$	$\frac{1}{2}$	$-\frac{1}{2}$	$\frac{1}{2}$	C_{2e}	0	0	$2^{-\frac{1}{2}}$	$2^{-\frac{1}{2}}$
$C_{3\gamma}^{-1}$	$\frac{1}{2}$	$-\frac{1}{2}$	$\frac{1}{2}$	$\frac{1}{2}$	C_{2f}	0	0	$-2^{-\frac{1}{2}}$	$2^{-\frac{1}{2}}$
$C_{3\delta}^{-1}$	$\frac{1}{2}$	$-\frac{1}{2}$	$-\frac{1}{2}$	$-\frac{1}{2}$	C_{3z}	$\frac{1}{2}$	0	0	$\frac{1}{2}3^{\frac{1}{2}}$
C_{2x}	0	1	0	0	C_{3z}^{-1}	$\frac{1}{2}$	0	0	$-\frac{1}{2}3^{\frac{1}{2}}$
C_{2y}	0	0	1	0	C_{6z}	$\frac{1}{2}3^{\frac{1}{2}}$	0	0	$\frac{1}{2}$
C_{2z}	0	0	0	1	C_{6z}^{-1}	$\frac{1}{2}3^{\frac{1}{2}}$	0	0	$-\frac{1}{2}$
C_{4x}	$2^{-\frac{1}{2}}$	$2^{-\frac{1}{2}}$	0	0	C_{2A}	0	$\frac{1}{2}$	$\frac{1}{2}3^{\frac{1}{2}}$	0
C_{4y}	$2^{-\frac{1}{2}}$	0	$2^{-\frac{1}{2}}$	0	C_{2B}	0	$-\frac{1}{2}$	$\frac{1}{2}3^{\frac{1}{2}}$	0
C_{4z}	$2^{-\frac{1}{2}}$	0	0	$2^{-\frac{1}{2}}$	C_{2C}	0	$-\frac{1}{2}3^{\frac{1}{2}}$	$\frac{1}{2}$	0
C_{4x}^{-1}	$2^{-\frac{1}{2}}$	$-2^{-\frac{1}{2}}$	0	0	C_{2D}	0	$\frac{1}{2}3^{\frac{1}{2}}$	$\frac{1}{2}$	0

so that

$$\frac{u_1}{u_0} = \frac{R_{23} - R_{32}}{\operatorname{Tr} \boldsymbol{R} + 1}, \qquad \frac{u_2}{u_0} = \frac{R_{31} - R_{13}}{\operatorname{Tr} \boldsymbol{R} + 1}, \qquad \frac{u_3}{u_0} = \frac{R_{12} - R_{21}}{\operatorname{Tr} \boldsymbol{R} + 1}. \tag{8.12}$$

Thus, if $u_0 \neq 0$, then (8.12) determines the ratios u_m/u_0, while (8.11) determines u_0^2. The sign of u_0 is not specified because both $\boldsymbol{u}$ and $-\boldsymbol{u}$ correspond to the same rotation $\boldsymbol{R}$.

If $u_0 = 0$, as it is for all the rotations through π of table 1.5, eqs. (8.9) and (8.10) give

$$u_j^2 = \tfrac{1}{2}(R_{jj} + 1), \qquad j = 1, 2, 3. \tag{8.13}$$

In every case inspection of table 1.5 shows that at least one u_j is zero. If only one u_j is zero, and u_k and u_l are the two non-zero coefficients, then the sign of $u_k u_l$ is given according to eq. (8.9) by

$$2u_k u_l = R_{kl}, \qquad k \neq l.$$

Again the overall sign of $\boldsymbol{u}$ remains undetermined.

Table 8.1 gives u_0, u_1, u_2 and u_3 for every proper rotation that is listed in table 1.5, the overall sign of $\boldsymbol{u}$ for rotations other than those through π being given by a convention that is described at the end of § 8.5.

It is sometimes convenient to specify a proper rotation in terms of the Eulerian angles θ, ϕ and ψ, as defined in fig. 8.1. Then

$$\left.\begin{aligned}
R_{11} &= -\sin\phi \sin\psi + \cos\theta \cos\phi \cos\psi, \\
R_{12} &= \cos\phi \sin\psi + \cos\theta \sin\phi \cos\psi, \\
R_{13} &= -\sin\theta \cos\psi, \\
R_{21} &= -\sin\phi \cos\psi - \cos\theta \cos\phi \sin\psi, \\
R_{22} &= \cos\phi \cos\psi - \cos\theta \sin\phi \sin\psi, \\
R_{23} &= \sin\theta \sin\psi, \\
R_{31} &= \sin\theta \cos\phi, \\
R_{32} &= \sin\theta \sin\phi, \\
R_{33} &= \cos\theta.
\end{aligned}\right| \tag{8.14}$$

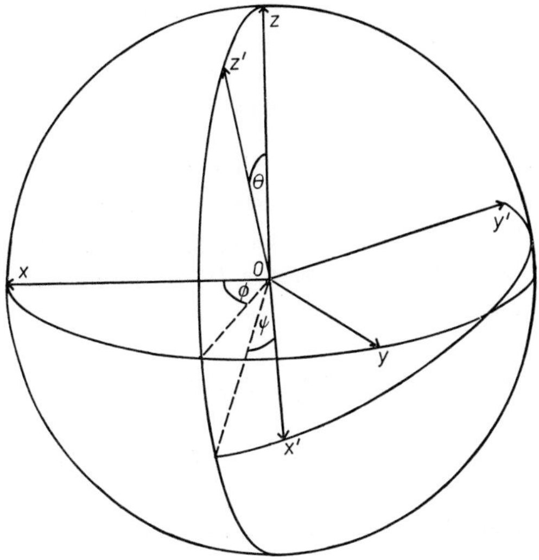

Fig. 8.1

Then eqs. (8.11) and (8.12) give

$$\left.\begin{aligned} u_0 &= \pm \cos\tfrac{1}{2}\theta \cos\tfrac{1}{2}(\phi+\psi),\\ u_1 &= \mp \sin\tfrac{1}{2}\theta \sin\tfrac{1}{2}(\phi-\psi),\\ u_2 &= \pm \sin\tfrac{1}{2}\theta \cos\tfrac{1}{2}(\phi-\psi),\\ u_3 &= \pm \cos\tfrac{1}{2}\theta \sin\tfrac{1}{2}(\phi+\psi). \end{aligned}\right| \qquad (8.15)$$

§ 3. TRANSFORMATION PROPERTIES OF SPINOR WAVE FUNCTIONS

A spinor function may be defined as a two-component quantity which transforms under a coordinate transformation T as

$$O(T)\, f(\boldsymbol{r}', i) = \sum_{j=1}^{2} u(\boldsymbol{R}_p)_{ij}\, f(\boldsymbol{r}, j), \qquad i = 1, 2, \qquad (8.16)$$

where

$$\boldsymbol{f}(\boldsymbol{r}) = \begin{pmatrix} f(\boldsymbol{r}, 1) \\ f(\boldsymbol{r}, 2) \end{pmatrix} \quad \text{and} \quad O(T)\,\boldsymbol{f}(\boldsymbol{r}') = \begin{pmatrix} O(T)\, f(\boldsymbol{r}', 1) \\ O(T)\, f(\boldsymbol{r}', 2) \end{pmatrix}$$

are the spinors defined relative to Ox, Oy and Oz and to Ox', Oy' and $O'z'$, respectively. Here T is the transformation

$$\boldsymbol{r}' = \{\boldsymbol{R} \mid \boldsymbol{t}\}\, \boldsymbol{r}\,,$$

and $\boldsymbol{R}_p$ is the proper part of $\boldsymbol{R}$, that is, $\boldsymbol{R}_p=\boldsymbol{R}$ if $\boldsymbol{R}$ is a proper rotation and $\boldsymbol{R}_p=-\boldsymbol{R}$ if $\boldsymbol{R}$ is an improper rotation. The functions $O(T)\, f(\boldsymbol{r}', i)$ may be thought of as being obtained from $f(\boldsymbol{r}, j)$ by acting on $f(\boldsymbol{r}, j)$ with the operator $O(T)$. Similarly, an operator $O(\bar{T})$ may be defined by

$$O(\bar{T})\, f(\boldsymbol{r}', i) = -\sum_{j=1}^{2} u(\boldsymbol{R}_p)_{ij}\, f(\boldsymbol{r}, j), \qquad i = 1, 2. \tag{8.17}$$

The transformation laws (8.16) and (8.17) for spinor functions are the analogues of the transformation law (3.1) for scalar functions. Not only does the functional form of each component of the spinor change, but its value changes as well. The operators $O(T)$ and $O(\bar{T})$ are the analogues of the operator $P(T)$ for scalar functions, and they play a very similar role. As will be shown in §4, it is necessary for each T to include *both* $O(T)$ and $O(\bar{T})$ in discussions of spin–orbit coupling.

As $O(T)$ and $O(\bar{T})$ differ only in the treatment of the rotational part of the transformation, it is possible to write $O(T)=O([\boldsymbol{R} \mid \boldsymbol{t}])$ and $O(\bar{T})=O([\bar{\boldsymbol{R}} \mid \boldsymbol{t}])$. It is then convenient to consider that for every transformation $\{\boldsymbol{R} \mid \boldsymbol{t}\}$ there correspond two 'generalized transformations' $[\boldsymbol{R} \mid \boldsymbol{t}]$ and $[\bar{\boldsymbol{R}} \mid \boldsymbol{t}]$ that are defined to be isomorphic to the operators $O([\boldsymbol{R} \mid \boldsymbol{t}])$ and $O([\bar{\boldsymbol{R}} \mid \boldsymbol{t}])$ respectively. These generalized transformations are distinguished from the ordinary transformations by writing them in square brackets. The product of two generalized transformations is then, by eqs. (8.16) and (8.17), given by

$$[\boldsymbol{R} \mid \boldsymbol{t}]\,[\boldsymbol{R}' \mid \boldsymbol{t}'] = [\bar{\boldsymbol{R}} \mid \boldsymbol{t}]\,[\bar{\boldsymbol{R}}' \mid \boldsymbol{t}'] = \begin{cases} [\boldsymbol{R}\boldsymbol{R}' \mid \boldsymbol{t} + \boldsymbol{R}\boldsymbol{t}'], & \text{if} \quad \boldsymbol{u}(\boldsymbol{R}_p)\, \boldsymbol{u}(\boldsymbol{R}'_p) = +\, \boldsymbol{u}(\boldsymbol{R}_p \boldsymbol{R}'_p), \\ [\overline{\boldsymbol{R}\boldsymbol{R}'} \mid \boldsymbol{t} + \boldsymbol{R}\boldsymbol{t}'], & \text{if} \quad \boldsymbol{u}(\boldsymbol{R}_p)\, \boldsymbol{u}(\boldsymbol{R}'_p) = -\, \boldsymbol{u}(\boldsymbol{R}_p \boldsymbol{R}'_p), \end{cases} \tag{8.18}$$

or by

$$[\bar{R} \mid t]\,[R' \mid t'] = [R \mid t]\,[\bar{R}' \mid t']$$
$$= \begin{cases} [RR' \mid t + Rt'], & \text{if } \; u(R_p)\,u(R'_p) = -\,u(R_pR'_p), \\ [\overline{RR'} \mid t + Rt'], & \text{if } \; u(R_p)\,u(R'_p) = +\,u(R_pR'_p), \end{cases} \tag{8.19}$$

whichever is appropriate. It is often convenient to write $[R(C_{ni}) \mid t]$ as $[C_{ni} \mid t]$ and $[\bar{R}(C_{ni}) \mid t]$ as $[\bar{C}_{ni} \mid t]$, with a similar notation for improper rotations.

If a set of ordinary transformations form a group, then the corresponding set of generalized transformations also form a group with twice as many elements. The group of ordinary transformations is then called a *single* group, and the corresponding group of generalized transformations is called a *double* group. The corresponding set of operators $O(T)$ and $O(\bar{T})$ then form a group that is isomorphic to the double group. The double group corresponding to a single group $\mathscr{G}$ will be denoted by $\mathscr{G}^D$.

The generalized rotations $[R \mid 0]$ and $[\bar{R} \mid 0]$ are only uniquely defined when a sign has been arbitrarily specified for $u(R_p)$, as for example in table 8.1. However, one convention that is always observed is that $u(E)$ is the 2×2 unit matrix, so that $[E \mid 0] = [E \mid 0] = [R(E) \mid 0]$ is the identity element of the double group.

It is important to note that, in general, the single group is *not* isomorphic to a subgroup of the double group. This is a consequence of the fact that, according to (8.18) and (8.19), the product of two generalized transformations does not necessarily correspond to the product of the corresponding ordinary transformations. It is sufficient to demonstrate this explicitly for a single group consisting only of proper rotations, so consider the point group C_2 with elements $\{E \mid 0\}$ and $\{C_{2x} \mid 0\}$. From table 8.1,

$$u(E) = \begin{pmatrix} 1 & 0 \\ 0 & 1 \end{pmatrix}, \qquad u(C_{2x}) = \begin{pmatrix} 0 & \mathrm{i} \\ \mathrm{i} & 0 \end{pmatrix}$$

so that the multiplication table of the double group with elements $[E \mid 0]$, $[\bar{E} \mid 0]$, $[C_{2x} \mid 0]$ and $[\bar{C}_{2x} \mid 0]$ is as given in table 8.2. As

TABLE 8.2

	$[E \mid 0]$	$[\bar{E} \mid 0]$	$[C_{2x} \mid 0]$	$[\bar{C}_{2x} \mid 0]$
$[E \mid 0]$	$[E \mid 0]$	$[\bar{E} \mid 0]$	$[C_{2x} \mid 0]$	$[\bar{C}_{2x} \mid 0]$
$[\bar{E} \mid 0]$	$[\bar{E} \mid 0]$	$[E \mid 0]$	$[\bar{C}_{2x} \mid 0]$	$[C_{2x} \mid 0]$
$[C_{2x} \mid 0]$	$[C_{2x} \mid 0]$	$[\bar{C}_{2x} \mid 0]$	$[\bar{E} \mid 0]$	$[E \mid 0]$
$[\bar{C}_{2x} \mid 0]$	$[\bar{C}_{2x} \mid 0]$	$[C_{2x} \mid 0]$	$[E \mid 0]$	$[\bar{E} \mid 0]$

$[C_{2x} \mid 0]^2$ is not equal to the identity $[E \mid 0]$, $[E \mid 0]$ and $[C_{2x} \mid 0]$ do not form a subgroup.

The inner product of two spinor functions

$$\boldsymbol{f}(\boldsymbol{r}) = \begin{pmatrix} f(\boldsymbol{r}, 1) \\ f(\boldsymbol{r}, 2) \end{pmatrix} \quad \text{and} \quad \boldsymbol{g}(\boldsymbol{r}) = \begin{pmatrix} g(\boldsymbol{r}, 1) \\ g(\boldsymbol{r}, 2) \end{pmatrix}$$

is defined, in a generalization of eq. (3.7), by

$$(\boldsymbol{f}, \boldsymbol{g}) = \sum_{j=1}^{2} \iiint f(\boldsymbol{r}, j)^* \, g(\boldsymbol{r}, j) \, \mathrm{d}x \, \mathrm{d}y \, \mathrm{d}z \,. \tag{8.20}$$

Because of the unitary property of the matrix $\boldsymbol{u}$, it follows that

$$(O(T)\,\boldsymbol{f}, O(T)\,\boldsymbol{g}) = (O(\bar{T})\,\boldsymbol{f}, O(\bar{T})\,\boldsymbol{g}) = (\boldsymbol{f}, \boldsymbol{g}),$$

for any two spinors $\boldsymbol{f}$ and $\boldsymbol{g}$ and any transformation T. The operators $O(T)$ and $O(\bar{T})$ may be described therefore as unitary operators.

This formulation of the theory for electron spin is entirely equivalent to the more usual Pauli theory. The physical interpretation of the components $f(\boldsymbol{r}, j)$ of a spinor function is that $|f(\boldsymbol{r}, 1)|^2$ and $|f(\boldsymbol{r}, 2)|^2$ are the probabilities that a measurement of the component of spin angular momentum about the axis Oz gives $+\frac{1}{2}\hbar$ and $-\frac{1}{2}\hbar$ respectively, while a measurement of position gives $\boldsymbol{r}$. It follows in this formulation that $P_1 = |O(T)f(\boldsymbol{r}', 1)|^2$ and $P_2 = |O(T)f(\boldsymbol{r}', 2)|^2$

must be the probabilities that a measurement of the spin angular momentum about $O'z'$ gives $+\frac{1}{2}\hbar$ and $-\frac{1}{2}\hbar$ respectively, while a measurement of position relative to the axes $O'x'$, $O'y'$ and $O'z'$ gives $\boldsymbol{r}'$.

For example, if the orientation of the axes $O'x'$, $O'y'$ and $O'z'$ relative to Ox, Oy and Oz is given by the Eulerian angles θ, ϕ and ψ of fig. 8.1, then

$$P_1 = |O(T)\, f(\boldsymbol{r}', 1)|^2 = |(u_0 + \mathrm{i}u_3)\, f(\boldsymbol{r}, 1) + (u_2 + \mathrm{i}u_1)\, f(\boldsymbol{r}, 2)|^2,$$

where u_0, u_1, u_2 and u_3 are defined in eq. (8.6). It then follows from (8.15) that

$$\begin{aligned} P_1 = |f(\boldsymbol{r}, 1)|^2 (\cos \tfrac{1}{2}\theta)^2 + |f(\boldsymbol{r}, 2)|^2 (\sin \tfrac{1}{2}\theta)^2 \\ + \mathrm{Re}\,\{f(\boldsymbol{r}, 1)\, f(\boldsymbol{r}, 2)^*\} \sin\theta \cos\phi \\ - \mathrm{Im}\,\{f(\boldsymbol{r}, 1)\, f(\boldsymbol{r}, 2)^*\} \sin\theta \sin\phi\,. \end{aligned} \tag{8.21}$$

The equivalence of this formulation with the Pauli theory is demonstrated by showing that precisely the same result is obtained in the Pauli theory. In this theory the probability P_1 that a measurement of spin angular momentum about $O'z'$ gives $+\frac{1}{2}\hbar$ while a measurement of position relative to the axes Ox, Oy, and Oz gives $\boldsymbol{r}$ is given by

$$P_1 = |\tilde{\boldsymbol{\Phi}}^* \boldsymbol{f}(\boldsymbol{r})|^2 = |\Phi_1^* f(\boldsymbol{r}, 1) + \Phi_2^* f(\boldsymbol{r}, 2)|^2\,. \tag{8.22}$$

Here $\boldsymbol{\Phi}$, whose components are Φ_1 and Φ_2, is the normalized eigenspinor of $S_{z'}$ corresponding to the eigenvalue $+\frac{1}{2}\hbar$, where $S_{z'}$ is the operator for spin angular momentum about $O'z'$. However $S_{z'}$ is given by

$$S_{z'} = \tfrac{1}{2}\hbar\, \{R_{31}\boldsymbol{\sigma}_1 + R_{32}\boldsymbol{\sigma}_2 + R_{33}\boldsymbol{\sigma}_3\},$$

so that $\boldsymbol{\Phi}$ satisfies the eigenvalue equation

$$\tfrac{1}{2}\hbar \begin{pmatrix} R_{33} & R_{31} - \mathrm{i}R_{32} \\ R_{31} + \mathrm{i}R_{32} & -R_{33} \end{pmatrix} \begin{pmatrix} \Phi_1 \\ \Phi_2 \end{pmatrix} = \tfrac{1}{2}\hbar \begin{pmatrix} \Phi_1 \\ \Phi_2 \end{pmatrix},$$

giving

$$\left.\begin{aligned} \Phi_1 &= (R_{31} - \mathrm{i}R_{32})\, \{2(1 - R_{33})\}^{-\frac{1}{2}}, \\ \Phi_2 &= (1 - R_{33})\, \{2(1 - R_{33})\}^{-\frac{1}{2}}. \end{aligned}\right| \tag{8.23}$$

(The fact that $\boldsymbol{R}$ is an orthogonal matrix has been used, giving $R_{31}^2+R_{32}^2+R_{33}^2=1$.) Substituting (8.14) into (8.23), and then (8.23) into (8.22), gives the same expression for P_1 as (8.21).

§ 4. TRANSFORMATION PROPERTIES OF THE HAMILTONIAN OPERATOR

The relativistic electron theory of Dirac shows that the Hamiltonian operator $\boldsymbol{H}(\boldsymbol{r})$ with the spin–orbit coupling term included is

$$\boldsymbol{H}(\boldsymbol{r}) = \boldsymbol{H}_0(\boldsymbol{r}) + \boldsymbol{H}_1(\boldsymbol{r}), \tag{8.24}$$

where

$$\boldsymbol{H}_0(\boldsymbol{r}) = \left\{ -\frac{\hbar^2}{2m}\left(\frac{\partial^2}{\partial x^2} + \frac{\partial^2}{\partial y^2} + \frac{\partial^2}{\partial z^2}\right) + V(\boldsymbol{r}) \right\} \boldsymbol{E},$$

and

$$\boldsymbol{H}_1(\boldsymbol{r}) = (\hbar^2/4\mathrm{i}m^2c^2)\,\{\mathrm{grad}\, V(\boldsymbol{r}) \wedge \mathrm{grad}\}\cdot\boldsymbol{\sigma},$$

as shown, for example, in the book by SCHIFF [1955]. Here $\boldsymbol{E}$ is the 2×2 unit matrix and $\boldsymbol{\sigma}=(\boldsymbol{\sigma}_1, \boldsymbol{\sigma}_2, \boldsymbol{\sigma}_3)$, $\boldsymbol{\sigma}_1, \boldsymbol{\sigma}_2$ and $\boldsymbol{\sigma}_3$ being the Pauli spin matrices (8.3). $\boldsymbol{H}(\boldsymbol{r})$ is therefore a matrix operator which acts on spinor functions, the Schrödinger equation then being of the form

$$\boldsymbol{H}(\boldsymbol{r}) \begin{pmatrix} \phi(\boldsymbol{r}, 1) \\ \phi(\boldsymbol{r}, 2) \end{pmatrix} = E \begin{pmatrix} \phi(\boldsymbol{r}, 1) \\ \phi(\boldsymbol{r}, 2) \end{pmatrix}.$$

The behaviour of the matrix Hamiltonian operator under transformations can be found by a simple generalization of the argument given in ch. 3 §3.1. The only difference is that, instead of considering any *scalar* function $f(\boldsymbol{r})$, one considers any *spinor* function with components $f(\boldsymbol{r}, i)$. The spinor with components $g(\boldsymbol{r}, j)$ may then be defined by

$$g(\boldsymbol{r}, j) = \sum_{i=1}^{2} H(\boldsymbol{r})_{ji}\, f(\boldsymbol{r}, i),$$

by analogy with eq. (3.9). The conclusions that replace eq. (3.10) are that

$$H(r) = O(T)\, u(R_p^{-1})\, H(Tr)\, u(R_p)\, O(T)^{-1}$$

and

$$H(r) = O(\bar{T})\, u(R_p^{-1})\, H(Tr)\, u(R_p)\, O(\bar{T})^{-1},$$

R_p being the proper rotational part of T, as in eqs. (8.16) and (8.17).

The group of the Schrödinger equation is then defined to be the group of operators $O(T)$ and $O(\bar{T})$ which commute with $H(r)$. As noted in §3, this group is a *double* group, which in general has no subgroup isomorphic to the corresponding single group. Clearly

$$O(T)\, H(r) = H(r)\, O(T)$$

and

$$O(\bar{T})\, H(r) = H(r)\, O(\bar{T})$$

if and only if

$$H(r) = u(R_p^{-1})\, H(Tr)\, u(R_p), \tag{8.25}$$

which is the analogue of eq. (3.11).

In the case of the spin–orbit coupling Hamiltonian defined by eq. (8.24), it will be shown that for any transformation T such that $H_0(r) = H_0(Tr)$, eq. (8.25) is satisfied. Thus the single group from which the group of the Schrödinger equation is formed is the group of all transformations leaving $H_0(r)$ invariant. In particular, for electrons in a crystalline solid, this single group is the space group of the crystal $\mathscr{G}$, and the group of the Schrödinger equation is the double space group $\mathscr{G}^D$.

The rest of this section is devoted to a proof of this assertion. As it is obvious that if $H_0(r) = H_0(Tr)$ then $H_0(r) = u(R_p^{-1})\, H_0(Tr)\, u(R_p)$, all that is necessary is to show then that $H_1(r) = u(R_p^{-1})\, H_1(Tr)\, u(R_p)$, which can be done as follows. One may write the scalar triple product as

$$(\text{grad}\, V(r) \wedge \text{grad}) \cdot \sigma = \begin{vmatrix} \partial V/\partial x & \partial V/\partial y & \partial V/\partial z \\ \partial/\partial x & \partial/\partial y & \partial/\partial z \\ \sigma_1 & \sigma_2 & \sigma_3 \end{vmatrix}.$$

If T is a transformation with a proper rotational part, so that

$\boldsymbol{r}' = \{\boldsymbol{R}_p \mid \boldsymbol{t}\}\, \boldsymbol{r}$, then as $\boldsymbol{R}_p^{-1} = \tilde{\boldsymbol{R}}_p$,

$$\partial/\partial x' = R_{11}\partial/\partial x + R_{12}\partial/\partial y + R_{13}\partial/\partial z\,, \tag{8.26}$$

so that

$$\partial V(\boldsymbol{r}')/\partial x' = R_{11}\partial V(\boldsymbol{r})/\partial x + R_{12}\partial V(\boldsymbol{r})/\partial y + R_{13}\partial V(\boldsymbol{r})/\partial z\,, \tag{8.27}$$

as $V(\boldsymbol{r}') = V(\boldsymbol{r})$. Moreover, by eqs. (8.2) and (8.5) with $y = z = 0$,

$$\boldsymbol{u}(\boldsymbol{R}_p)\, \boldsymbol{\sigma}_1 \boldsymbol{u}(\boldsymbol{R}_p^{-1}) = R_{11}\boldsymbol{\sigma}_1 + R_{21}\boldsymbol{\sigma}_2 + R_{31}\boldsymbol{\sigma}_3\,,$$

so that

$$\boldsymbol{u}(\boldsymbol{R}_p^{-1})\, \boldsymbol{\sigma}_1 \boldsymbol{u}(\boldsymbol{R}_p) = R_{11}\boldsymbol{\sigma}_1 + R_{12}\boldsymbol{\sigma}_2 + R_{13}\boldsymbol{\sigma}_3\,. \tag{8.28}$$

Each of the expressions (8.26), (8.27) and (8.28) involve the *same* matrix elements of $\boldsymbol{R}_p$, and there are similar expressions for $\partial/\partial y'$, and so on. The required result then follows by simple determinant manipulation. If the rotational part of the transformation T is improper, so that $\boldsymbol{r}' = \{-\boldsymbol{R}_p \mid \boldsymbol{t}\}\, \boldsymbol{r}$, there is then merely a change of sign in the right-hand side of eqs. (8.26) and (8.27), leaving the value of the determinant unchanged.

§ 5. ENERGY EIGENFUNCTIONS AND BASIS FUNCTIONS

It was shown in §4 that the group of the Schrödinger equation when the Hamiltonian operator $\boldsymbol{H}(\boldsymbol{r})$ is spin dependent is the double group of operators $O(T)$ and $O(\bar{T})$, each of which commutes with $\boldsymbol{H}(\boldsymbol{r})$. The theory developed in ch. 3 §§4–6 then applies, with only the following simple modifications.

(a) The group of operators $P(T)$ is replaced by the double group of operators $O(T)$ and $O(\bar{T})$.

(b) Every scalar wave function is replaced by a spinor wave function. In particular, the basis functions of the group of the Schrödinger equation are spinor functions.

As an example, consider the analogue of eq. (3.13), which is

$$O(T)\, \boldsymbol{\phi}_n(\boldsymbol{r}) = \sum_{m=1}^{l} \Gamma(T)_{mn}\, \boldsymbol{\phi}_m(\boldsymbol{r})\,.$$

In terms of components this is

$$O(T)\,\phi_n(\boldsymbol{r}, i) = \sum_{m=1}^{l} \Gamma(T)_{mn}\,\phi_m(\boldsymbol{r}, i), \qquad (8.29)$$

for $i = 1, 2$. Similar expressions hold when T is replaced by $\bar{T}$. It will be seen that the labels m and n now apply to spinors, considered as single entities, and a further index is required to specify their components. Similarly, the projection operator $\mathscr{P}^p_{mn}$, originally defined in eq. (3.22), becomes

$$\mathscr{P}^p_{mn} = (l_p/g) \sum_T \{\Gamma^p(T)^*_{mn}\, O(T) + \Gamma^p(\bar{T})^*_{mn}\, O(\bar{T})\}, \qquad (8.30)$$

l_p being the dimension of the unitary irreducible representation $\boldsymbol{\Gamma}^p$ of the double group of transformations of order g, the summation being over all elements T of the corresponding single group. This projection operator now acts on spinor functions of course. Similarly, the matrix element theorems given in ch. 3 §6.1 now apply to the matrix elements $(\boldsymbol{\phi}^p_m, \boldsymbol{\psi}^q_n)$ and $(\boldsymbol{\phi}^p_m, \boldsymbol{H}\boldsymbol{\psi}^q_n)$; $\boldsymbol{\phi}^p_m(\boldsymbol{r})$ and $\boldsymbol{\psi}^q_n(\boldsymbol{r})$ being spinor basis functions and the inner product being as defined in eq. (8.20).

§ 6. GENERAL PROPERTIES OF IRREDUCIBLE REPRESENTATIONS OF DOUBLE GROUPS

As $[\bar{E} \mid \boldsymbol{0}]$ commutes with every element of the double group, Schur's lemma and the fact that $[\bar{E} \mid \boldsymbol{0}]^2 = [E \mid \boldsymbol{0}]$ imply that

$$\boldsymbol{\Gamma}([\bar{E} \mid \boldsymbol{0}]) = \pm\,\boldsymbol{\Gamma}([E \mid \boldsymbol{0}]), \qquad (8.31)$$

$\boldsymbol{\Gamma}([E \mid \boldsymbol{0}])$ being a unit matrix of some dimension. Moreover, for any $[\bar{\boldsymbol{R}} \mid \boldsymbol{t}]$, $[\bar{\boldsymbol{R}} \mid \boldsymbol{t}] = [\bar{E} \mid \boldsymbol{0}]\,[\boldsymbol{R} \mid \boldsymbol{t}]$, so that

$$\boldsymbol{\Gamma}([\bar{\boldsymbol{R}} \mid \boldsymbol{t}]) = \pm\,\boldsymbol{\Gamma}([\boldsymbol{R} \mid \boldsymbol{t}]), \qquad (8.32)$$

the sign here being the same as the sign in eq. (8.31). Thus, every irreducible representation of a double group is of one of the following two types.

(a) The first type of irreducible representation is that for which $\boldsymbol{\Gamma}([\bar{\boldsymbol{R}} \mid \boldsymbol{t}]) = +\boldsymbol{\Gamma}([\boldsymbol{R} \mid \boldsymbol{t}])$ for every transformation $[\boldsymbol{R} \mid \boldsymbol{t}]$ of the group. Any irreducible representation of the corresponding single group $\boldsymbol{\Gamma}(\{\boldsymbol{R} \mid \boldsymbol{t}\})$ gives a representation of the double group of this type with

$$\boldsymbol{\Gamma}([\bar{\boldsymbol{R}} \mid \boldsymbol{t}]) = \boldsymbol{\Gamma}([\boldsymbol{R} \mid \boldsymbol{t}]) = \boldsymbol{\Gamma}(\{\boldsymbol{R} \mid \boldsymbol{t}\}).$$

It follows from the second theorem of ch. 2 §6 that this constitutes an *irreducible* representation of the double group. Moreover, it is clear that all the irreducible representations of the double group of this type can be formed this way.

(b) The second type of irreducible representation is that for which $\boldsymbol{\Gamma}([\bar{\boldsymbol{R}} \mid \boldsymbol{t}]) = -\boldsymbol{\Gamma}([\boldsymbol{R} \mid \boldsymbol{t}])$. This is called an '*extra*' irreducible representation of the double group.

Spinor functions only transform as basis functions for the rows of the *extra* representations. This is most easily seen by noting that the projection operator (8.30) is identically zero if $\boldsymbol{\Gamma}^p$ is not an extra representation, as $\{O(T)+O(\bar{T})\}$ is zero for any T by eqs. (8.16) and (8.17). However, if $\boldsymbol{\Gamma}^p$ is an extra representation, then

$$\mathscr{P}^p_{mn} = (2l_p/g) \sum_T \Gamma^p(T)^*_{mn}\, O(T),$$

where the summation is over all the elements T of the *single* group.

§ 7. IRREDUCIBLE REPRESENTATIONS OF THE DOUBLE GROUP OF PURE PRIMITIVE TRANSLATIONS $\mathscr{T}^D$

The double space group of a crystal contains as a subgroup the double group of pure primitive translations, which will be denoted by $\mathscr{T}^D$. The development given in ch. 4 for single groups shows that it is worthwhile to study this group on its own, before considering the effects of rotations.

The double group $\mathscr{T}^D$ is Abelian, so that all its irreducible representations are one-dimensional. The imposition of the Born cyclic boundary conditions of ch. 4 §2 is equivalent to the imposition of the set of conditions that

$$O([E \mid a_j])^{N_j} = O([E \mid 0])$$

for $j=1$, 2, and 3, these being the analogues of (4.3). By a simple extension of the argument given in ch. 4 § 3, it follows that the extra irreducible representations of $\mathscr{T}^D$ are given by

$$\Gamma^k([E \mid t_n]) = -\Gamma^k([\bar{E} \mid t_n]) = (\exp(-ik\cdot t_n)), \qquad (8.33)$$

k being as defined in eq. (4.8).

It follows from eqs. (8.16) and (8.17) that

$$\left.\begin{aligned} O([E \mid t_n])\, f(r, i) &= f(r - t_n, i), \\ O([\bar{E} \mid t_n])\, f(r, i) &= -f(r - t_n, i), \end{aligned}\right|$$

for $i=1$ and 2. This shows again that spinor functions can only transform as basis functions of the extra representations of $\mathscr{T}^D$, and that if $\phi_k(r, i)$ are the components of a basis spinor for the irreducible representation of (8.33), then

$$O([E \mid t_n])\, \phi_k(r, i) = \phi_k(r - t_n, i) = \exp(-ik\cdot t_n)\, \phi_k(r, i),$$

so that

$$\phi_k(r, i) = \exp(ik\cdot r)\, u_k(r, i), \qquad (8.34)$$

$i=1$ and 2, where the functions $u_k(r, i)$ have the periodicity of the crystal lattice.

Eq. (8.34) is the extension of the theorem of BLOCH [1928] for spinors. It shows that a spinor which is an energy eigenfunction has components that are Bloch functions. All the consequences of Bloch's theorem that are described in ch. 4 §§ 4, 5 therefore still apply, the most important being the concept of the Brillouin zone.

§ 8. DOUBLE SYMMORPHIC SPACE GROUPS

§ 8.1. *Irreducible representations of double point groups*

It was shown in ch. 5 § 1 that the basis functions of a single symmorphic space group $\mathscr{G}$ are determined from a knowledge of the irreducible representations of the point groups $\mathscr{G}_o(\boldsymbol{k})$. As will be described in § 8.2, a similar result holds for double groups, the basis functions of the double symmorphic space group $\mathscr{G}^D$ being determined by the irreducible representations of the double point groups $\mathscr{G}_o^D(\boldsymbol{k})$.

Double point groups were investigated first by BETHE [1929], and then more thoroughly by OPECHOWSKI [1940]. The construction of the extra representations of the double point groups is helped by the use of the following simple rules derived by Opechowski.

(a) If a set of proper or improper rotations $\{\boldsymbol{R} \mid \boldsymbol{0}\}$ through $2\pi/n$ form a class in the single group, then the sets of generalized rotations $[\boldsymbol{R} \mid \boldsymbol{0}]$ and $[\bar{\boldsymbol{R}} \mid \boldsymbol{0}]$ form two separate classes in the double group.

(b) There is one exception to (a), namely that if $n=2$, then $[\boldsymbol{R} \mid \boldsymbol{0}]$ and $[\bar{\boldsymbol{R}} \mid \boldsymbol{0}]$ lie in the same class in the double group if, and only if, there is in the single group a proper or improper rotation through π about an axis perpendicular to the axis of $\{\boldsymbol{R} \mid \boldsymbol{0}\}$.

The characters of the extra representations must satisfy the relations

$$\chi([\bar{\boldsymbol{R}} \mid \boldsymbol{0}]) = -\chi([\boldsymbol{R} \mid \boldsymbol{0}])$$

for every $\boldsymbol{R}$ of the group. As the characters of all the elements in a class are equal, then in the exceptional case (b), $\chi([C_{2i} \mid \boldsymbol{0}])=0$. The character tables of the extra representations for all the double crystallographic point groups are given in appendix 2.

Spinor basis functions formed from spherical harmonics that transform as the extra irreducible representations of the double point groups O_h^D, T_d^D, O^D, D_{4h}^D, D_{3d}^D, C_{4v}^D, C_{3v}^D and C_{2v}^D have been tabulated by ONODERA and OKAZAKI [1966], and those for O_h^D and D_{3h}^D have been tabulated by TELEMAN and GLODEANU [1967].

§ 8.2. *Irreducible representations of a double symmorphic space group*

The construction of the basis functions of a double symmorphic space group $\mathscr{G}^D$ follows the same line as that given for the corresponding single group $\mathscr{G}$ in ch. 5 §1. The point groups $\mathscr{G}_o(\boldsymbol{k})$, symmetry points, axes and planes, and the concept of the star of $\boldsymbol{k}$ may be defined exactly as before. The only modification is in the fundamental theorem, which is now as follows.

Fundamental theorem on the spinor basis functions of the irreducible representations of a double symmorphic space group $\mathscr{G}^D$. Suppose that $\boldsymbol{k}$ is an allowed $\boldsymbol{k}$-vector, and $\phi^p_{\boldsymbol{k}s}(\boldsymbol{r}, j)$ are Bloch functions which are the $j=1$ and 2 components of a spinor which transforms as the sth row of the *extra* unitary irreducible representation $\boldsymbol{\Gamma}^p$ of the double point group $\mathscr{G}^D_o(\boldsymbol{k})$ corresponding to $\mathscr{G}_o(\boldsymbol{k})$, this representation being of dimension l. Suppose that the star of $\boldsymbol{k}$ consists of the $M(\boldsymbol{k})$ wave vectors $\boldsymbol{k}_1\,(=\boldsymbol{k}), \boldsymbol{k}_2, \ldots$, associated with which are $M(\boldsymbol{k})$ rotations $\boldsymbol{R}_i$, as defined in eq. (5.1). Then the set of $lM(\boldsymbol{k})$ spinors with components

$$O([\boldsymbol{R}_i \mid \boldsymbol{0}])\, \phi^p_{\boldsymbol{k}s}(\boldsymbol{r}, j), \qquad j = 1, 2,$$
$$i = 1, 2, \ldots, M(\boldsymbol{k}), \qquad s = 1, 2, \ldots, l,$$

form a basis for a $lM(\boldsymbol{k})$-dimensional extra unitary irreducible representation of the double space group $\mathscr{G}^D$. Moreover, *all* the extra irreducible representations of $\mathscr{G}^D$ may be obtained in this way, by working through all the extra irreducible representations of $\mathscr{G}^D_o(\boldsymbol{k})$ for all the allowed $\boldsymbol{k}$-vectors that are in different stars.

This theorem implies that the extra irreducible representations of $\mathscr{G}^D$ can be specified by an allowed $\boldsymbol{k}$-vector and a label p specifying an extra irreducible representation of $\mathscr{G}^D_o(\boldsymbol{k})$, and so may be denoted by $\boldsymbol{\Gamma}^{\boldsymbol{k}p}$. An explicit expression for the matrix elements of $\boldsymbol{\Gamma}^{\boldsymbol{k}p}$, obtained by the same argument as that leading to eq. (5.3), is

$$\Gamma^{\boldsymbol{k}p}([\boldsymbol{R} \mid \boldsymbol{t}_n])_{j,t;\,i,s} = \begin{cases} 0, & \text{if } \boldsymbol{R}_j^{-1}\boldsymbol{R}\boldsymbol{R}_i \text{ is not a member of } \mathscr{G}_o(\boldsymbol{k}), \\ \Gamma^p([\boldsymbol{R}_j \mid \boldsymbol{0}]^{-1}\,[\boldsymbol{R} \mid \boldsymbol{0}]\,[\boldsymbol{R}_i \mid \boldsymbol{0}])_{ts} \exp(-\,\mathrm{i}\boldsymbol{R}_j\boldsymbol{k}\cdot\boldsymbol{t}_n), & \text{otherwise}. \end{cases}$$

The proof of the theorem is almost identical to the proof of the corresponding theorem for single space groups that is given in ch. 5 §1.

The consequences of the theorem as regards degeneracies of energy levels, the symmetry of $E(\boldsymbol{k})$, and matrix elements follow exactly as in ch. 5 §§2.1, 2.2, the only difference being that scalar functions are to be replaced everywhere by spinor functions. In particular, $E(\boldsymbol{k})$ retains the symmetry of the point group $\mathscr{G}_o$. Similarly, continuity and compatibility may be discussed in exactly the same way as in ch. 5 §5.1.

The construction of symmetrized plane wave spinors for all 73 double symmorphic space groups may be simplified by use of the tables given by LUEHRMANN [1968].

§ 8.3. *Time-reversal symmetry for double symmorphic space groups*

The spin–orbit coupling Hamiltonian $\boldsymbol{H}(\boldsymbol{r})$ of eq. (8.24) is not real, so that the complex conjugate $\phi_{\boldsymbol{k}}(\boldsymbol{r})^*$ of a spinor eigenfunction $\phi_{\boldsymbol{k}}(\boldsymbol{r})$ is not, in general, another eigenfunction. However, it is easily verified that

$$\boldsymbol{\sigma}_2^{-1}\boldsymbol{H}(\boldsymbol{r})\,\boldsymbol{\sigma}_2 = \boldsymbol{H}(\boldsymbol{r})^*, \tag{8.35}$$

so that if

$$\boldsymbol{H}(\boldsymbol{r})\,\phi_{\boldsymbol{k}}(\boldsymbol{r}) = E(\boldsymbol{k})\,\phi_{\boldsymbol{k}}(\boldsymbol{r}),$$

then taking the complex conjugate and using (8.35) gives

$$\boldsymbol{H}(\boldsymbol{r})\,\{\boldsymbol{\sigma}_2\phi_{\boldsymbol{k}}(\boldsymbol{r})^*\} = E(\boldsymbol{k})\,\{\boldsymbol{\sigma}_2\phi_{\boldsymbol{k}}(\boldsymbol{r})^*\}.$$

Thus $\boldsymbol{\sigma}_2\phi_{\boldsymbol{k}}(\boldsymbol{r})^*$ is an eigenspinor with the same energy eigenvalue.

Although it is possible to carry out a non-group-theoretical argument similar to that of the first part of ch. 5 §7.1, an account of which may be found for example in the book by KITTEL [1963], this is more complicated with spin–orbit coupling present. However, the group theoretical argument is no more difficult, and of course gives *all* the extra degeneracies, so only this approach will be given here. The general conclusion is that time-reversal symmetry plays a much more important role in the theory of double space groups than in the theory of single space groups. The original study of this problem was made by ELLIOTT [1954].

WIGNER [1932, 1959] has shown that for double groups there are again three cases, which are as follows.
(a) $\boldsymbol{\Gamma}$ is equivalent to a representation by real matrices.
(b) $\boldsymbol{\Gamma}$ is not equivalent to $\boldsymbol{\Gamma}^*$.
(c) $\boldsymbol{\Gamma}$ is equivalent to $\boldsymbol{\Gamma}^*$, but is not equivalent to a representation by real matrices.
Although these are the same cases as in ch. 5 §7.1, the degeneracies corresponding to them are quite different when the Hamiltonian $\boldsymbol{H}(\boldsymbol{r})$ is spin-dependent. In this situation there is now in case (a) an extra degeneracy, and the representation $\boldsymbol{\Gamma}$ occurs doubled. In case (b) there is still an extra degeneracy, the eigenvalues corresponding $\boldsymbol{\Gamma}$ and $\boldsymbol{\Gamma}^*$ being equal, while in case (c) there is now no extra degeneracy. In cases (a) and (b), if $\boldsymbol{\phi}(\boldsymbol{r})$ is an eigenspinor, $\boldsymbol{\sigma}_2\boldsymbol{\phi}(\boldsymbol{r})^*$ is a linearly independent eigenspinor with the same energy eigenvalue. (A shorter proof of these results based directly on the Dirac equation has been given by JOHNSTON [1958].)

The condition (5.26) of FROBENIUS and SCHUR [1906] remains unchanged. In terms of the irreducible representations of a double symmorphic space group the criterion analogous to (5.27) is

$$\sum \chi^p([\boldsymbol{Q} \mid \boldsymbol{0}]^2) = \begin{cases} g_o(\boldsymbol{k}), & \text{case (a)}, \\ 0, & \text{case (b)}, \\ -\,g_o(\boldsymbol{k}), & \text{case (c)}, \end{cases} \tag{8.36}$$

where the summation is over all rotations $\boldsymbol{Q}$ of the point group $\mathscr{G}_o$ which transform $\boldsymbol{k}$ into a vector equivalent to $-\boldsymbol{k}$, $g_o(\boldsymbol{k})$ is the order of $\mathscr{G}_o(\boldsymbol{k})$, and $\chi^p([\boldsymbol{Q} \mid \boldsymbol{0}]^2)$ denotes the character of $[\boldsymbol{Q} \mid \boldsymbol{0}]^2$ in the irreducible representation $\boldsymbol{\Gamma}^p$ of $\mathscr{G}_o^D(\boldsymbol{k})$. Although $[\boldsymbol{Q} \mid \boldsymbol{0}]$ may not be a member of $\mathscr{G}_o^D(\boldsymbol{k})$, $[\boldsymbol{Q} \mid \boldsymbol{0}]^2$ is necessarily a member. The proof of (8.36) from the criterion (5.26) is essentially the same as in the corresponding proof for single symmorphic space groups that is given in ch. 5 §7.1.

In terms of the degeneracy of $E(\boldsymbol{k})$, in the sense of the last paragraph but one of ch. 5 §2.1, there are again three cases.
(1) If $-\boldsymbol{k}$ is not in the star of $\boldsymbol{k}$ nor equivalent to $\boldsymbol{k}$, then an extra

degeneracy following from cases (a) and (b) above merely corresponds to $E(\boldsymbol{k})=E(-\boldsymbol{k})$.

(2) If $-\boldsymbol{k}$ is in the star of $\boldsymbol{k}$, then an extra degeneracy following from cases (a) and (b) above causes an extra degeneracy of $E(\boldsymbol{k})$, and the two irreducible representations of $\mathscr{G}_o^D(\boldsymbol{k})$ that are involved are said to 'stick together'.

(3) If $-\boldsymbol{k}$ is equivalent to $\boldsymbol{k}$, any extra degeneracy takes the same form as in case (2).

The criteria for deciding which irreducible representations of $\mathscr{G}_o^D(\boldsymbol{k})$ of type (b) stick together in cases (2) and (3) are very similar to the corresponding criteria (5.28) and (5.30) for single groups. In fact for case (2), if $[\boldsymbol{R}_- \mid \boldsymbol{0}]$ is a generalized rotation such that $\boldsymbol{R}_-\boldsymbol{k}=-\boldsymbol{k}$, and if $\boldsymbol{\Gamma}^p$ and $\boldsymbol{\Gamma}^q$ are two irreducible representations of $\mathscr{G}_o^D(\boldsymbol{k})$ of type (b) that stick together, and χ^p and χ^q denote their characters, then

$$\chi^p([\boldsymbol{R} \mid \boldsymbol{0}])^* = \chi^q([\boldsymbol{R}_- \mid \boldsymbol{0}]^{-1}\,[\boldsymbol{R} \mid \boldsymbol{0}]\,[\boldsymbol{R}_- \mid \boldsymbol{0}]), \tag{8.37}$$

for *every* $[\boldsymbol{R} \mid \boldsymbol{0}]$ of $\mathscr{G}_o^D(\boldsymbol{k})$. The proof of (8.37) is very similar to the proof of (5.28), the only new feature being that if the spinors $\boldsymbol{\phi}_{\boldsymbol{k}s}^p(\boldsymbol{r})$ form a basis for $\boldsymbol{\Gamma}^{\boldsymbol{k}p}$, then the spinors $\boldsymbol{\sigma}_2\boldsymbol{\phi}_{\boldsymbol{k}s}^p(\boldsymbol{r})^*$ form a basis for $(\boldsymbol{\Gamma}^{\boldsymbol{k}p})^*$. This is a consequence of the fact that for any 2×2 unitary matrix $\boldsymbol{u}$, $\boldsymbol{u}^*=\boldsymbol{\sigma}_2^{-1}\boldsymbol{u}\boldsymbol{\sigma}_2$, so that for any spinor $\boldsymbol{f}(\boldsymbol{r})$, $\{O(T)\boldsymbol{f}(\boldsymbol{r})\}^*=\boldsymbol{\sigma}_2^{-1}O(T)\,\boldsymbol{\sigma}_2\boldsymbol{f}(\boldsymbol{r})^*$. Then if

$$O(T)\,\boldsymbol{f}_n(\boldsymbol{r}) = \sum_m \Gamma(T)_{mn}\,\boldsymbol{f}_m(\boldsymbol{r}),$$

taking the complex conjugate gives

$$O(T)\,\{\boldsymbol{\sigma}_2\boldsymbol{f}_n(\boldsymbol{r})^*\} = \sum_m \Gamma(T)_{mn}^*\,\{\boldsymbol{\sigma}_2\boldsymbol{f}_m(\boldsymbol{r})^*\}.$$

For case (3), the irreducible representations $\boldsymbol{\Gamma}^p$ and $\boldsymbol{\Gamma}^q$ of $\mathscr{G}_o^D(\boldsymbol{k})$ stick together if

$$\chi^p([\boldsymbol{R} \mid \boldsymbol{0}])^* = \chi^q([\boldsymbol{R} \mid \boldsymbol{0}]) \tag{8.38}$$

for every $[\boldsymbol{R} \mid \boldsymbol{0}]$ of $\mathscr{G}_o^D(\boldsymbol{k})$.

It is very interesting to see what these results imply if $\boldsymbol{k}$ is a general point of the Brillouin zone. There are two cases to be considered.

(α) If $\mathscr{G}_o$ contains the inversion operator I, as in the cubic space groups O_h^5 and O_h^9, then the only rotation $\boldsymbol{Q}$ for eq. (8.36) is I. As $[I \mid 0]^2 = [E \mid 0]$, this means that case (a) has occurred. Moreover, this situation is an example of case (2), so there must be a two-fold degeneracy at every general point of the Brillouin zone if $\mathscr{G}_o$ contains I. As the $E(\boldsymbol{k})$ are continuous functions of $\boldsymbol{k}$, this implies that there is at least a two-fold degeneracy everywhere in the Brillouin zone, although at non-general points this degeneracy may also be present for other reasons.

(β) If $\mathscr{G}_o$ does not contain I, as for example for the space group T_d^2 of the zinc blende structure, there is no rotation $\boldsymbol{Q}$ for eq. (8.36). This is then an example of cases (b) and (1). The spinors $\boldsymbol{\phi}_{\boldsymbol{k}}(\boldsymbol{r})$ and $\boldsymbol{\sigma}_2\boldsymbol{\phi}_{\boldsymbol{k}}(\boldsymbol{r})^*$ are degenerate, but correspond to wave vectors $\boldsymbol{k}$ and $-\boldsymbol{k}$ respectively, so that no two-fold degeneracy exists at a general point $\boldsymbol{k}$ in this case.

§ 8.4. *Splitting of degeneracies by spin-orbit coupling*

The introduction of the spin–orbit coupling term into a Hamiltonian operator reduces the symmetry of the Hamiltonian, and therefore, in general, causes degeneracies of energy levels to be split. In order to determine exactly how the degeneracies are split, it is necessary to relate the eigenfunctions of the spin-independent Hamiltonian to those of the spin–orbit coupling Hamiltonian. This implies relating the basis functions of the single groups to those of the corresponding double groups.

Suppose therefore that $\psi_s^p(\boldsymbol{r})$, for $s = 1, 2, \ldots, l$, is a set of scalar basis functions for an irreducible representation $\boldsymbol{\Gamma}^p$ of dimension l of a single group. Define a set of spinor functions $\boldsymbol{\phi}_{sm}^p(\boldsymbol{r})$ to have components given by

$$\phi_{sm}^p(\boldsymbol{r}, i) = \psi_s^p(\boldsymbol{r})\, \delta_{mi}, \qquad s = 1, 2, \ldots, l, \qquad m = 1, 2, \qquad i = 1, 2. \tag{8.39}$$

Then

$$\boldsymbol{\phi}^p_{s1}(\boldsymbol{r}) = \begin{pmatrix} \psi^p_s(\boldsymbol{r}) \\ 0 \end{pmatrix} \quad \text{and} \quad \boldsymbol{\phi}^p_{s2}(\boldsymbol{r}) = \begin{pmatrix} 0 \\ \psi^p_s(\boldsymbol{r}) \end{pmatrix}.$$

The set of spinor functions $\boldsymbol{\phi}^p_{sm}(\boldsymbol{r})$, for $s=1, 2, \ldots, l$, $m=1, 2$, then form a basis for the $2l$-dimensional direct product representation $\boldsymbol{\Gamma}^p \otimes \boldsymbol{u}$ of the corresponding double group. Here $\boldsymbol{u}(T)$ is defined to be $\boldsymbol{u}(\boldsymbol{R}_p)$, where $\boldsymbol{R}_p$ is the proper rotational part of T.

This may be proved as follows. The matrices $\boldsymbol{\Gamma}^p$ and $\boldsymbol{u}$ each form representations of the double group, so that as shown in ch. 3 §7.2, their direct product also forms a representation. Then by the definitions (8.16) and (8.39),

$$\begin{aligned} O(T)\, \phi^p_{sm}(\boldsymbol{r}, i) &= \sum_{j=1}^{2} u(T)_{ij}\, \phi^p_{sm}(T^{-1}\boldsymbol{r}, j) \\ &= u(T)_{im}\, P(T)\, \psi^p_s(\boldsymbol{r}) \\ &= \sum_{t=1}^{l} \Gamma^p(T)_{ts}\, \psi^p_t(\boldsymbol{r})\, u(T)_{im}. \end{aligned}$$

However, by (3.34) and (8.39)

$$\sum_{t=1}^{l} \sum_{n=1}^{2} \{\boldsymbol{\Gamma}^p(T) \otimes \boldsymbol{u}(T)\}_{tn,\, sm}\, \phi^p_{tn}(\boldsymbol{r}, i) = \sum_{t=1}^{l} \Gamma^p(T)_{ts}\, u(T)_{im}\, \psi^p_t(\boldsymbol{r}).$$

Thus

$$O(T)\, \boldsymbol{\phi}^p_{sm}(\boldsymbol{r}) = \sum_{t=1}^{l} \sum_{n=1}^{2} \{\boldsymbol{\Gamma}^p(T) \otimes \boldsymbol{u}(T)\}_{tn,\, sm}\, \boldsymbol{\phi}^p_{tn}(\boldsymbol{r}),$$

as required. A similar result holds for T replaced by $\bar{T}$.

Now suppose that the functions $\psi^p_{\boldsymbol{k}s}(\boldsymbol{r})$ are eigenfunctions of the spin-independent Hamiltonian having wavevector $\boldsymbol{k}$ and belonging to a l-fold degenerate energy eigenvalue $E(\boldsymbol{k})$, that in turn corresponds to the l-dimensional irreducible representation $\boldsymbol{\Gamma}^p$ of $\mathscr{G}_o(\boldsymbol{k})$. Let $\phi^p_{\boldsymbol{k}sm}(\boldsymbol{r}, i) = \psi^p_{\boldsymbol{k}s}(\boldsymbol{r})\, \delta_{mi}$. Then the spinors $\boldsymbol{\phi}^p_{\boldsymbol{k}sm}(\boldsymbol{r})$ are eigenspinors of a $2l$-fold degenerate energy eigenvalue $E(\boldsymbol{k})$ of the spin-independent matrix Hamiltonian operator $\boldsymbol{H}_0(\boldsymbol{r})$ of (8.24), as they form a basis for the $2l$-dimensional representation $\boldsymbol{\Gamma}^p \otimes \boldsymbol{u}$ of $\mathscr{G}^D_o(\boldsymbol{k})$.

If the representation $\boldsymbol{\Gamma}^p \otimes \boldsymbol{u}$ is reducible, as it often is, and $\boldsymbol{\Gamma}^q$, $\boldsymbol{\Gamma}^r, \ldots$ are the irreducible representations of the double point group $\mathscr{G}_o^D(\boldsymbol{k})$ that appear in its reduction, then spinors $\boldsymbol{\phi}_{\boldsymbol{k}sm}^p(\boldsymbol{r})$ can be rearranged so as to form bases for $\boldsymbol{\Gamma}^q, \boldsymbol{\Gamma}^r, \ldots$. Thus, in the absence of spin–orbit coupling, the energy levels at $\boldsymbol{k}$ corresponding to the irreducible representations $\boldsymbol{\Gamma}^q, \boldsymbol{\Gamma}^r, \ldots$ stick together, this being an example of an 'accidental' degeneracy, caused here by the particular form of $\boldsymbol{H}_0(\boldsymbol{r})$. When spin–orbit coupling is taken into account, the energy eigenvalues, corresponding to $\boldsymbol{\Gamma}^q, \boldsymbol{\Gamma}^r, \ldots$ will in general be different, and the degeneracy will be split. Thus the reduction series

$$\boldsymbol{\Gamma}^p \otimes \boldsymbol{u} = n_q \boldsymbol{\Gamma}^q \oplus n_r \boldsymbol{\Gamma}^r \oplus \cdots \tag{8.40}$$

shows immediately how the splitting takes place. Of course, it has to be remembered that some of the irreducible representations of the right-hand side of (8.40) may occur doubled or stick together because of time-reversal symmetry.

The number of times n_q that $\boldsymbol{\Gamma}^q$ appears in $\boldsymbol{\Gamma}^p \otimes \boldsymbol{u}$ is given, according to eq. (3.36), by

$$\begin{aligned} n_q = \{1/2g_o(\boldsymbol{k})\} \sum \chi^p([\boldsymbol{R} \mid \boldsymbol{0}]) \\ \times \operatorname{Tr}\{\boldsymbol{u}(\boldsymbol{R})\} \{\chi^q([\boldsymbol{R} \mid \boldsymbol{0}] - \chi^q([\bar{\boldsymbol{R}} \mid \boldsymbol{0}])\}^* \end{aligned} \tag{8.41}$$

the sum being over all rotations $\{\boldsymbol{R} \mid \boldsymbol{0}\}$ of $\mathscr{G}_o(\boldsymbol{k})$, where the fact has been used that $\chi^p([\boldsymbol{R} \mid \boldsymbol{0}]) = \chi^p([\bar{\boldsymbol{R}} \mid \boldsymbol{0}])$. Thus $n_q = 0$ for all $\boldsymbol{\Gamma}^q$ that are not extra representations of $\mathscr{G}_o^D(\boldsymbol{k})$, demonstrating again that spinor functions only transform according to the extra representations. If $\boldsymbol{\Gamma}^q$ is an extra representation, then (8.41) simplifies to give

$$n_q = \{1/g_o(\boldsymbol{k})\} \sum \chi^p([\boldsymbol{R} \mid \boldsymbol{0}]) \operatorname{Tr}\{\boldsymbol{u}(\boldsymbol{R})\} \chi^q([\boldsymbol{R} \mid \boldsymbol{0}])^*,$$

the sum again being over all rotations $\{\boldsymbol{R} \mid \boldsymbol{0}\}$ of $\mathscr{G}_o(\boldsymbol{k})$. The splitting caused by spin–orbit coupling therefore is very easily determined, particularly as $\operatorname{Tr}\{\boldsymbol{u}(\boldsymbol{R})\} = 2u_0$, where u_0 is defined in (8.7) and given for many rotations in table 8.1. This argument does not tell the *ordering* of the energy levels after the splitting has taken place. This requires explicit calculation.

An example will be considered in §8.5. The series (8.40) has been tabulated in the literature for the more common space groups. In such tabulations the matrices $\boldsymbol{u}$ are often replaced by the equivalent (but not necessarily identical) matrices denoted by $\boldsymbol{D}_{\frac{1}{2}}$, as for example in the papers of ELLIOTT [1954] and DRESSELHAUS [1955].

§ 8.5. *The double groups corresponding to the space groups O_h^9 and O_h^5*

The character tables for the extra irreducible representations of all the double point groups are given in appendix 2. For the double point groups $\mathscr{G}_o^D(\boldsymbol{k})$ belonging to the space groups O_h^1, O_h^5, O_h^9 and T_d^2 the notations of ELLIOTT [1954] and PARMENTER [1955] have been used to label the irreducible representations. The appropriate point group for a particular $\boldsymbol{k}$ for the space groups O_h^9 and O_h^5 may be found from tables 5.1 and 5.2. (PARMENTER [1955] has pointed out that the character tables given by ELLIOTT [1954] for the axes Λ and F are incorrect. The character table quoted by ELLIOTT [1954] for the axis Q is also incorrect.) The assignment of these labels to the irreducible representations of $\mathscr{G}_o^D(\boldsymbol{k})$ implies that a particular choice has been made of the sign of the matrices $\boldsymbol{u}(\boldsymbol{R}_p)$ in eqs. (8.16) and (8.17). This matter is examined in detail at the end of this subsection.

For the symmetry points K and U and the symmetry axes Σ, S and Z of the space group O_h^5 and for the symmetry axes Σ, D and G of the space group O_h^9, the groups $\mathscr{G}_o^D(\boldsymbol{k})$ have only one extra irreducible representation. This representation has dimension two, and is such that $\chi([\boldsymbol{R} \mid \boldsymbol{0}])=0$ for every rotation $\{\boldsymbol{R} \mid \boldsymbol{0}\}$ other than the identity transformation. The series (8.40) and the compatibility relations involving these axes are therefore trivial. The only non-trivial compatibility relations are between the axes Δ, Λ, F and Q and their end points. These are given in table 8.3.

Time-reversal symmetry causes the irreducible representations L_4^+, L_5^+, and L_4^-, L_5^-, and Λ_4, Λ_5, and F_4, F_5 and Q_3, Q_4 to stick

TABLE 8.3

Compatibility relations between symmetry points and symmetry axes for the extra representations for the body-centred cubic space group O_h^9 and the face-centred cubic space group O_h^5

Γ_6^+	Γ_7^+	Γ_8^+	Γ_6^-	Γ_7^-	Γ_8^-
Δ_6	Δ_7	$\Delta_6\Delta_7$	Δ_6	Δ_7	$\Delta_6\Delta_7$
Λ_6	Λ_6	$\Lambda_4\Lambda_5\Lambda_6$	Λ_6	Λ_6	$\Lambda_4\Lambda_5\Lambda_6$

H_6^+	H_7^+	H_8^+	H_6^-	H_7^-	H_8^-
Δ_6	Δ_7	$\Delta_6\Delta_7$	Δ_6	Δ_7	$\Delta_6\Delta_7$
F_6	F_6	$F_4F_5F_6$	F_6	F_6	$F_4F_5F_6$

P_6	P_7	P_8
Λ_6	Λ_6	$\Lambda_4\Lambda_5\Lambda_6$
F_6	F_6	$F_4F_5F_6$

L_4^+	L_5^+	L_4^-	L_5^-	L_6
Λ_4	Λ_5	Λ_5	Λ_4	Λ_6
Q_3	Q_4	Q_3	Q_4	Q_3Q_4

X_6^+	X_7^+	X_6^-	X_7^-
Δ_6	Δ_7	Δ_6	Δ_7

W_6	W_7
Q_3Q_4	Q_3Q_4

together in pairs. [All these representations are of the type (b) considered in § 8.3, and the point L is an example of case (3), while Λ, F and Q are examples of case (2).] The 'even' and 'odd' extra representations at every point on a symmetry plane also stick together. As noted in § 8.3, there is a two-fold degeneracy at all general points because of time-reversal symmetry, so there is a two-fold degeneracy for every point $\boldsymbol{k}$ of the Brillouin zone.

TABLE 8.4

Compatibility table connecting representations of the single groups with the extra representations of the double groups of O_h^9 and O_h^5

Γ_i	Γ_1	Γ_2	Γ_{12}	$\Gamma_{15'}$	$\Gamma_{25'}$	$\Gamma_{1'}$	$\Gamma_{2'}$	$\Gamma_{12'}$	Γ_{15}	Γ_{25}
$\Gamma_i \otimes \boldsymbol{u}$	Γ_6^+	Γ_7^+	Γ_8^+	$\Gamma_6^+ \oplus \Gamma_8^+$	$\Gamma_7^+ \oplus \Gamma_8^+$	Γ_6^-	Γ_7^-	Γ_8^-	$\Gamma_6^- \oplus \Gamma_8^-$	$\Gamma_7^- \oplus \Gamma_8^-$

H_i	H_1	H_2	H_{12}	$H_{15'}$	$H_{25'}$	$H_{1'}$	$H_{2'}$	$H_{12'}$	H_{15}	H_{25}
$H_i \otimes \boldsymbol{u}$	H_6^+	H_7^+	H_8^+	$H_6^+ \oplus H_8^+$	$H_7^+ \otimes H_8^+$	H_6^-	H_7^-	H_8^-	$H_6^- \oplus H_8^-$	$H_7^- \oplus H_8^-$

P_i	P_1	P_2	P_3	P_4	P_5
$P_i \otimes \boldsymbol{u}$	P_6	P_7	P_8	$P_7 \oplus P_8$	$P_6 \oplus P_8$

N_i	N_1	N_2	N_3	N_4	$N_{1'}$	$N_{2'}$	$N_{3'}$	$N_{4'}$
$N_i \otimes \boldsymbol{u}$	N_5	N_5	N_5	N_5	$N_{5'}$	$N_{5'}$	$N_{5'}$	$N_{5'}$

L_i	L_1	L_2	L_3	$L_{1'}$	$L_{2'}$	$L_{3'}$
$L_i \otimes \boldsymbol{u}$	L_6^+	L_6^+	$L_4^+ \oplus L_5^+ \oplus L_6^+$	L_6^-	L_6^-	$L_4^- \oplus L_5^- \oplus L_6^-$

table 8.4 (continued)

W_i	W_1	W_2	$W_{1'}$	$W_{2'}$	W_3
$W_i \otimes \boldsymbol{u}$	W_6	W_6	W_7	W_7	$W_6 \oplus W_7$

X_i	X_1	X_2	X_3	X_4	X_5	$X_{1'}$	$X_{2'}$	$X_{3'}$	$X_{4'}$	$X_{5'}$
$X_i \otimes \boldsymbol{u}$	X_6^+	X_7^+	X_7^+	X_6^+	$X_6^+ \oplus X_7^+$	X_6^-	X_7^-	X_7^-	X_6^-	$X_6^- \oplus X_7^-$

Δ_i	Δ_1	Δ_2	$\Delta_{1'}$	$\Delta_{2'}$	Δ_5
$\Delta_i \otimes \boldsymbol{u}$	Δ_6	Δ_7	Δ_6	Δ_7	$\Delta_6 \oplus \Delta_7$

Λ_i	Λ_1	Λ_2	Λ_3
$\Lambda_i \otimes \boldsymbol{u}$	Λ_6	Λ_6	$\Lambda_4 \oplus \Lambda_5 \oplus \Lambda_6$

F_i	F_1	F_2	F_3
$F_i \otimes \boldsymbol{u}$	F_6	F_6	$F_4 \oplus F_5 \oplus F_6$

Q_i	Q_1	Q_2
$Q_i \otimes \boldsymbol{u}$	$Q_3 \oplus Q_4$	$Q_3 \oplus Q_4$

The non-trivial series (8.40) are given in table 8.4. As an example of its use, consider the energy levels at H which in the absence of spin–orbit coupling correspond to the three-dimensional representation $H_{15'}$. When the two possible linearly independent spin functions that may be attached to every orbital eigenfunction are taken into account, these levels are 6-fold degenerate. As $H_{15'} \otimes \boldsymbol{u} = H_6^+ \oplus H_8^+$, and H_6^+ and H_8^+ have dimensions 2 and 4 respectively, the introduction of spin–orbit coupling splits the 6-fold degenerate level into a 2-fold degenerate level and a 4-fold degenerate level. This argument cannot predict which of these is the highest.

The use of the convention of ELLIOTT [1954] for the labelling of the extra irreducible representations of the groups $\mathscr{G}_o^D(\boldsymbol{k})$ does imply that a particular choice has been made of the sign of the matrices $\boldsymbol{u}(\boldsymbol{R}_p)$ of eqs. (8.16) and (8.17). That is, it implies that a choice has been made as to which of the two generalized transformations corresponding to $\{\boldsymbol{R} \mid \boldsymbol{0}\}$ should be regarded as $[\boldsymbol{R} \mid \boldsymbol{0}]$. To see this, consider a pair of extra irreducible representations $\boldsymbol{\Gamma}^1$ and $\boldsymbol{\Gamma}^2$ for which

$$\begin{aligned} \chi^1([\boldsymbol{R} \mid \boldsymbol{0}]) &= c\,, & \chi^1([\bar{\boldsymbol{R}} \mid \boldsymbol{0}]) &= -c\,, \\ \chi^2([\boldsymbol{R} \mid \boldsymbol{0}]) &= -c\,, & \chi^2([\bar{\boldsymbol{R}} \mid \boldsymbol{0}]) &= c\,, \end{aligned}$$

where c is a non-zero constant. Then it is clear that if $[\boldsymbol{R} \mid \boldsymbol{0}]$ and $[\bar{\boldsymbol{R}} \mid \boldsymbol{0}]$ were interchanged, the assignment of labels would have to be interchanged as well. Thus a specification that $\chi^1([\boldsymbol{R} \mid \boldsymbol{0}]) = c$, rather than $-c$, uniquely distinguishes $[\boldsymbol{R} \mid \boldsymbol{0}]$ from $[\bar{\boldsymbol{R}} \mid \boldsymbol{0}]$.

It is possible to ensure in the following simple way that the conventions adopted here, by the choice of the signs of the matrices $\boldsymbol{u}(\boldsymbol{R}_p)$ in table 8.1, are consistent with the conventions of ELLIOTT [1954]. First consider the generalized rotations corresponding to the elements of the cubic point group O_h. ELLIOTT [1954] has stated that, with his conventions, the series (8.40), for the case when $\boldsymbol{\Gamma}^p$ is Γ_1, is $\Gamma_1 \otimes \boldsymbol{u} = \Gamma_6^+$. Thus, by eq. (3.35), for every $[\boldsymbol{R} \mid \boldsymbol{0}]$ of $\mathscr{G}_o^D(\boldsymbol{k})$,

$$\chi_1([\boldsymbol{R} \mid \boldsymbol{0}])\,\mathrm{Tr}\,\{\boldsymbol{u}(\boldsymbol{R}_p)\} = \chi_6^+([\boldsymbol{R} \mid \boldsymbol{0}]) \tag{8.42}$$

The character systems of χ_1 and χ_6^+ are given in the papers of BOUCKAERT *et al.* [1936] and ELLIOTT [1954] respectively, and are reproduced in tables A.1 and A.25 respectively. Thus the sign of

$\mathrm{Tr}\{\boldsymbol{u}(R_p)\}=2u_0$ is determined. This procedure gives the sign of $\boldsymbol{u}(R_p)$ for all rotations, except those through π, for which $u_0=0$. The signs of the matrices $\boldsymbol{u}(R_p)$ given in table 8.1 are chosen so that eq. (8.42) is satisfied. In the same way the signs of the matrices $\boldsymbol{u}(R_p)$ for the members of D_{6h} not in O_h are such that the equation $\Gamma_1^+ \otimes \boldsymbol{u} = \Gamma_7^+$ of ELLIOTT [1954] is satisfied.

Although this procedure distinguishes $[C_{ni} \mid \mathit{0}]$ and $[IC_{ni} \mid \mathit{0}]$ from $[\bar{C}_{ni} \mid \mathit{0}]$ and $[I\bar{C}_{ni} \mid \mathit{0}]$ if $n \neq 2$, it does not do so if $n=2$. In fact, there is *no* way of making this distinction if $n=2$, which must imply that the distinction is unobservable. That this is so can be seen from considering in the following two cases.

(a) If Γ is an extra irreducible representation of $\mathscr{G}_o^D(\boldsymbol{k})$ such that $\chi([C_{2i} \mid \mathit{0}])=0$ for all rotations through π, then the distinction between $[C_{2i} \mid \mathit{0}]$ and $[\bar{C}_{2i} \mid \mathit{0}]$ does not effect the labelling of the irreducible representation, nor its basis functions, nor the way spin–orbit coupling splits degeneracies. The same is true if $\chi([IC_{2i} \mid \mathit{0}])=0$ for all reflections of $\mathscr{G}_o^D(\boldsymbol{k})$. This case applies to all the irreducible representations of the groups $\mathscr{G}_o^D(\boldsymbol{k})$ for the space groups O_h^9 and O_h^5, except for L_4^+, L_5^+, L_4^-, L_5^-, Λ_4, Λ_5, F_4, F_5, Q_3, Q_4, and representations for symmetry planes.

(b) If $\chi([C_{2i} \mid \mathit{0}]) \neq 0$ or $\chi([IC_{2i} \mid \mathit{0}]) \neq 0$, as in the exceptional cases just mentioned, the choice between $[C_{2i} \mid \mathit{0}]$ or $[IC_{2i} \mid \mathit{0}]$ and $[\bar{C}_{2i} \mid \mathit{0}]$ or $[I\bar{C}_{2i} \mid \mathit{0}]$ is coupled to the choice of the labelling of the irreducible representations. For example, with L_4^+ and L_5^+, if $[C_{2i} \mid \mathit{0}]$ is replaced by $[\bar{C}_{2i} \mid \mathit{0}]$, then L_4^+ and L_5^+ are interchanged. However, the energy eigenvalues corresponding to L_4^+ and L_5^+ are always equal because of time-reversal symmetry, and L_4^+ and L_5^+ always occur together in the series of the form (8.40), as shown in table 8.4. Therefore the energy eigenvalues corresponding to L_4^+ and L_5^+ are indistinguishable, and choice of the labelling is quite arbitrary. The same is true of all the other exceptional pairs mentioned above.

The conventions of PARMENTER [1955] and DRESSELHAUS [1955] are also consistent with the present conventions.

§ 9. DOUBLE NON-SYMMORPHIC SPACE GROUPS

§ 9.1. *Irreducible representations of a double non-symmorphic space group*

The theory for double non-symmorphic space groups can be regarded as a generalization of the theories for both single non-symmorphic and double symmorphic space groups, and as such combines all the features of the two theories.

The construction of the basis functions follows a parallel line to that of ch. 7 § 1, the groups $\mathscr{G}(\boldsymbol{k})$, symmetry points, axes and planes, and the star of $\boldsymbol{k}$ being defined exactly as before. The fundamental theorem requires only minor modification, and is now as follows.

Fundamental theorem on the spinor basis functions of irreducible representations of a double non-symmorphic space group $\mathscr{G}^D$. Suppose that $\boldsymbol{k}$ is an allowed $\boldsymbol{k}$-vector, and $\phi^p_{\boldsymbol{k}s}(\boldsymbol{r}, j)$ are Bloch functions that are the $j=1$ and $j=2$ components of a spinor which transforms as the sth row of the *extra* irreducible representation $\boldsymbol{\Gamma}^p$ of the double group $\mathscr{G}^D(\boldsymbol{k})$ that satisfies

$$\boldsymbol{\Gamma}^p([E \mid \boldsymbol{t}_n]) = \exp(-\mathrm{i}\boldsymbol{k}\cdot\boldsymbol{t}_n)\, \boldsymbol{\Gamma}^p([E \mid \boldsymbol{0}]) \tag{8.43}$$

for every generalized primitive translation $[\boldsymbol{E} \mid \boldsymbol{t}_n]$, this representation being of dimension l. Then the set of $lM(\boldsymbol{k})$ spinors with components

$$O([\boldsymbol{R}_i \mid \boldsymbol{t}_i])\, \phi^p_{\boldsymbol{k}s}(\boldsymbol{r}, j), \qquad j = 1, 2,$$
$$i = 1, 2, \ldots, M(\boldsymbol{k}), \qquad s = 1, 2, \ldots, l,$$

form a basis for a $lM(\boldsymbol{k})$-dimensional extra unitary irreducible representation of the double space group $\mathscr{G}^D$. Moreover, all the extra irreducible representations of $\mathscr{G}^D$ may be obtained in this way, by working through all the extra irreducible representations of $\mathscr{G}^D(\boldsymbol{k})$ that satisfy (8.43) for all the allowed values of $\boldsymbol{k}$ that are in different stars.

An explicit expression for the matrix elements of the irreducible representation $\boldsymbol{\Gamma}^{\boldsymbol{k}p}$ that is formed in this way is

$$\Gamma^{\boldsymbol{k}p}([\boldsymbol{R} \mid \boldsymbol{t}])_{j,t;i,s} = \begin{cases} 0, & \text{if } \boldsymbol{R}_j^{-1}\boldsymbol{R}\boldsymbol{R}_i\boldsymbol{k} \text{ is not equivalent to } \boldsymbol{k}, \\ \Gamma^p([\boldsymbol{R}_j \mid \boldsymbol{t}_j]^{-1}[\boldsymbol{R} \mid \boldsymbol{t}][\boldsymbol{R}_i \mid \boldsymbol{t}_i])_{ts}, & \text{otherwise}. \end{cases}$$

As in the case of symmorphic space groups, the consequences of the theorem are the same for double groups as for single groups, the only modification being that scalar functions are to be replaced everywhere by spinor functions.

A development similar to that of ch. 7 §2 can be made for the double group $\mathscr{G}^D(\boldsymbol{k})$. The group $\mathscr{T}(\boldsymbol{k})$ may be defined as the group of all generalized primitive translations $[\boldsymbol{E} \mid \boldsymbol{t}_n]$ such that $\exp(-\mathrm{i}\boldsymbol{k}\cdot\boldsymbol{t}_n)=1$. This group is isomorphic to the group of corresponding ordinary transformations $\{\boldsymbol{E} \mid \boldsymbol{t}_n\}$ that was also denoted by $\mathscr{T}(\boldsymbol{k})$ in ch. 7 §2. (The new group $\mathscr{T}(\boldsymbol{k})$ does not contain any translations with rotational part $\bar{\boldsymbol{E}}$, by its very definition.) The group $\mathscr{T}(\boldsymbol{k})$ is then an invariant subgroup of $\mathscr{G}^D(\boldsymbol{k})$, so that it is possible to form the factor group $\mathscr{G}^D(\boldsymbol{k})/\mathscr{T}(\boldsymbol{k})$. This has as its elements cosets of the form $[\boldsymbol{R} \mid \boldsymbol{t}]\,\mathscr{T}(\boldsymbol{k})$ and $[\bar{\boldsymbol{R}} \mid \boldsymbol{t}]\,\mathscr{T}(\boldsymbol{k})$, the identity element being $[\boldsymbol{E} \mid \boldsymbol{0}]\,\mathscr{T}(\boldsymbol{k})$.

The following theorem then holds.

Theorem. Let $\boldsymbol{\Gamma}'$ be an extra irreducible representation of $\mathscr{G}^D(\boldsymbol{k})/\mathscr{T}(\boldsymbol{k})$ that satisfies

$$\boldsymbol{\Gamma}'([\boldsymbol{E} \mid \boldsymbol{t}_n]\,\mathscr{T}(\boldsymbol{k})) = \exp(-\mathrm{i}\boldsymbol{k}\cdot\boldsymbol{t}_n)\,\boldsymbol{\Gamma}'([\boldsymbol{E} \mid \boldsymbol{0}]\,\mathscr{T}(\boldsymbol{k})) \tag{8.44}$$

for every $\boldsymbol{t}_n$. Then the set of matrices $\boldsymbol{\Gamma}([\boldsymbol{R} \mid \boldsymbol{t}])$ defined by

$$\boldsymbol{\Gamma}([\boldsymbol{R} \mid \boldsymbol{t}]) = \boldsymbol{\Gamma}'([\boldsymbol{R} \mid \boldsymbol{t}]\,\mathscr{T}(\boldsymbol{k}))$$

for every $[\boldsymbol{R} \mid \boldsymbol{t}]$ of $\mathscr{G}^D(\boldsymbol{k})$ form an extra irreducible representation of $\mathscr{G}^D(\boldsymbol{k})$ which satisfies (8.43). Moreover, all such extra irreducible representations of $\mathscr{G}^D(\boldsymbol{k})$ can be constructed in this way.

This theorem shows that $\mathscr{G}^D(\boldsymbol{k})/\mathscr{T}(\boldsymbol{k})$ replaces $\mathscr{G}(\boldsymbol{k})/\mathscr{T}(\boldsymbol{k})$ in the theory of double non-symmorphic space groups.

The splitting of degeneracies by spin–orbit coupling may be treated in the same way as in §8.4, the splitting being given by the reduction

$$\boldsymbol{\Gamma}^p \otimes \boldsymbol{u} = n_{1p}\boldsymbol{\Gamma}^1 \oplus n_{2p}\boldsymbol{\Gamma}^2 \oplus \cdots, \tag{8.45}$$

where $\boldsymbol{\Gamma}^p$ now denotes an irreducible representation of $\mathscr{G}(\boldsymbol{k})$ and $\boldsymbol{\Gamma}^1, \boldsymbol{\Gamma}^2, \ldots$ are extra irreducible representations of $\mathscr{G}^D(\boldsymbol{k})$.

§ 9.2. *Time-reversal symmetry for double non-symmorphic space groups*

The theory of time-reversal symmetry for double non-symmorphic space groups is very similar to the corresponding theory for double symmorphic space groups. The irreducible representations of $\mathscr{G}^D(\boldsymbol{k})$ belong to one of the three types (a), (b) or (c), with exactly the same consequences as were described in §8.3. (It should be noted that these consequences are not the same as those mentioned in ch. 5 §7.1 and ch. 7 §8.) The criterion replacing (8.36) is

$$\sum \chi^p([\boldsymbol{Q} \mid \boldsymbol{t}]^2) = \begin{cases} g_o/M(\boldsymbol{k}), & \text{case (a)}, \\ 0, & \text{case (b)}, \\ -\,g_o/M(\boldsymbol{k}), & \text{case (c)}, \end{cases} \tag{8.46}$$

where the summation is over all the generalized transformations $[\boldsymbol{Q} \mid \boldsymbol{t}]$ that correspond to the coset representatives $\{\boldsymbol{Q} \mid \boldsymbol{t}\}$ of $\mathscr{G}/\mathscr{T}$ that have the property that $\boldsymbol{Qk}$ is equivalent to $-\boldsymbol{k}$. In (8.46) g_o is the order of $\mathscr{G}_o$, $M(\boldsymbol{k})$ is the number of vectors in the star of $\boldsymbol{k}$, and $\chi^p([\boldsymbol{Q} \mid \boldsymbol{t}]^2)$ denotes the character of $[\boldsymbol{Q} \mid \boldsymbol{t}]^2$ in the extra irreducible representation $\boldsymbol{\Gamma}^p$ of $\mathscr{G}^D(\boldsymbol{k})$.

The analysis of an extra degeneracy corresponding to cases (a) and (b) in terms of the three possible cases (1), (2) or (3) for $\boldsymbol{k}$ is also similar to that given in §8.3. The condition for case (2) that replaces (8.37) is that

$$\chi^p([\boldsymbol{R} \mid \boldsymbol{t}])^* = \chi^q([\boldsymbol{R}_- \mid \boldsymbol{t}_-]^{-1}\,[\boldsymbol{R} \mid \boldsymbol{t}]\,[\boldsymbol{R}_- \mid \boldsymbol{t}_-]), \tag{8.47}$$

for every $[\boldsymbol{R} \mid \boldsymbol{t}]$ of $\mathscr{G}^D(\boldsymbol{k})$, where $[\boldsymbol{R}_- \mid \boldsymbol{t}_-]$ is the coset representative

TABLE 8.5

Character table of the extra representations of $\mathscr{G}^D(\boldsymbol{k})/\mathscr{T}(\boldsymbol{k})$ for A of D_{6h}^4

	A_4	A_5	A_6
$[E \mid \boldsymbol{0}]$	2	2	4
$[\bar{E} \mid \boldsymbol{0}]$	-2	-2	-4
$[E \mid \boldsymbol{a}_1]$	-2	-2	-4
$[\bar{E} \mid \boldsymbol{a}_1]$	2	2	4
$[C_{6z}, C_{6z}^{-1} \mid \boldsymbol{\tau}, \boldsymbol{\tau}+\boldsymbol{a}_1]$	0	0	0
$[\bar{C}_{6z}, \bar{C}_{6z}^{-1} \mid \boldsymbol{\tau}, \boldsymbol{\tau}+\boldsymbol{a}_1]$	0	0	0
$[C_{3z}, C_{3z}^{-1} \mid \boldsymbol{0}]$	-2	-2	2
$[\bar{C}_{3z}, \bar{C}_{3z}^{-1} \mid \boldsymbol{0}]$	2	2	-2
$[C_{3z}, C_{3z}^{-1} \mid \boldsymbol{a}_1]$	2	2	-2
$[\bar{C}_{3z}, \bar{C}_{3z}^{-1} \mid \boldsymbol{a}_1]$	-2	-2	2
$[C_{2z}, \bar{C}_{2z} \mid \boldsymbol{\tau}, \boldsymbol{\tau}+\boldsymbol{a}_1]$	0	0	0
$[C_{2x}, C_{2A}, C_{2B} \mid \boldsymbol{\tau}], [\bar{C}_{2x}, \bar{C}_{2A}, \bar{C}_{2B} \mid \boldsymbol{\tau}+\boldsymbol{a}_1]$	2i	-2i	0
$[\bar{C}_{2x}, \bar{C}_{2A}, \bar{C}_{2B} \mid \boldsymbol{\tau}], [C_{2x}, C_{2A}, C_{2B} \mid \boldsymbol{\tau}+\boldsymbol{a}_1]$	-2i	2i	0
$[C_{2y}, C_{2C}, C_{2D}, \bar{C}_{2y}, \bar{C}_{2C}, \bar{C}_{2D} \mid \boldsymbol{0}, \boldsymbol{a}_1]$	0	0	0
$[I \mid \boldsymbol{\tau}, \boldsymbol{\tau}+\boldsymbol{a}_1]$	0	0	0
$[\bar{I} \mid \boldsymbol{\tau}, \boldsymbol{\tau}+\boldsymbol{a}_1]$	0	0	0
$[IC_{6z}, IC_{6z}^{-1} \mid \boldsymbol{0}, \boldsymbol{a}_1]$	0	0	0
$[\overline{IC}_{6z}, \overline{IC}_{6z}^{-1} \mid \boldsymbol{0}, \boldsymbol{a}_1]$	0	0	0
$[IC_{3z}, IC_{3z}^{-1} \mid \boldsymbol{\tau}, \boldsymbol{\tau}+\boldsymbol{a}_1]$	0	0	0
$[\overline{IC}_{3z}, \overline{IC}_{3z}^{-1} \mid \boldsymbol{\tau}, \boldsymbol{\tau}+\boldsymbol{a}_1]$	0	0	0
$[IC_{2z}, \overline{IC}_{2z} \mid \boldsymbol{0}, \boldsymbol{a}_1]$	0	0	0
$[IC_{2x}, IC_{2A}, IC_{2B}, \overline{IC}_{2x}, \overline{IC}_{2A}, \overline{IC}_{2B} \mid \boldsymbol{0}]$	0	0	0
$[IC_{2x}, IC_{2A}, IC_{2B}, \overline{IC}_{2x}, \overline{IC}_{2A}, \overline{IC}_{2B} \mid \boldsymbol{a}_1]$	0	0	0
$[IC_{2y}, IC_{2C}, IC_{2D}, \overline{IC}_{2y}, \overline{IC}_{2C}, \overline{IC}_{2D} \mid \boldsymbol{\tau}, \boldsymbol{\tau}+\boldsymbol{a}_1]$	0	0	0

of the decomposition of $\mathscr{G}^D$ into left cosets with respect to $\mathscr{G}^D(\boldsymbol{k})$ such that $\boldsymbol{R}_{-}\boldsymbol{k}=-\boldsymbol{k}$. Similarly, the condition for case (3) that replaces (8.38) is that

$$\chi^p([\boldsymbol{R} \mid \boldsymbol{t}])^* = \chi^q([\boldsymbol{R} \mid \boldsymbol{t}]), \tag{8.48}$$

for every $[\boldsymbol{R} \mid \boldsymbol{t}]$ of $\mathscr{G}^D(\boldsymbol{k})$. It is sufficient to test (8.47) and (8.48) with only those elements $[\boldsymbol{R} \mid \boldsymbol{t}]$ of $\mathscr{G}^D(\boldsymbol{k})$ that are coset representatives of $\mathscr{G}^D(\boldsymbol{k})/\mathscr{T}(\boldsymbol{k})$ but are not pure translations.

§ 9.3. *The double group corresponding to the space group D_{6h}^4*

ELLIOTT [1954] has given the character tables for the extra representations of the groups $\mathscr{G}^D(\boldsymbol{k})/\mathscr{T}(\boldsymbol{k})$ for the space group D_{6h}^4. As a typical example, the table for the point A is translated into the notation used elsewhere in this book and reproduced in table 8.5. The representations A_4 and A_5 are both of type (b), and therefore stick together because of time-reversal symmetry. The series (8.45) is given in table 8.6. (The corresponding series for the point Γ as given by ELLIOTT [1954] contains a minor error that has been corrected by CORNWELL [1966].)

A very interesting situation occurs at points on the hexagonal face of the Brillouin zone. Let $\boldsymbol{k}=(k_y, k_y, \pi/c)$. The relevant charac-

TABLE 8.6

Compatibility table connecting representations of the single group $\mathscr{G}(\boldsymbol{k})/\mathscr{T}(\boldsymbol{k})$ with the extra representations of the double group $\mathscr{G}^D(\boldsymbol{k})/\mathscr{T}(\boldsymbol{k})$ for the point A of D_{6h}^4

A_i	A_1	A_2	A_3
$A_i \otimes \boldsymbol{u}$	A_6	A_6	$A_4 \oplus A_5 \oplus A_6$

ters of the single group $\mathscr{G}(\boldsymbol{k})$ for this point were derived in ch. 7 §4.4, and the effects of time-reversal symmetry for this point were described in ch. 7 §8. It follows by essentially the same argument as that given in ch. 7 §4.4 that the relevant extra irreducible representations of $\mathscr{G}^D(\boldsymbol{k})$ are one-dimensional. The only elements of $\mathscr{G}^D(\boldsymbol{k})$ whose characters are not given immediately by (8.43) are $[IC_{2z} \mid \boldsymbol{0}]$ and $[\bar{I}\bar{C}_{2z} \mid \boldsymbol{0}]$. The character table for the two relevant extra irreducible representations $\boldsymbol{\Gamma}^{2+}$ and $\boldsymbol{\Gamma}^{2-}$ for these elements and $[E \mid \boldsymbol{0}]$ and $[\bar{E} \mid \boldsymbol{0}]$ is given in table 8.7. [The entries $\pm$i for

TABLE 8.7

	$\boldsymbol{\Gamma}^{2+}$	$\boldsymbol{\Gamma}^{2-}$
$[E \mid \boldsymbol{0}]$	1	1
$[\bar{E} \mid \boldsymbol{0}]$	-1	-1
$[IC_{2z} \mid \boldsymbol{0}]$	i	$-$i
$[\overline{IC}_{2z} \mid \boldsymbol{0}]$	$-$i	i

$[IC_{2z} \mid \boldsymbol{0}]$ and $[\bar{I}\bar{C}_{2z} \mid \boldsymbol{0}]$ arise from the fact that $[IC_{2z} \mid \boldsymbol{0}]^2 = [\bar{I}\bar{C}_{2z} \mid \boldsymbol{0}]^2 = [\bar{E} \mid \boldsymbol{0}]$, by eq. (8.18).] The only operations $[\boldsymbol{Q} \mid \boldsymbol{t}]$ of eq. (8.46) for this $\boldsymbol{k}$ are $[I \mid \boldsymbol{\tau}]$ and $[C_{2z} \mid \boldsymbol{\tau}]$, whose squares are $[E \mid \boldsymbol{0}]$ and $[\bar{E} \mid \boldsymbol{a}_1]$, which both have character $+1$ for both representations $\boldsymbol{\Gamma}^{2+}$ and $\boldsymbol{\Gamma}^{2-}$. Thus $\boldsymbol{\Gamma}^{2+}$ and $\boldsymbol{\Gamma}^{2-}$ are both of type (a), and therefore occur doubled. In the absence of spin–orbit coupling, but taking the spin degeneracy into account, there is a four-fold degeneracy at all points on the hexagonal face of the Brillouin zone. The above argument shows that for points that are not on symmetry axes, this degeneracy is split by spin–orbit coupling to give two two-fold degenerate levels. It has been pointed out by COHEN and FALICOV

[1960] that this means that the 'double zone' description of ch. 7 § 8 can no longer be used, and that the topology of the Fermi surface may then be considerably altered, thereby giving an explanation of the observed magnetoresistance of certain metals.

There is a two-fold degeneracy at all general points of the Brillouin zone. This follows because the only $[\boldsymbol{Q} \mid \boldsymbol{t}]$ of eq. (8.46) for such a point is $[I \mid \boldsymbol{\tau}]$, and $[I \mid \boldsymbol{\tau}]^2 = [E \mid \boldsymbol{0}]$. This gives an example of case (a) and case (2) in which the only relevant extra irreducible representation of $\mathscr{G}^D(\boldsymbol{k})$ occurs doubled. As the $E(\boldsymbol{k})$ are continuous functions of $\boldsymbol{k}$, this implies that there is at least a two-fold degeneracy everywhere in the Brillouin zone.

APPENDIX 1

THE CRYSTALLOGRAPHIC POINT GROUPS

The following is a list of the 32 crystallographic point groups, arranged in roughly decreasing order of complexity. For each group the following details are given.

(a) The group elements. These are arranged in classes (as defined in ch. 1 § 2.4). The notation for the rotations is given in ch. 1 §§ 3.2, 3.3.

(b) The character table (as described in ch. 2 § 7). If the irreducible representations have been assigned labels that are in very common use in the solid-state literature, these are indicated in the left-hand columns. Other less commonly used notations are defined in the references given in appendix 3.

(c) Matrices for the irreducible representations of more than one dimension. These matrices are only unique up to a similarity transformation (as discussed in ch. 2 § 2). For one-dimensional representations the characters themselves are the matrix elements.

Each point group is defined relative to certain axes, but the orientation of these axes in space is quite arbitrary. Point groups that differ in the orientation of these defining axes are isomorphic (as discussed in ch. 5 § 4.1).

For a point group that is isomorphic to a group $\mathscr{G}_o(\boldsymbol{k})$, as defined in ch. 5 § 1, for some point $\boldsymbol{k}$ that is a symmetry point or is on a symmetry axis of the body-centred cubic space group O_h^9 or the face-centred cubic space group O_h^5, the group elements are chosen to be those of $\mathscr{G}_o(\boldsymbol{k})$. The corresponding $\boldsymbol{k}$-vector is as defined in tables 5.1 and 5.2. As described in ch. 5 § 3, it is possible for two or more points in different stars to have point groups $\mathscr{G}_o(\boldsymbol{k})$ that are isomorphic. The group elements for each $\mathscr{G}_o(\boldsymbol{k})$ are listed when this occurs.

More information on the 32 crystallographic point groups may be found in the book by KOSTER *et al.* [1964], which is wholly devoted to this subject, and in the articles by ALTMANN [1962, 1963].

The notation for the point groups is that of SCHÖNFLIESS [1923].

(1) O_h

(a) Classes [for $\mathscr{G}_o(\boldsymbol{k})$ of Γ and H]: $\mathscr{C}_1 = E$; $\mathscr{C}_2 = C_{3\alpha}, C_{3\beta}, C_{3\gamma}, C_{3\delta}, C_{3\alpha}^{-1}, C_{3\beta}^{-1}, C_{3\gamma}^{-1}, C_{3\delta}^{-1}$; $\mathscr{C}_3 = C_{2x}, C_{2y}, C_{2z}$; $\mathscr{C}_4 = C_{4x}, C_{4y}, C_{4z}, C_{4x}^{-1}, C_{4y}^{-1}, C_{4z}^{-1}$; $\mathscr{C}_5 = C_{2a}, C_{2b}, C_{2c}, C_{2d}, C_{2e}, C_{2f}$; $\mathscr{C}_6 = I$; $\mathscr{C}_7 = IC_{3\alpha}, IC_{3\beta}, IC_{3\gamma}, IC_{3\delta}, IC_{3\alpha}^{-1}, IC_{3\beta}^{-1}, IC_{3\gamma}^{-1}, IC_{3\delta}^{-1}$; $\mathscr{C}_8 = IC_{2x}, IC_{2y}, IC_{2z}$; $\mathscr{C}_9 = IC_{4x}, IC_{4y}, IC_{4z}, IC_{4x}^{-1}, IC_{4y}^{-1}, IC_{4z}^{-1}$; $\mathscr{C}_{10} = IC_{2a}, IC_{2b}, IC_{2c}, IC_{2d}, IC_{2e}, IC_{2f}$.

TABLE A.1

		$\mathscr{C}_1$	$\mathscr{C}_2$	$\mathscr{C}_3$	$\mathscr{C}_4$	$\mathscr{C}_5$	$\mathscr{C}_6$	$\mathscr{C}_7$	$\mathscr{C}_8$	$\mathscr{C}_9$	$\mathscr{C}_{10}$
$\Gamma_1 H_1 R_1$	$\boldsymbol{\Gamma}^1$	1	1	1	1	1	1	1	1	1	1
$\Gamma_2 H_2 R_2$	$\boldsymbol{\Gamma}^2$	1	1	1	−1	−1	1	1	1	−1	−1
$\Gamma_{12} H_{12} R_{12}$	$\boldsymbol{\Gamma}^3$	2	−1	2	0	0	2	−1	2	0	0
$\Gamma_{15'} H_{15'} R_{15'}$	$\boldsymbol{\Gamma}^4$	3	0	−1	1	−1	3	0	−1	1	−1
$\Gamma_{25'} H_{25'} R_{25'}$	$\boldsymbol{\Gamma}^5$	3	0	−1	−1	1	3	0	−1	−1	1
$\Gamma_{1'} H_{1'} R_{1'}$	$\boldsymbol{\Gamma}^6$	1	1	1	1	1	−1	−1	−1	−1	−1
$\Gamma_{2'} H_{2'} R_{2'}$	$\boldsymbol{\Gamma}^7$	1	1	1	−1	−1	−1	−1	−1	1	1
$\Gamma_{12'} H_{12'} R_{12'}$	$\boldsymbol{\Gamma}^8$	2	−1	2	0	0	−2	1	−2	0	0
$\Gamma_{15} H_{15} R_{15}$	$\boldsymbol{\Gamma}^9$	3	0	−1	1	−1	−3	0	1	−1	1
$\Gamma_{25} H_{25} R_{25}$	$\boldsymbol{\Gamma}^{10}$	3	0	−1	−1	1	−3	0	1	1	−1

(b) The character table is given in table A.1. The notations in the first column are those of BOUCKAERT *et al.* [1936].

(c) Matrices for irreducible representations of more than one dimension: for any proper rotation C_{na} of O_h,

$$\boldsymbol{\Gamma}^3(C_{na})=\boldsymbol{\Gamma}^8(C_{na})=\boldsymbol{\Gamma}''(C_{na}),\quad \boldsymbol{\Gamma}^3(IC_{na})=-\boldsymbol{\Gamma}^8(IC_{na})=\boldsymbol{\Gamma}''(C_{na}),$$

$$\boldsymbol{\Gamma}^4(C_{na})=\boldsymbol{\Gamma}^9(C_{na})=\boldsymbol{R}(C_{na}),\quad \boldsymbol{\Gamma}^4(IC_{na})=-\boldsymbol{\Gamma}^9(IC_{na})=\boldsymbol{R}(C_{na}),$$

$$\boldsymbol{\Gamma}^5(C_{na})=\boldsymbol{\Gamma}^{10}(C_{na})=\boldsymbol{\Gamma}'(C_{na}),$$

$$\boldsymbol{\Gamma}^5(IC_{na})=-\boldsymbol{\Gamma}^{10}(IC_{na})=\boldsymbol{\Gamma}'(C_{na}),$$

TABLE A.2

$$\boldsymbol{\Gamma}'(E)=\begin{pmatrix}1&0&0\\0&1&0\\0&0&1\end{pmatrix},\qquad \boldsymbol{\Gamma}'(C_{3\alpha})=\begin{pmatrix}0&-1&0\\0&0&1\\-1&0&0\end{pmatrix},$$

$$\boldsymbol{\Gamma}'(C_{3\beta})=\begin{pmatrix}0&1&0\\0&0&-1\\-1&0&0\end{pmatrix},\qquad \boldsymbol{\Gamma}'(C_{3\gamma})=\begin{pmatrix}0&-1&0\\0&0&-1\\1&0&0\end{pmatrix},$$

$$\boldsymbol{\Gamma}'(C_{3\delta})=\begin{pmatrix}0&1&0\\0&0&1\\1&0&0\end{pmatrix},\qquad \boldsymbol{\Gamma}'(C_{3\alpha}^{-1})=\begin{pmatrix}0&0&-1\\-1&0&0\\0&1&0\end{pmatrix},$$

$$\boldsymbol{\Gamma}'(C_{3\beta}^{-1})=\begin{pmatrix}0&0&-1\\1&0&0\\0&-1&0\end{pmatrix},\qquad \boldsymbol{\Gamma}'(C_{3\gamma}^{-1})=\begin{pmatrix}0&0&1\\-1&0&0\\0&-1&0\end{pmatrix},$$

$$\boldsymbol{\Gamma}'(C_{3\delta}^{-1})=\begin{pmatrix}0&0&1\\1&0&0\\0&1&0\end{pmatrix},\qquad \boldsymbol{\Gamma}'(C_{2x})=\begin{pmatrix}-1&0&0\\0&1&0\\0&0&-1\end{pmatrix},$$

$$\boldsymbol{\Gamma}'(C_{2y})=\begin{pmatrix}-1&0&0\\0&-1&0\\0&0&1\end{pmatrix},\qquad \boldsymbol{\Gamma}'(C_{2z})=\begin{pmatrix}1&0&0\\0&-1&0\\0&0&-1\end{pmatrix},$$

$$\boldsymbol{\Gamma}'(C_{4x})=\begin{pmatrix}0&0&1\\0&-1&0\\-1&0&0\end{pmatrix},\qquad \boldsymbol{\Gamma}'(C_{4y})=\begin{pmatrix}0&-1&0\\1&0&0\\0&0&-1\end{pmatrix},$$

$$\boldsymbol{\Gamma}'(C_{4z})=\begin{pmatrix}-1&0&0\\0&0&1\\0&-1&0\end{pmatrix},\qquad \boldsymbol{\Gamma}'(C_{4x}^{-1})=\begin{pmatrix}0&0&-1\\0&-1&0\\1&0&0\end{pmatrix},$$

$$\boldsymbol{\Gamma}'(C_{4y}^{-1})=\begin{pmatrix}0&1&0\\-1&0&0\\0&0&-1\end{pmatrix},\qquad \boldsymbol{\Gamma}'(C_{4z}^{-1})=\begin{pmatrix}-1&0&0\\0&0&-1\\0&1&0\end{pmatrix},$$

$$\boldsymbol{\Gamma}'(C_{2a})=\begin{pmatrix}1&0&0\\0&0&-1\\0&-1&0\end{pmatrix},\qquad \boldsymbol{\Gamma}'(C_{2b})=\begin{pmatrix}1&0&0\\0&0&1\\0&1&0\end{pmatrix},$$

$$\boldsymbol{\Gamma}'(C_{2c})=\begin{pmatrix}0&-1&0\\-1&0&0\\0&0&1\end{pmatrix},\qquad \boldsymbol{\Gamma}'(C_{2d})=\begin{pmatrix}0&1&0\\1&0&0\\0&0&1\end{pmatrix},$$

$$\boldsymbol{\Gamma}'(C_{2e})=\begin{pmatrix}0&0&-1\\0&1&0\\-1&0&0\end{pmatrix},\qquad \boldsymbol{\Gamma}'(C_{2f})=\begin{pmatrix}0&0&1\\0&1&0\\1&0&0\end{pmatrix},$$

TABLE A.3

$\boldsymbol{\Gamma}''(E)$	$= \boldsymbol{\Gamma}''(C_{2x})$	$= \boldsymbol{\Gamma}''(C_{2y})$	$= \boldsymbol{\Gamma}''(C_{2z})$	$= \begin{pmatrix} 1 & 0 \\ 0 & 1 \end{pmatrix}$
$\boldsymbol{\Gamma}''(C_{3\alpha})$	$= \boldsymbol{\Gamma}''(C_{3\beta})$	$= \boldsymbol{\Gamma}''(C_{3\gamma})$	$= \boldsymbol{\Gamma}''(C_{3\delta})$	$= \begin{pmatrix} -\frac{1}{2} & -\frac{1}{2}\sqrt{3} \\ \frac{1}{2}\sqrt{3} & -\frac{1}{2} \end{pmatrix}$
$\boldsymbol{\Gamma}''(C_{3\alpha}^{-1})$	$= \boldsymbol{\Gamma}''(C_{3\beta}^{-1})$	$= \boldsymbol{\Gamma}''(C_{3\gamma}^{-1})$	$= \boldsymbol{\Gamma}''(C_{3\delta}^{-1})$	$= \begin{pmatrix} -\frac{1}{2} & \frac{1}{2}\sqrt{3} \\ -\frac{1}{2}\sqrt{3} & -\frac{1}{2} \end{pmatrix}$
$\boldsymbol{\Gamma}''(C_{4x})$	$= \boldsymbol{\Gamma}''(C_{4x}^{-1})$	$= \boldsymbol{\Gamma}''(C_{2e})$	$= \boldsymbol{\Gamma}''(C_{2f})$	$= \begin{pmatrix} \frac{1}{2} & -\frac{1}{2}\sqrt{3} \\ -\frac{1}{2}\sqrt{3} & -\frac{1}{2} \end{pmatrix}$
$\boldsymbol{\Gamma}''(C_{4y})$	$= \boldsymbol{\Gamma}''(C_{4y}^{-1})$	$= \boldsymbol{\Gamma}''(C_{2c})$	$= \boldsymbol{\Gamma}''(C_{2d})$	$= \begin{pmatrix} \frac{1}{2} & \frac{1}{2}\sqrt{3} \\ \frac{1}{2}\sqrt{3} & -\frac{1}{2} \end{pmatrix}$
$\boldsymbol{\Gamma}''(C_{4z})$	$= \boldsymbol{\Gamma}''(C_{4z}^{-1})$	$= \boldsymbol{\Gamma}''(C_{2a})$	$= \boldsymbol{\Gamma}''(C_{2b})$	$= \begin{pmatrix} -1 & 0 \\ 0 & 1 \end{pmatrix}$

TABLE A.4

		$\mathscr{C}_1$	$\mathscr{C}_2$	$\mathscr{C}_3$	$\mathscr{C}_4$	$\mathscr{C}_5$	$\mathscr{C}_6$	$\mathscr{C}_7$	$\mathscr{C}_8$	$\mathscr{C}_9$	$\mathscr{C}_{10}$	$\mathscr{C}_{11}$	$\mathscr{C}_{12}$
Γ_1^+	$\boldsymbol{\Gamma}^1$	1	1	1	1	1	1	1	1	1	1	1	1
Γ_2^+	$\boldsymbol{\Gamma}^2$	1	1	1	1	−1	−1	1	1	1	1	−1	−1
Γ_3^+	$\boldsymbol{\Gamma}^3$	1	−1	1	−1	1	−1	1	−1	1	−1	1	−1
Γ_4^+	$\boldsymbol{\Gamma}^4$	1	−1	1	−1	−1	1	1	−1	1	−1	−1	1
Γ_5^+	$\boldsymbol{\Gamma}^5$	2	−1	−1	2	0	0	2	−1	−1	2	0	0
Γ_6^+	$\boldsymbol{\Gamma}^6$	2	1	−1	−2	0	0	2	1	−1	−2	0	0
Γ_1^-	$\boldsymbol{\Gamma}^7$	1	1	1	1	1	1	−1	−1	−1	−1	−1	−1
Γ_2^-	$\boldsymbol{\Gamma}^8$	1	1	1	1	−1	−1	−1	−1	−1	−1	1	1
Γ_3^-	$\boldsymbol{\Gamma}^9$	1	−1	1	−1	1	−1	−1	1	−1	1	−1	1
Γ_4^-	$\boldsymbol{\Gamma}^{10}$	1	−1	1	−1	−1	1	−1	1	−1	1	1	−1
Γ_5^-	$\boldsymbol{\Gamma}^{11}$	2	−1	−1	2	0	0	−2	1	1	−2	0	0
Γ_6^-	$\boldsymbol{\Gamma}^{12}$	2	1	−1	−2	0	0	−2	−1	1	2	0	0

where the matrices $\boldsymbol{\Gamma}'(C_{na})$, $\boldsymbol{\Gamma}''(C_{na})$ and $\boldsymbol{R}(C_{na})$ are given in tables A.2, A.3 and 1.5 respectively.

(2) D_{6h}

(a) Classes: $\mathscr{C}_1 = E$; $\mathscr{C}_2 = C_{6z}, C_{6z}^{-1}$; $\mathscr{C}_3 = C_{3z}, C_{3z}^{-1}$;

$\mathscr{C}_4 = C_{2z}$; $\mathscr{C}_5 = C_{2x}, C_{2A}, C_{2B}$; $\mathscr{C}_6 = C_{2y}, C_{2C}, C_{2D}$;
$\mathscr{C}_7 = I$; $\mathscr{C}_8 = IC_{6z}, IC_{6z}^{-1}$; $\mathscr{C}_9 = IC_{3z}, IC_{3z}^{-1}$; $\mathscr{C}_{10} = IC_{2z}$;
$\mathscr{C}_{11} = IC_{2x}, IC_{2A}, IC_{2B}$; $\mathscr{C}_{12} = IC_{2y}, IC_{2C}, IC_{2D}$.

(b) The character table is given in table A.4. The notation in the first column is that of HERRING [1942].

TABLE A.5

$$D'(E) = D'(C_{2z}) = \begin{pmatrix} 1 & 0 \\ 0 & 1 \end{pmatrix}, \quad D'(C_{6z}) = D'(C_{3z}^{-1}) = \begin{pmatrix} -\frac{1}{2} & -\frac{1}{2}\sqrt{3} \\ \frac{1}{2}\sqrt{3} & -\frac{1}{2} \end{pmatrix}$$

$$D'(C_{6z}^{-1}) = D'(C_{3z}) = \begin{pmatrix} -\frac{1}{2} & \frac{1}{2}\sqrt{3} \\ -\frac{1}{2}\sqrt{3} & -\frac{1}{2} \end{pmatrix}, \quad D'(C_{2x}) = D'(C_{2y}) = \begin{pmatrix} -1 & 0 \\ 0 & 1 \end{pmatrix}$$

$$D'(C_{2A}) = D'(C_{2C}) = \begin{pmatrix} \frac{1}{2} & -\frac{1}{2}\sqrt{3} \\ -\frac{1}{2}\sqrt{3} & -\frac{1}{2} \end{pmatrix}, \quad D'(C_{2B}) = D'(C_{2D}) = \begin{pmatrix} \frac{1}{2} & \frac{1}{2}\sqrt{3} \\ \frac{1}{2}\sqrt{3} & -\frac{1}{2} \end{pmatrix}$$

(c) Matrices for irreducible representations of more than one dimension: for any proper rotation C_{na} of D_{6h},

$\Gamma^5(C_{na}) = \Gamma^{11}(C_{na}) = D'(C_{na})$,
$\Gamma^5(IC_{na}) = -\Gamma^{11}(IC_{na}) = D'(C_{na})$,
$\Gamma^6(C_{na}) = \Gamma^{12}(C_{na}) = D''(C_{na})$,
$\Gamma^6(IC_{na}) = -\Gamma^{12}(IC_{na}) = D''(C_{na})$,

TABLE A.6

$$D''(E) = \begin{pmatrix} 1 & 0 \\ 0 & 1 \end{pmatrix}, \quad D''(C_{6z}) = \begin{pmatrix} \frac{1}{2} & \frac{1}{2}\sqrt{3} \\ -\frac{1}{2}\sqrt{3} & \frac{1}{2} \end{pmatrix},$$

$$D''(C_{6z}^{-1}) = \begin{pmatrix} \frac{1}{2} & -\frac{1}{2}\sqrt{3} \\ \frac{1}{2}\sqrt{3} & \frac{1}{2} \end{pmatrix}, \quad D''(C_{3z}) = \begin{pmatrix} -\frac{1}{2} & \frac{1}{2}\sqrt{3} \\ -\frac{1}{2}\sqrt{3} & -\frac{1}{2} \end{pmatrix},$$

$$D''(C_{3z}^{-1}) = \begin{pmatrix} -\frac{1}{2} & -\frac{1}{2}\sqrt{3} \\ \frac{1}{2}\sqrt{3} & -\frac{1}{2} \end{pmatrix}, \quad D''(C_{2z}) = \begin{pmatrix} -1 & 0 \\ 0 & -1 \end{pmatrix},$$

$$D''(C_{2x}) = \begin{pmatrix} 1 & 0 \\ 0 & -1 \end{pmatrix}, \quad D''(C_{2A}) = \begin{pmatrix} -\frac{1}{2} & \frac{1}{2}\sqrt{3} \\ \frac{1}{2}\sqrt{3} & \frac{1}{2} \end{pmatrix},$$

$$D''(C_{2B}) = \begin{pmatrix} -\frac{1}{2} & -\frac{1}{2}\sqrt{3} \\ -\frac{1}{2}\sqrt{3} & \frac{1}{2} \end{pmatrix}, \quad D''(C_{2y}) = \begin{pmatrix} -1 & 0 \\ 0 & 1 \end{pmatrix},$$

$$D''(C_{2C}) = \begin{pmatrix} \frac{1}{2} & -\frac{1}{2}\sqrt{3} \\ -\frac{1}{2}\sqrt{3} & -\frac{1}{2} \end{pmatrix}, \quad D''(C_{2D}) = \begin{pmatrix} \frac{1}{2} & \frac{1}{2}\sqrt{3} \\ \frac{1}{2}\sqrt{3} & -\frac{1}{2} \end{pmatrix},$$

where the matrices $\boldsymbol{D}'(C_{na})$ and $\boldsymbol{D}''(C_{na})$ are given in tables A.5 and A.6 respectively.

(3) T_d

(a) Classes [for $\mathscr{G}_o(k)$ of P]: $\mathscr{C}_1=E$; $\mathscr{C}_2=C_{3\alpha}, C_{3\beta}, C_{3\gamma}, C_{3\delta}, C_{3\alpha}^{-1}, C_{3\beta}^{-1}, C_{3\gamma}^{-1}, C_{3\delta}^{-1}$; $\mathscr{C}_3=C_{2x}, C_{2y}, C_{2z}$; $\mathscr{C}_4=IC_{4x}, IC_{4y}, IC_{4z}, IC_{4x}^{-1}, IC_{4y}^{-1}, IC_{4z}^{-1}$; $\mathscr{C}_5=IC_{2a}, IC_{2b}, IC_{2c}, IC_{2d}, IC_{2e}, IC_{2f}$.

TABLE A.7

			$\mathscr{C}_1$	$\mathscr{C}_2$	$\mathscr{C}_3$	$\mathscr{C}_4$	$\mathscr{C}_5$
P_1	Γ_1	$\boldsymbol{\Gamma}^1$	1	1	1	1	1
P_2	Γ_2	$\boldsymbol{\Gamma}^2$	1	1	1	−1	−1
P_3	Γ_{12}	$\boldsymbol{\Gamma}^3$	2	−1	2	0	0
P_4	Γ_{15}	$\boldsymbol{\Gamma}^4$	3	0	−1	−1	1
P_5	Γ_{25}	$\boldsymbol{\Gamma}^5$	3	0	−1	1	−1

(b) The character table is given in table A.7. The notations of the first and second columns are those of BOUCKAERT *et al.* [1936] and PARMENTER [1955] respectively.

(c) Matrices for irreducible representations of more than one dimension: for proper rotations C_{na} of T_d,
$\boldsymbol{\Gamma}^3(C_{na})=\boldsymbol{\Gamma}''(C_{na})$, $\boldsymbol{\Gamma}^4(C_{na})=\boldsymbol{R}(C_{na})$, $\boldsymbol{\Gamma}^5(C_{na})=\boldsymbol{\Gamma}'(C_{na})$,
and for the improper rotations IC_{na} of T_d,
$\boldsymbol{\Gamma}^3(IC_{na})=-\boldsymbol{\Gamma}''(C_{na})$, $\boldsymbol{\Gamma}^4(IC_{na})=-\boldsymbol{R}(C_{na})$,
$\boldsymbol{\Gamma}^5(IC_{na})=-\boldsymbol{\Gamma}'(C_{na})$,
where the matrices $\boldsymbol{\Gamma}'(C_{na})$, $\boldsymbol{\Gamma}''(C_{na})$ and $\boldsymbol{R}(C_{na})$ are given in tables A.2, A.3 and 1.5 respectively.

(4) O

(a) Classes: $\mathscr{C}_1=E$; $\mathscr{C}_2=C_{3\alpha}, C_{3\beta}, C_{3\gamma}, C_{3\delta}, C_{3\alpha}^{-1}, C_{3\beta}^{-1}, C_{3\gamma}^{-1}$,

$C_{3\delta}^{-1}$; $\mathscr{C}_3 = C_{2x}, C_{2y}, C_{2z}$; $\mathscr{C}_4 = C_{4x}, C_{4y}, C_{4z}, C_{4x}^{-1}, C_{4y}^{-1}, C_{4z}^{-1}$; $\mathscr{C}_5 = C_{2a}, C_{2b}, C_{2c}, C_{2d}, C_{2e}, C_{2f}$.

(b) The character table is given in table A.7.

(c) Matrices for irreducible representations of more than one dimension: for all rotations C_{na} of O,

$\boldsymbol{\Gamma}^3(C_{na}) = \boldsymbol{\Gamma}''(C_{na})$, $\boldsymbol{\Gamma}^4(C_{na}) = \boldsymbol{\Gamma}'(C_{na})$, $\boldsymbol{\Gamma}^5(C_{na}) = \boldsymbol{R}(C_{na})$,

where the matrices $\boldsymbol{\Gamma}'(C_{na})$, $\boldsymbol{\Gamma}''(C_{na})$ and $\boldsymbol{R}(C_{na})$ are given in tables A.2, A.3 and 1.5 respectively.

(5) T_h

(a) Classes: $\mathscr{C}_1 = E$; $\mathscr{C}_2 = C_{3\alpha}, C_{3\beta}, C_{3\gamma}, C_{3\delta}$; $\mathscr{C}_3 = C_{3\alpha}^{-1}, C_{3\beta}^{-1}, C_{3\gamma}^{-1}, C_{3\delta}^{-1}$; $\mathscr{C}_4 = C_{2x}, C_{2y}, C_{2z}$; $\mathscr{C}_5 = I$; $\mathscr{C}_6 = IC_{3\alpha}, IC_{3\beta}, IC_{3\gamma}, IC_{3\delta}$; $\mathscr{C}_7 = IC_{3\alpha}^{-1}, IC_{3\beta}^{-1}, IC_{3\gamma}^{-1}, IC_{3\delta}^{-1}$; $\mathscr{C}_8 = IC_{2x}, IC_{2y}, IC_{2z}$.

(b) The character table is given in table A.8.

(c) Matrices for irreducible representations of more than one dimension: for any proper rotation C_{na} of T_h,

TABLE A.8

	$\mathscr{C}_1$	$\mathscr{C}_2$	$\mathscr{C}_3$	$\mathscr{C}_4$	$\mathscr{C}_5$	$\mathscr{C}_6$	$\mathscr{C}_7$	$\mathscr{C}_8$
$\boldsymbol{\Gamma}^1$	1	1	1	1	1	1	1	1
$\boldsymbol{\Gamma}^2$	1	ϕ	ϕ^2	1	1	ϕ	ϕ^2	1
$\boldsymbol{\Gamma}^3$	1	ϕ^2	ϕ	1	1	ϕ^2	ϕ	1
$\boldsymbol{\Gamma}^4$	3	0	0	-1	3	0	0	-1
$\boldsymbol{\Gamma}^5$	1	1	1	1	-1	-1	-1	-1
$\boldsymbol{\Gamma}^6$	1	ϕ	ϕ^2	1	-1	$-\phi$	$-\phi^2$	-1
$\boldsymbol{\Gamma}^7$	1	ϕ^2	ϕ	1	-1	$-\phi^2$	$-\phi$	-1
$\boldsymbol{\Gamma}^8$	3	0	0	-1	-3	0	0	1

$\phi = \exp(\frac{2}{3}\pi i)$

TABLE A.9

		$\mathscr{C}_1$	$\mathscr{C}_2$	$\mathscr{C}_3$	$\mathscr{C}_4$	$\mathscr{C}_5$	$\mathscr{C}_6$	$\mathscr{C}_7$	$\mathscr{C}_8$	$\mathscr{C}_9$	$\mathscr{C}_{10}$
X_1, M_1	$\boldsymbol{\Gamma}^1$	1	1	1	1	1	1	1	1	1	1
X_2, M_2	$\boldsymbol{\Gamma}^2$	1	1	1	−1	−1	1	1	1	−1	−1
X_3, M_3	$\boldsymbol{\Gamma}^3$	1	−1	1	−1	1	1	−1	1	−1	1
X_4, M_4	$\boldsymbol{\Gamma}^4$	1	−1	1	1	−1	1	−1	1	1	−1
X_5, M_5	$\boldsymbol{\Gamma}^5$	2	0	−2	0	0	2	0	−2	0	0
$X_{1'}, M_{1'}$	$\boldsymbol{\Gamma}^6$	1	1	1	1	1	−1	−1	−1	−1	−1
$X_{2'}, M_{2'}$	$\boldsymbol{\Gamma}^7$	1	1	1	−1	−1	−1	−1	−1	1	1
$X_{3'}, M_{3'}$	$\boldsymbol{\Gamma}^8$	1	−1	1	−1	1	−1	1	−1	1	−1
$X_{4'}, M_{4'}$	$\boldsymbol{\Gamma}^9$	1	−1	1	1	−1	−1	1	−1	−1	1
$X_{5'}, M_{5'}$	$\boldsymbol{\Gamma}^{10}$	2	0	−2	0	0	−2	0	2	0	0

$\boldsymbol{\Gamma}^4(C_{na}) = \boldsymbol{\Gamma}^8(C_{na}) = \boldsymbol{\Gamma}'(C_{na})$, $\boldsymbol{\Gamma}^4(IC_{na}) = -\boldsymbol{\Gamma}^8(IC_{na}) = \boldsymbol{\Gamma}'(C_{na})$,
where the matrices $\boldsymbol{\Gamma}'(C_{na})$ are contained in table A.2.

(6) D_{4h}

(a) Classes [for $\mathscr{G}_o(\boldsymbol{k})$ of X]; $\mathscr{C}_1 = E$; $\mathscr{C}_2 = C_{2x}, C_{2y}$; $\mathscr{C}_3 = C_{2z}$; $\mathscr{C}_4 = C_{4z}, C_{4z}^{-1}$; $\mathscr{C}_5 = C_{2a}, C_{2b}$; $\mathscr{C}_6 = I$; $\mathscr{C}_7 = IC_{2x}, IC_{2y}$; $\mathscr{C}_8 = IC_{2z}$; $\mathscr{C}_9 = IC_{4z}, IC_{4z}^{-1}$; $\mathscr{C}_{10} = IC_{2a}, IC_{2b}$.

(b) The character table is given in table A.9. The notations in the first column are those of BOUCKAERT *et al.* [1936].

(c) Matrices for irreducible representations of more than one dimension: for any proper rotation C_{na} of D_{4h},
$\boldsymbol{\Gamma}^5(C_{na}) = \boldsymbol{\Gamma}^{10}(C_{na}) = \boldsymbol{D}(C_{na})$,
$\boldsymbol{\Gamma}^5(IC_{na}) = -\boldsymbol{\Gamma}^{10}(IC_{na}) = \boldsymbol{D}(C_{na})$,
where the matrices $\boldsymbol{D}(C_{na})$ are given in table A.10.

(7) D_{3h}

(a) Classes: $\mathscr{C}_1 = E$; $\mathscr{C}_2 = C_{3z}, C_{3z}^{-1}$; $\mathscr{C}_3 = C_{2x}, C_{2A}, C_{2B}$; $\mathscr{C}_4 = IC_{2z}$; $\mathscr{C}_5 = IC_{6z}, IC_{6z}^{-1}$; $\mathscr{C}_6 = IC_{2y}, IC_{2C}, IC_{2D}$.

TABLE A.10

$\boldsymbol{D}(E) = \begin{pmatrix} 1 & 0 \\ 0 & 1 \end{pmatrix}$, $\boldsymbol{D}(C_{2x}) = \begin{pmatrix} 1 & 0 \\ 0 & -1 \end{pmatrix}$, $\boldsymbol{D}(C_{2y}) = \begin{pmatrix} -1 & 0 \\ 0 & 1 \end{pmatrix}$

$\boldsymbol{D}(C_{2z}) = \begin{pmatrix} -1 & 0 \\ 0 & -1 \end{pmatrix}$, $\boldsymbol{D}(C_{4z}) = \begin{pmatrix} 0 & 1 \\ -1 & 0 \end{pmatrix}$, $\boldsymbol{D}(C_{4z}^{-1}) = \begin{pmatrix} 0 & -1 \\ 1 & 0 \end{pmatrix}$

$\boldsymbol{D}(C_{2a}) = \begin{pmatrix} 0 & 1 \\ 1 & 0 \end{pmatrix}$, $\boldsymbol{D}(C_{2b}) = \begin{pmatrix} 0 & -1 \\ -1 & 0 \end{pmatrix}$

(b) The character table is given in table A.11.

(c) Matrices for irreducible representations of more than one dimension: for any proper rotation C_{na} of D_{3h},
$\boldsymbol{\Gamma}^3(C_{na}) = \boldsymbol{\Gamma}^6(C_{na}) = \boldsymbol{D}'(C_{na})$,
and for any improper rotation IC_{na} of D_{3h},
$\boldsymbol{\Gamma}^3(IC_{na}) = -\boldsymbol{\Gamma}^6(IC_{na}) = \boldsymbol{D}'(C_{na})$,
where the matrices $\boldsymbol{D}'(C_{na})$ are contained in table A.5.

(8) D_{3d}

(a) Classes [for $\mathscr{G}_o(\boldsymbol{k})$ of L]: $\mathscr{C}_1 = E$; $\mathscr{C}_2 = C_{3\delta}, C_{3\delta}^{-1}$; $\mathscr{C}_3 = C_{2b}, C_{2d}, C_{2f}$; $\mathscr{C}_4 = I$; $\mathscr{C}_5 = IC_{3\delta}, IC_{3\delta}^{-1}$; $\mathscr{C}_6 = IC_{2b}, IC_{2d}, IC_{2f}$.

(b) The character table is given in table A.11. The notation in the first column is that of BOUCKAERT *et al.* [1936].

TABLE A.11

		$\mathscr{C}_1$	$\mathscr{C}_2$	$\mathscr{C}_3$	$\mathscr{C}_4$	$\mathscr{C}_5$	$\mathscr{C}_6$
L_1	$\boldsymbol{\Gamma}^1$	1	1	1	1	1	1
L_2	$\boldsymbol{\Gamma}^2$	1	1	−1	1	1	−1
L_3	$\boldsymbol{\Gamma}^3$	2	−1	0	2	−1	0
$L_{1'}$	$\boldsymbol{\Gamma}^4$	1	1	1	−1	−1	−1
$L_{2'}$	$\boldsymbol{\Gamma}^5$	1	1	−1	−1	−1	1
$L_{3'}$	$\boldsymbol{\Gamma}^6$	2	−1	0	−2	1	0

(c) Matrices for irreducible representations of more than one dimension: for any proper rotation C_{na} of D_{3d},

$$\boldsymbol{\Gamma}^3(C_{na})=\boldsymbol{\Gamma}^6(C_{na})=\boldsymbol{\Gamma}''(C_{na}),\quad \boldsymbol{\Gamma}^3(IC_{na})=-\boldsymbol{\Gamma}^6(IC_{na})=\boldsymbol{\Gamma}''(C_{na}),$$

where the matrices $\boldsymbol{\Gamma}''(C_{na})$ are contained in table A.3.

(9) C_{6v}

(a) Classes: $\mathscr{C}_1=E$; $\mathscr{C}_2=C_{3z}, C_{3z}^{-1}$; $\mathscr{C}_3=IC_{2x}, IC_{2A}, IC_{2B}$; $\mathscr{C}_4=C_{2z}$; $\mathscr{C}_5=C_{6z}, C_{6z}^{-1}$; $\mathscr{C}_6=IC_{2y}, IC_{2C}, IC_{2D}$.

(b) The character table is given in table A.11.

(c) Matrices for irreducible representations of more than one dimension: for any proper rotation C_{na} of C_{6v},

$$\boldsymbol{\Gamma}^6(C_{na})=\boldsymbol{D}''(C_{na}),\quad \boldsymbol{\Gamma}^3(C_{na})=\boldsymbol{D}'(C_{na}),$$

and for any improper rotation IC_{na} of C_{6v},

$$\boldsymbol{\Gamma}^6(IC_{na})=\boldsymbol{D}''(C_{na}),\quad \boldsymbol{\Gamma}^3(IC_{na})=\boldsymbol{D}'(C_{na}),$$

TABLE A.12

	$\mathscr{C}_1$	$\mathscr{C}_2$	$\mathscr{C}_3$	$\mathscr{C}_4$	$\mathscr{C}_5$	$\mathscr{C}_6$	$\mathscr{C}_7$	$\mathscr{C}_8$	$\mathscr{C}_9$	$\mathscr{C}_{10}$	$\mathscr{C}_{11}$	$\mathscr{C}_{12}$
$\boldsymbol{\Gamma}^1$	1	1	1	1	1	1	1	1	1	1	1	1
$\boldsymbol{\Gamma}^2$	1	ω	ω^2	-1	$-\omega$	$-\omega^2$	1	ω	ω^2	-1	$-\omega$	$-\omega^2$
$\boldsymbol{\Gamma}^3$	1	ω^2	$-\omega$	1	ω^2	$-\omega$	1	ω^2	$-\omega$	1	ω^2	$-\omega$
$\boldsymbol{\Gamma}^4$	1	-1	1	-1	1	-1	1	-1	1	-1	1	-1
$\boldsymbol{\Gamma}^5$	1	$-\omega$	ω^2	1	$-\omega$	ω^2	1	$-\omega$	ω^2	1	$-\omega$	ω^2
$\boldsymbol{\Gamma}^6$	1	$-\omega^2$	$-\omega$	-1	ω^2	ω	1	$-\omega^2$	$-\omega$	-1	ω^2	ω
$\boldsymbol{\Gamma}^7$	1	1	1	1	1	1	-1	-1	-1	-1	-1	-1
$\boldsymbol{\Gamma}^8$	1	ω	ω^2	-1	$-\omega$	$-\omega^2$	-1	$-\omega$	$-\omega^2$	1	ω	ω^2
$\boldsymbol{\Gamma}^9$	1	ω^2	$-\omega$	1	ω^2	$-\omega$	-1	$-\omega^2$	ω	-1	$-\omega^2$	ω
$\boldsymbol{\Gamma}^{10}$	1	-1	1	-1	1	-1	-1	1	-1	1	-1	1
$\boldsymbol{\Gamma}^{11}$	1	$-\omega$	ω^2	1	$-\omega$	ω^2	-1	ω	$-\omega^2$	-1	ω	$-\omega^2$
$\boldsymbol{\Gamma}^{12}$	1	$-\omega^2$	$-\omega$	-1	ω^2	ω	-1	ω^2	ω	1	$-\omega^2$	$-\omega$

$\omega=\exp(\frac{1}{3}\pi\mathrm{i})$

where the matrices $\boldsymbol{D}'(C_{na})$ and $\boldsymbol{D}''(C_{na})$ are contained in tables A.5 and A.6 respectively.

(10) C_{6h}

(a) Classes: $\mathscr{C}_1=E$; $\mathscr{C}_2=C_{6z}$; $\mathscr{C}_3=C_{3z}$; $\mathscr{C}_4=C_{2z}$; $\mathscr{C}_5=C_{3z}^{-1}$; $\mathscr{C}_6=C_{6z}^{-1}$; $\mathscr{C}_7=I$; $\mathscr{C}_8=IC_{6z}$; $\mathscr{C}_9=IC_{3z}$; $\mathscr{C}_{10}=IC_{2z}$; $\mathscr{C}_{11}=IC_{3z}^{-1}$; $\mathscr{C}_{12}=IC_{6z}^{-1}$.

(b) The character table is given in table A.12.

(11) D_6

(a) Classes: $\mathscr{C}_1=E$; $\mathscr{C}_2=C_{3z}, C_{3z}^{-1}$; $\mathscr{C}_3=C_{2x}, C_{2A}, C_{2B}$; $\mathscr{C}_4=C_{2z}$; $\mathscr{C}_5=C_{6z}, C_{6z}^{-1}$; $\mathscr{C}_6=C_{2y}, C_{2C}, C_{2D}$.

(b) The character table is given in table A.11.

(c) Matrices for irreducible representations of more than one dimension: for any rotation C_{na} of D_6,

$\boldsymbol{\Gamma}^6(C_{na})=\boldsymbol{D}''(C_{na})$, $\boldsymbol{\Gamma}^3(C_{na})=\boldsymbol{D}'(C_{na})$,

where the matrices $\boldsymbol{D}'(C_{na})$ and $\boldsymbol{D}''(C_{na})$ are given in tables A.5 and A.6 respectively.

(12) T

(a) Classes: $\mathscr{C}_1=E$; $\mathscr{C}_2=C_{3\alpha}, C_{3\beta}, C_{3\gamma}, C_{3\delta}$; $\mathscr{C}_3=C_{3\alpha}^{-1}, C_{3\beta}^{-1}, C_{3\gamma}^{-1}, C_{3\delta}^{-1}$; $\mathscr{C}_4=C_{2x}, C_{2y}, C_{2z}$.

(b) The character table is given in table A.13.

(c) Matrices for irreducible representations of more than one

TABLE A.13

	$\mathscr{C}_1$	$\mathscr{C}_2$	$\mathscr{C}_3$	$\mathscr{C}_4$
$\boldsymbol{\Gamma}^1$	1	1	1	1
$\boldsymbol{\Gamma}^2$	1	ϕ	ϕ^2	1
$\boldsymbol{\Gamma}^3$	1	ϕ^2	ϕ	1
$\boldsymbol{\Gamma}^4$	3	0	0	-1

$\phi=\exp(\frac{2}{3}\pi i)$

dimension: for any rotation C_{na} of T,
$\boldsymbol{\Gamma}^4(C_{na})=\boldsymbol{\Gamma}'(C_{na})$,
where the matrices $\boldsymbol{\Gamma}'(C_{na})$ are contained in table A.2.

(13) D_{2h} or V_h

(a) Classes [for $\mathscr{G}_o(\boldsymbol{k})$ of N]: $\mathscr{C}_1=E$; $\mathscr{C}_2=C_{2x}$; $\mathscr{C}_3=C_{2e}$; $\mathscr{C}_4=C_{2f}$; $\mathscr{C}_5=I$; $\mathscr{C}_6=IC_{2x}$; $\mathscr{C}_7=IC_{2e}$; $\mathscr{C}_8=IC_{2f}$.

(b) The character table is given in table A.14. The notation in the first column is that of BOUCKAERT *et al.* [1936].

TABLE A.14

		$\mathscr{C}_1$	$\mathscr{C}_2$	$\mathscr{C}_3$	$\mathscr{C}_4$	$\mathscr{C}_5$	$\mathscr{C}_6$	$\mathscr{C}_7$	$\mathscr{C}_8$
N_1	$\boldsymbol{\Gamma}^1$	1	1	1	1	1	1	1	1
N_2	$\boldsymbol{\Gamma}^2$	1	−1	1	−1	1	−1	1	−1
N_3	$\boldsymbol{\Gamma}^3$	1	−1	−1	1	1	−1	−1	1
N_4	$\boldsymbol{\Gamma}^4$	1	1	−1	−1	1	1	−1	−1
$N_{2'}$	$\boldsymbol{\Gamma}^5$	1	1	1	1	−1	−1	−1	−1
$N_{1'}$	$\boldsymbol{\Gamma}^6$	1	−1	1	−1	−1	1	−1	1
$N_{4'}$	$\boldsymbol{\Gamma}^7$	1	−1	−1	1	−1	1	1	−1
$N_{3'}$	$\boldsymbol{\Gamma}^8$	1	1	−1	−1	−1	−1	1	1

(14) C_{4v}

(a) Classes [for $\mathscr{G}_o(\boldsymbol{k})$ of Δ]: $\mathscr{C}_1=E$; $\mathscr{C}_2=C_{2z}$; $\mathscr{C}_3=C_{4z}$, C_{4z}^{-1}; $\mathscr{C}_4=IC_{2x}$, IC_{2y}; $\mathscr{C}_5=IC_{2a}$, IC_{2b}.

(b) The character table is given in table A.15. The notations in the third column are those of BOUCKAERT *et al.* [1936].

(c) Matrices for irreducible representations of more than one dimension: for any proper rotation C_{na} of C_{4v},
$\boldsymbol{\Gamma}^5(C_{na})=\boldsymbol{D}(C_{na})$,

TABLE A.15

				$\mathscr{C}_1$	$\mathscr{C}_2$	$\mathscr{C}_3$	$\mathscr{C}_4$	$\mathscr{C}_5$
X_1	W_1	Δ_1, T_1	Γ^1	1	1	1	1	1
X_4	$W_{2'}$	Δ_2, T_2	Γ^2	1	1	-1	1	-1
X_3	W_2	$\Delta_{1'}, T_{1'}$	Γ^3	1	1	1	-1	-1
X_2	$W_{1'}$	$\Delta_{2'}, T_{2'}$	Γ^4	1	1	-1	-1	1
X_5	W_3	Δ_5, T_5	Γ^5	2	-2	0	0	0

and for any improper rotation IC_{na} of C_{4v},
$\Gamma^5(IC_{na}) = -D(C_{na})$,
where the matrices $D(C_{na})$ are contained in table A.10.

(15) D_4

(a) Classes: $\mathscr{C}_1 = E$; $\mathscr{C}_2 = C_{2y}$; $\mathscr{C}_3 = C_{4y}, C_{4y}^{-1}$; $\mathscr{C}_4 = C_{2x}, C_{2z}$; $\mathscr{C}_5 = C_{2c}, C_{2d}$.

(b) The character table is given in table A.15.

(c) Matrices for irreducible representations of more than one dimension: for any rotation C_{na} of D_4,
$\Gamma^5(C_{na}) = S(C_{na})$,
where the matrices $S(C_{na})$ are given in table A.16.

(16) D_{2d} or V_d

(a) Classes [for $\mathscr{G}_o(k)$ of W]: $\mathscr{C}_1 = E$; $\mathscr{C}_2 = C_{2y}$; $\mathscr{C}_3 = IC_{4y}, IC_{4y}^{-1}$; $\mathscr{C}_4 = IC_{2x}, IC_{2z}$; $\mathscr{C}_5 = C_{2c}, C_{2d}$.

(b) The character table is given in table A.15. The notations in the first and second columns are those of PARMENTER [1955] and BOUCKAERT *et al.* [1936] respectively.

(c) Matrices for irreducible representations of more than one dimension: for any proper rotation C_{na} of D_{2d},
$\Gamma^5(C_{na}) = S(C_{na})$,
and for any improper rotation IC_{na} of D_{2d},

TABLE A.16

$$S(E) = \begin{pmatrix} 1 & 0 \\ 0 & 1 \end{pmatrix}, \quad S(C_{2y}) = \begin{pmatrix} -1 & 0 \\ 0 & -1 \end{pmatrix}, \quad S(C_{2c}) = \begin{pmatrix} 0 & 1 \\ 1 & 0 \end{pmatrix}$$

$$S(C_{2d}) = \begin{pmatrix} 0 & -1 \\ -1 & 0 \end{pmatrix}, \quad S(C_{4y}) = \begin{pmatrix} 0 & -1 \\ 1 & 0 \end{pmatrix}, \quad S(C_{4y}^{-1}) = \begin{pmatrix} 0 & 1 \\ -1 & 0 \end{pmatrix}$$

$$S(C_{2x}) = \begin{pmatrix} 1 & 0 \\ 0 & -1 \end{pmatrix}, \quad S(C_{2z}) = \begin{pmatrix} -1 & 0 \\ 0 & 1 \end{pmatrix}$$

$\Gamma^5(IC_{na}) = -S(C_{na})$,
where the matrices $S(C_{na})$ are given in table A.16.

(17) C_{4h}
(a) Classes: $\mathscr{C}_1 = E$; $\mathscr{C}_2 = C_{4z}$; $\mathscr{C}_3 = C_{2z}$; $\mathscr{C}_4 = C_{4z}^{-1}$;
$C_5 = I$; $\mathscr{C}_6 = IC_{4z}$; $\mathscr{C}_7 = IC_{2z}$; $\mathscr{C}_8 = IC_{4z}^{-1}$.
(b) The character table is given in table A.17.

(18) C_{3h}
(a) Classes: $\mathscr{C}_1 = E$; $\mathscr{C}_2 = IC_{6z}$; $\mathscr{C}_3 = C_{3z}$; $\mathscr{C}_4 = IC_{2z}$;
$\mathscr{C}_5 = C_{3z}^{-1}$; $\mathscr{C}_6 = IC_{6z}^{-1}$.
(b) The character table is given in table A.18.

TABLE A.17

	$\mathscr{C}_1$	$\mathscr{C}_2$	$\mathscr{C}_3$	$\mathscr{C}_4$	$\mathscr{C}_5$	$\mathscr{C}_6$	$\mathscr{C}_7$	$\mathscr{C}_8$
Γ^1	1	1	1	1	1	1	1	1
Γ^2	1	−1	1	−1	1	−1	1	−1
Γ^3	1	i	−1	−i	1	i	−1	−i
Γ^4	1	−i	−1	i	1	−i	−1	i
Γ^5	1	1	1	1	−1	−1	−1	−1
Γ^6	1	−1	1	−1	−1	1	−1	1
Γ^7	1	i	−1	−i	−1	−i	1	i
Γ^8	1	−i	−1	i	−1	i	1	−i

TABLE A.18

	$\mathscr{C}_1$	$\mathscr{C}_2$	$\mathscr{C}_3$	$\mathscr{C}_4$	$\mathscr{C}_5$	$\mathscr{C}_6$
$\boldsymbol{\Gamma}^1$	1	1	1	1	1	1
$\boldsymbol{\Gamma}^2$	1	ω	ω^2	-1	$-\omega$	$-\omega^2$
$\boldsymbol{\Gamma}^3$	1	ω^2	$-\omega$	1	ω^2	$-\omega$
$\boldsymbol{\Gamma}^4$	1	-1	1	-1	1	-1
$\boldsymbol{\Gamma}^5$	1	$-\omega$	ω^2	1	$-\omega$	ω^2
$\boldsymbol{\Gamma}^6$	1	$-\omega^2$	$-\omega$	-1	ω^2	ω

$\omega = \exp(\frac{1}{3}\pi i)$

(19) C_{3v}

(a) Classes [for $\mathscr{G}_o(\boldsymbol{k})$ of Λ]: $\mathscr{C}_1 = E$; $\mathscr{C}_2 = C_{3\delta}, C_{3\delta}^{-1}$; $\mathscr{C}_3 = IC_{2b}, IC_{2d}, IC_{2f}$.
Classes [for $\mathscr{G}_o(\boldsymbol{k})$ of F]: $\mathscr{C}_1 = E$; $\mathscr{C}_2 = C_{3\alpha}, C_{3\alpha}^{-1}$; $\mathscr{C}_3 = IC_{2b}, IC_{2c}, IC_{2e}$.

(b) The character table is given in table A.19. The notations in the first column are those of BOUCKAERT *et al.* [1936].

(c) Matrices for irreducible representations of more than one dimension: for any proper rotation C_{na} of C_{3v},
$\boldsymbol{\Gamma}^3(C_{na}) = \boldsymbol{\Gamma}''(C_{na})$,
and for any improper rotation IC_{na} of C_{3v},
$\boldsymbol{\Gamma}^3(IC_{na}) = \boldsymbol{\Gamma}''(C_{na})$,
where the matrices $\boldsymbol{\Gamma}''(C_{na})$ are contained in table A.3.

(20) D_3

(a) Classes: $\mathscr{C}_1 = E$; $\mathscr{C}_2 = C_{3z}, C_{3z}^{-1}$; $\mathscr{C}_3 = C_{2x}, C_{2A}, C_{2B}$.

(b) The character table is given in table A.19.

(c) Matrices for irreducible representations of more than one dimension: for any rotation C_{na} of D_3,
$\boldsymbol{\Gamma}^3(C_{na}) = \boldsymbol{D}'(C_{na})$,
where the matrices $\boldsymbol{D}'(C_{na})$ are contained in table A.5.

TABLE A.19

		$\mathscr{C}_1$	$\mathscr{C}_2$	$\mathscr{C}_3$
F_1, Λ_1	$\boldsymbol{\Gamma}^1$	1	1	1
F_2, Λ_2	$\boldsymbol{\Gamma}^2$	1	1	-1
F_3, Λ_3	$\boldsymbol{\Gamma}^3$	2	-1	0

(21) C_{3i} or S_6

(a) Classes: $\mathscr{C}_1=E$; $\mathscr{C}_2=IC_{3z}^{-1}$; $\mathscr{C}_3=C_{3z}$; $\mathscr{C}_4=I$; $\mathscr{C}_5=C_{3z}^{-1}$; $\mathscr{C}_6=IC_{3z}$.

(b) The character table is given in table A.18.

(22) C_6

(a) Classes: $\mathscr{C}_1=E$; $\mathscr{C}_2=C_{6z}$; $\mathscr{C}_3=C_{3z}$; $\mathscr{C}_4=C_{2z}$; $\mathscr{C}_5=C_{3z}^{-1}$; $\mathscr{C}_6=C_{6z}^{-1}$.

(b) The character table is given in table A.18.

(23) C_{2v}

(a) Classes [for $\mathscr{G}_o(\boldsymbol{k})$ for Σ and K]: $\mathscr{C}_1=E$; $\mathscr{C}_2=C_{2e}$; $\mathscr{C}_3=IC_{2x}$; $\mathscr{C}_4=IC_{2f}$.
Classes [for $\mathscr{G}_o(\boldsymbol{k})$ for D]: $\mathscr{C}_1=E$; $\mathscr{C}_2=C_{2x}$; $\mathscr{C}_3=IC_{2e}$; $\mathscr{C}_4=IC_{2f}$.
Classes [for $\mathscr{G}_o(\boldsymbol{k})$ for U and S]: $\mathscr{C}_1=E$; $\mathscr{C}_2=C_{2a}$; $\mathscr{C}_3=IC_{2z}$; $\mathscr{C}_4=IC_{2b}$.
Classes [for $\mathscr{G}_o(\boldsymbol{k})$ for Z]: $\mathscr{C}_1=E$; $\mathscr{C}_2=C_{2y}$; $\mathscr{C}_3=IC_{2z}$; $\mathscr{C}_4=IC_{2x}$.
Classes [for $\mathscr{G}_o(\boldsymbol{k})$ for G]: $\mathscr{C}_1=E$; $\mathscr{C}_2=C_{2f}$; $\mathscr{C}_3=IC_{2x}$; $\mathscr{C}_4=IC_{2e}$.

(b) The character table is given in table A.20. The notations in the first and second columns are those of PARMENTER [1955] and BOUCKAERT *et al.* [1936].

(24) C_{2h}

(a) Classes: $\mathscr{C}_1=E$; $\mathscr{C}_2=C_{2z}$; $\mathscr{C}_3=I$; $\mathscr{C}_4=IC_{2z}$.

TABLE A.20

			$\mathscr{C}_1$	$\mathscr{C}_2$	$\mathscr{C}_3$	$\mathscr{C}_4$
Δ_1	Σ_1 S_1 Z_1 G_1 K_1 U_1 D_1	Γ^1	1	1	1	1
Δ_2	Σ_2 S_2 Z_2 G_2 K_2 U_2 D_2	Γ^2	1	1	−1	−1
Δ_3	Σ_3 S_3 Z_3 G_3 K_3 U_3 D_3	Γ^3	1	−1	−1	1
Δ_4	Σ_4 S_4 Z_4 G_4 K_4 U_4 D_4	Γ^4	1	−1	1	−1

(b) The character table is given in table A.20.

(25) D_2 or V

(a) Classes: $\mathscr{C}_1=E$; $\mathscr{C}_2=C_{2x}$; $\mathscr{C}_3=C_{2y}$; $\mathscr{C}_4=C_{2z}$.

(b) The character table is given in table A.20.

(26) C_4

(a) Classes: $\mathscr{C}_1=E$; $\mathscr{C}_2=C_{4z}$; $\mathscr{C}_3=C_{2z}$; $\mathscr{C}_4=C_{4z}^{-1}$.

(b) The character table is given in table A.21.

(27) S_4

(a) Classes: $\mathscr{C}_1=E$; $\mathscr{C}_2=IC_{4y}$; $\mathscr{C}_3=C_{2y}$; $\mathscr{C}_4=IC_{4y}^{-1}$.

(b) The character table is given in table A.21. The notation in the first column is that of PARMENTER [1955].

TABLE A.21

		$\mathscr{C}_1$	$\mathscr{C}_2$	$\mathscr{C}_3$	$\mathscr{C}_4$
W_1	Γ^1	1	1	1	1
W_3	Γ^2	1	−1	1	−1
W_4	Γ^3	1	i	−1	−i
W_2	Γ^4	1	−i	−1	i

TABLE A.22

	$\mathscr{C}_1$	$\mathscr{C}_2$	$\mathscr{C}_3$
$\boldsymbol{\Gamma}^1$	1	1	1
$\boldsymbol{\Gamma}^2$	1	ϕ	ϕ^2
$\boldsymbol{\Gamma}^3$	1	ϕ^2	ϕ

$\phi = \exp(\frac{2}{3}\pi \mathrm{i})$

(28) C_3

(a) Classes: $\mathscr{C}_1 = E$; $\mathscr{C}_2 = C_{3z}$; $\mathscr{C}_3 = C_{3z}^{-1}$.

(b) The character table is given in table A.22.

(29) C_s or C_{1h}

(a) Classes: $\mathscr{C}_1 = E$; $\mathscr{C}_2 = IC_{2z}$.

(b) The character table is given in table A.23.

(30) C_2

(a) Classes [for $\mathscr{G}_o(\boldsymbol{k})$ of Q]: $\mathscr{C}_1 = E$; $\mathscr{C}_2 = C_{2d}$.

(b) The character table is given in table A.23. The notation in the first column is that of Bouckaert *et al.* [1936].

(31) C_i

(a) Classes: $\mathscr{C}_1 = E$; $\mathscr{C}_2 = I$.

(b) The character table is given in table A.23.

(32) C_1

(a) This group consists of E alone.

(b) $\chi(E) = 1$.

TABLE A.23

		$\mathscr{C}_1$	$\mathscr{C}_2$
Q_1	$\boldsymbol{\Gamma}^1$	1	1
Q_2	$\boldsymbol{\Gamma}^2$	1	-1

APPENDIX 2

THE DOUBLE CRYSTALLOGRAPHIC POINT GROUPS

The following is a list of the double crystallographic point groups, arranged in the same order as the single groups of appendix 1. For each group the following details are given.

(a) The group elements. These are arranged in classes (as defined in ch. 1 § 2.4). For brevity the following conventions have been used.

(1) The generalized rotations $[C_{ni} \mid 0]$ and $[\bar{C}_{ni} \mid 0]$ defined in ch. 8 § 3 are written simply as C_{ni} and $\bar{C}_{ni}$, with a similar convention for improper rotations. The product of two such rotations is defined in ch. 8 § 3, and, as noted there, it is not necessarily the same as the product of the corresponding elements of the single group. The notation for the corresponding elements of the single groups are given in ch. 1 §§ 3.2, 3.3.

(2) If the class $\mathscr{C}_i$ consists entirely of a set of 'unbarred' elements, and the class $\mathscr{C}_j$ consists of the corresponding 'barred' elements, then $\mathscr{C}_j$ is said to be equal to $\bar{\mathscr{C}}_i$. For example, if $\mathscr{C}_2 = C_{3\alpha}, C_{3\alpha}^{-1}$, and $\mathscr{C}_5 = \bar{C}_{3\alpha}, \bar{C}_{3\alpha}^{-1}$, then $\mathscr{C}_5 = \bar{\mathscr{C}}_2$.

(b) The character systems of the extra representations (as defined in ch. 2 § 7 and ch. 8 § 6). If $\chi(\bar{\mathscr{C}}_i)$ denotes the character of the class $\bar{\mathscr{C}}_i$ in an extra irreducible representation, then $\chi(\bar{\mathscr{C}}_i) = -\chi(\mathscr{C}_i)$. The tables have been shortened by omitting the characters of the classes $\bar{\mathscr{C}}_i$. If the irreducible representations have been assigned labels that are in very common use in the solid-state literature, this is indicated, usually in the left-hand column. Other less commonly used notations are defined in the references given in appendix 3.

For a point group that is isomorphic to a group $\mathscr{G}_o^D(\boldsymbol{k})$, as defined in ch. 8 § 8.2, for some point $\boldsymbol{k}$ that is a symmetry point or is on a symmetry axis of the body-centred cubic space group O_h^9 or the face-centred cubic space group O_h^5, the group elements are chosen to be those of $\mathscr{G}_o^D(\boldsymbol{k})$. The corresponding $\boldsymbol{k}$-vector is as defined in tables 5.1 and 5.2. As described in ch. 5 § 3, it is possible for two or more points in different stars to have point groups $\mathscr{G}_o(\boldsymbol{k})$, and hence $\mathscr{G}_o^D(\boldsymbol{k})$, that are isomorphic. The group elements for each $\mathscr{G}_o^D(\boldsymbol{k})$ are listed when this occurs.

Explicit matrices for the extra irreducible representations of O_h^D, T_d^D, O^D, D_{4h}^D, D_{3d}^D, C_{4v}^D, C_{3v}^D and C_{2v}^D have been given by ONODERA and OKAZAKI [1966].

(1) O_h^D

(a) Classes [for $\mathscr{G}_o^D(\boldsymbol{k})$ of Γ and H]: $\mathscr{C}_1=E$; $\mathscr{C}_2=C_{3\alpha}, C_{3\beta}, C_{3\gamma}, C_{3\delta}, C_{3\alpha}^{-1}, C_{3\beta}^{-1}, C_{3\gamma}^{-1}, C_{3\delta}^{-1}$; $\mathscr{C}_3=C_{2x}, C_{2y}, C_{2z}, \bar{C}_{2x}, \bar{C}_{2y}, \bar{C}_{2z}$; $\mathscr{C}_4=C_{4x}, C_{4y}, C_{4z}, C_{4x}^{-1}, C_{4y}^{-1}, C_{4z}^{-1}$; $\mathscr{C}_5=C_{2a}, C_{2b}, C_{2c}, C_{2d}, C_{2e}, C_{2f}, \bar{C}_{2a}, \bar{C}_{2b}, \bar{C}_{2c}, \bar{C}_{2d}, \bar{C}_{2e}, \bar{C}_{2f}$; $\mathscr{C}_6=I$; $\mathscr{C}_7=IC_{3\alpha}, IC_{3\beta}, IC_{3\gamma}, IC_{3\delta}, IC_{3\alpha}^{-1}, IC_{3\beta}^{-1}, IC_{3\gamma}^{-1}, IC_{3\delta}^{-1}$; $\mathscr{C}_8=IC_{2x}, IC_{2y}, IC_{2z}, \overline{IC}_{2x}, \overline{IC}_{2y}, \overline{IC}_{2z}$; $\mathscr{C}_9=IC_{4x}, IC_{4y}, IC_{4z}, IC_{4x}^{-1}, IC_{4y}^{-1}, IC_{4z}^{-1}$; $\mathscr{C}_{10}=IC_{2a}, IC_{2b}, IC_{2c}, IC_{2d}, IC_{2e}, IC_{2f}, \overline{IC}_{2a}, \overline{IC}_{2b}, \overline{IC}_{2c}, \overline{IC}_{2d}, \overline{IC}_{2e}, \overline{IC}_{2f}$; $\mathscr{C}_{11}=\bar{\mathscr{C}}_1$; $\mathscr{C}_{12}=\bar{\mathscr{C}}_2$; $\mathscr{C}_{13}=\bar{\mathscr{C}}_4$; $\mathscr{C}_{14}=\bar{\mathscr{C}}_6$; $\mathscr{C}_{15}=\bar{\mathscr{C}}_7$; $\mathscr{C}_{16}=\bar{\mathscr{C}}_9$.

(b) The character table is given in table A.25. The notations in the first row are those of ELLIOTT [1954].

(2) D_{6h}^D

(a) Classes: $\mathscr{C}_1=E$; $\mathscr{C}_2=C_{6z}, C_{6z}^{-1}$; $\mathscr{C}_3=C_{3z}, C_{3z}^{-1}$; $\mathscr{C}_4=C_{2z}, \bar{C}_{2z}$; $\mathscr{C}_5=C_{2x}, C_{2A}, C_{2B}, \bar{C}_{2x}, \bar{C}_{2A}, \bar{C}_{2B}$; $\mathscr{C}_6=C_{2y}, C_{2C}, C_{2D}, \bar{C}_{2y}, \bar{C}_{2C}, \bar{C}_{2D}$; $\mathscr{C}_7=I$; $\mathscr{C}_8=IC_{6z}, IC_{6z}^{-1}$; $\mathscr{C}_9=IC_{3z}, IC_{3z}^{-1}$; $\mathscr{C}_{10}=IC_{2z}, \overline{IC}_{2z}$; $\mathscr{C}_{11}=IC_{2x}, IC_{2A}, IC_{2B}, \overline{IC}_{2x}, \overline{IC}_{2A}, \overline{IC}_{2B}$; $\mathscr{C}_{12}=IC_{2y}, IC_{2C}, IC_{2D}, \overline{IC}_{2y}, \overline{IC}_{2C}, \overline{IC}_{2D}$; $\mathscr{C}_{13}=\bar{\mathscr{C}}_1$; $\mathscr{C}_{14}=\bar{\mathscr{C}}_2$; $\mathscr{C}_{15}=\bar{\mathscr{C}}_3$; $\mathscr{C}_{16}=\bar{\mathscr{C}}_7$; $\mathscr{C}_{17}=\bar{\mathscr{C}}_8$; $\mathscr{C}_{18}=\bar{\mathscr{C}}_9$.

TABLE A.25

	Γ_6^+, H_6^+	Γ_7^+, H_7^+	Γ_8^+, H_8^+	Γ_6^-, H_6^-	Γ_7^-, H_7^-	Γ_8^-, H_8^-
$\mathscr{C}_1$	2	2	4	2	2	4
$\mathscr{C}_2$	1	1	-1	1	1	-1
$\mathscr{C}_3$	0	0	0	0	0	0
$\mathscr{C}_4$	$\sqrt{2}$	$-\sqrt{2}$	0	$\sqrt{2}$	$-\sqrt{2}$	0
$\mathscr{C}_5$	0	0	0	0	0	0
$\mathscr{C}_6$	2	2	4	-2	-2	-4
$\mathscr{C}_7$	1	1	-1	-1	-1	1
$\mathscr{C}_8$	0	0	0	0	0	0
$\mathscr{C}_9$	$\sqrt{2}$	$-\sqrt{2}$	0	$-\sqrt{2}$	$\sqrt{2}$	0
$\mathscr{C}_{10}$	0	0	0	0	0	0

TABLE A.26

	Γ_7^+	Γ_8^+	Γ_9^+	Γ_7^-	Γ_8^-	Γ_9^-
$\mathscr{C}_1$	2	2	2	2	2	2
$\mathscr{C}_2$	$\sqrt{3}$	$-\sqrt{3}$	0	$\sqrt{3}$	$-\sqrt{3}$	0
$\mathscr{C}_3$	1	1	-2	1	1	-2
$\mathscr{C}_4$	0	0	0	0	0	0
$\mathscr{C}_5$	0	0	0	0	0	0
$\mathscr{C}_6$	0	0	0	0	0	0
$\mathscr{C}_7$	2	2	2	-2	-2	-2
$\mathscr{C}_8$	$\sqrt{3}$	$-\sqrt{3}$	0	$-\sqrt{3}$	$\sqrt{3}$	0
$\mathscr{C}_9$	1	1	-2	-1	-1	2
$\mathscr{C}_{10}$	0	0	0	0	0	0
$\mathscr{C}_{11}$	0	0	0	0	0	0
$\mathscr{C}_{12}$	0	0	0	0	0	0

(b) The character table is given in table A.26. The notation in the first row is that of ELLIOTT [1954].

(3) T_d^D

(a) Classes [for $\mathscr{G}_o^D(\boldsymbol{k})$ of P]: $\mathscr{C}_1=E$; $\mathscr{C}_2=C_{3\alpha}, C_{3\beta}, C_{3\gamma}, C_{3\delta}, C_{3\alpha}^{-1}, C_{3\beta}^{-1}, C_{3\gamma}^{-1}, C_{3\delta}^{-1}$; $\mathscr{C}_3=C_{2x}, C_{2y}, C_{2z}, \bar{C}_{2x}, \bar{C}_{2y}, \bar{C}_{2z}$; $\mathscr{C}_4=IC_{4x}, IC_{4y}, IC_{4z}, IC_{4x}^{-1}, IC_{4y}^{-1}, IC_{4z}^{-1}$; $\mathscr{C}_5=IC_{2a}, IC_{2b}, IC_{2c}, IC_{2d}, IC_{2e}, IC_{2f}, \overline{IC}_{2a}, \overline{IC}_{2b}, \overline{IC}_{2c}, \overline{IC}_{2d}, \overline{IC}_{2e}, \overline{IC}_{2f}$; $\mathscr{C}_6=\overline{\mathscr{C}}_1$; $\mathscr{C}_7=\overline{\mathscr{C}}_2$; $\mathscr{C}_8=\overline{\mathscr{C}}_4$.

(b) The character table is given in table A.27. The notations in the first and second columns are those of ELLIOTT [1954] and PARMENTER [1955] respectively.

TABLE A.27

		$\mathscr{C}_1$	$\mathscr{C}_2$	$\mathscr{C}_3$	$\mathscr{C}_4$	$\mathscr{C}_5$
P_6	Γ_6	2	1	0	$\sqrt{2}$	0
P_7	Γ_7	2	1	0	$-\sqrt{2}$	0
P_8	Γ_8	4	-1	0	0	0

(4) O^D

(a) Classes: $\mathscr{C}_1=E$; $\mathscr{C}_2=C_{3\alpha}, C_{3\beta}, C_{3\gamma}, C_{3\delta}, C_{3\alpha}^{-1}, C_{3\beta}^{-1}, C_{3\gamma}^{-1}, C_{3\delta}^{-1}$; $\mathscr{C}_3=C_{2x}, C_{2y}, C_{2z}, \bar{C}_{2x}, \bar{C}_{2y}, \bar{C}_{2z}$; $\mathscr{C}_4=C_{4x}, C_{4y}, C_{4z}, C_{4x}^{-1}, C_{4y}^{-1}, C_{4z}^{-1}$; $\mathscr{C}_5=C_{2a}, C_{2b}, C_{2c}, C_{2d}, C_{2e}, C_{2f}, \bar{C}_{2a}, \bar{C}_{2b}, \bar{C}_{2c}, \bar{C}_{2d}, \bar{C}_{2e}, \bar{C}_{2f}$; $\mathscr{C}_6=\overline{\mathscr{C}}_1$; $\mathscr{C}_7=\overline{\mathscr{C}}_2$; $\mathscr{C}_8=\overline{\mathscr{C}}_4$.

(b) The character table is given in table A.27.

(5) T_h^D

(a) Classes: $\mathscr{C}_1=E$; $\mathscr{C}_2=C_{3\alpha}, C_{3\beta}, C_{3\gamma}, C_{3\delta}$; $\mathscr{C}_3=C_{3\alpha}^{-1}, C_{3\beta}^{-1}, C_{3\gamma}^{-1}, C_{3\delta}^{-1}$; $\mathscr{C}_4=C_{2x}, C_{2y}, C_{2z}, \bar{C}_{2x}, \bar{C}_{2y}, \bar{C}_{2z}$;

$\mathscr{C}_5 = I$; $\mathscr{C}_6 = IC_{3\alpha}, IC_{3\beta}, IC_{3\gamma}, IC_{3\delta}$; $\mathscr{C}_7 = IC_{3\alpha}^{-1}, IC_{3\beta}^{-1}, IC_{3\gamma}^{-1}, IC_{3\delta}^{-1}$; $\mathscr{C}_8 = IC_{2x}, IC_{2y}, IC_{2z}, \overline{IC}_{2x}, \overline{IC}_{2y}, \overline{IC}_{2z}$; $\mathscr{C}_9 = \bar{\mathscr{C}}_1$; $\mathscr{C}_{10} = \bar{\mathscr{C}}_2$; $\mathscr{C}_{11} = \bar{\mathscr{C}}_3$; $\mathscr{C}_{12} = \bar{\mathscr{C}}_5$; $\mathscr{C}_{13} = \bar{\mathscr{C}}_6$; $\mathscr{C}_{14} = \bar{\mathscr{C}}_7$.

(b) The character table is given in table A.28.

TABLE A.28

$\mathscr{C}_1$	$\mathscr{C}_2$	$\mathscr{C}_3$	$\mathscr{C}_4$	$\mathscr{C}_5$	$\mathscr{C}_6$	$\mathscr{C}_7$	$\mathscr{C}_8$
2	1	1	0	2	1	1	0
2	ω	ω^2	0	2	ω	ω^2	0
2	$-\omega^2$	$-\omega$	0	2	$-\omega^2$	$-\omega$	0
2	1	1	0	-2	-1	-1	0
2	ω	ω^2	0	-2	$-\omega$	$-\omega^2$	0
2	$-\omega^2$	$-\omega$	0	-2	ω^2	ω	0

$\omega = \exp(\frac{1}{3}\pi i)$

(6) D_{4h}^D

(a) Classes [for $\mathscr{G}_o^D(\boldsymbol{k})$ of X]: $\mathscr{C}_1 = E$; $\mathscr{C}_2 = C_{2x}, C_{2y}, \bar{C}_{2x}, \bar{C}_{2y}$; $\mathscr{C}_3 = C_{2z}, \bar{C}_{2z}$; $\mathscr{C}_4 = C_{4z}, C_{4z}^{-1}$; $\mathscr{C}_5 = C_{2a}, C_{2b}, \bar{C}_{2a}, \bar{C}_{2b}$; $\mathscr{C}_6 = I$; $\mathscr{C}_7 = IC_{2x}, IC_{2y}, \overline{IC}_{2x}, \overline{IC}_{2y}$; $\mathscr{C}_8 = IC_{2z}, \overline{IC}_{2z}$; $\mathscr{C}_9 = IC_{4z}, IC_{4z}^{-1}$; $\mathscr{C}_{10} = IC_{2a}, IC_{2b}, \overline{IC}_{2a}, \overline{IC}_{2b}$; $\mathscr{C}_{11} = \bar{\mathscr{C}}_1$; $\mathscr{C}_{12} = \bar{\mathscr{C}}_4$; $\mathscr{C}_{13} = \bar{\mathscr{C}}_6$; $\mathscr{C}_{14} = \bar{\mathscr{C}}_9$.

(b) The character table is given in table A.29. The notations in the first column are those of ELLIOTT [1954].

(7) D_{3h}^D

(a) Classes: $\mathscr{C}_1 = E$; $\mathscr{C}_2 = C_{3z}, C_{3z}^{-1}$; $\mathscr{C}_3 = C_{2x}, C_{2A}, C_{2B}, \bar{C}_{2x}, \bar{C}_{2A}, \bar{C}_{2B}$; $\mathscr{C}_4 = IC_{2z}, \overline{IC}_{2z}$; $\mathscr{C}_5 = IC_{6z}, IC_{6z}^{-1}$;

TABLE A.29

	$\mathscr{C}_1$	$\mathscr{C}_2$	$\mathscr{C}_3$	$\mathscr{C}_4$	$\mathscr{C}_5$	$\mathscr{C}_6$	$\mathscr{C}_7$	$\mathscr{C}_8$	$\mathscr{C}_9$	$\mathscr{C}_{10}$
X_6^+, M_6^+	2	0	0	$\sqrt{2}$	0	2	0	0	$\sqrt{2}$	0
X_7^+, M_7^+	2	0	0	$-\sqrt{2}$	0	2	0	0	$-\sqrt{2}$	0
X_6^-, M_6^-	2	0	0	$\sqrt{2}$	0	-2	0	0	$-\sqrt{2}$	0
X_7^-, M_7^-	2	0	0	$-\sqrt{2}$	0	-2	0	0	$\sqrt{2}$	0

$\mathscr{C}_6 = IC_{2y}, IC_{2C}, IC_{2D}, \overline{IC}_{2y}, \overline{IC}_{2C}, \overline{IC}_{2D}$; $\mathscr{C}_7 = \overline{\mathscr{C}}_1$; $\mathscr{C}_8 = \overline{\mathscr{C}}_2$; $\mathscr{C}_9 = \overline{\mathscr{C}}_5$.

(b) The character table is given in table A.30.

(8) D_{3d}^D

(a) Classes [for $\mathscr{G}_o^D(\boldsymbol{k})$ of L]: $\mathscr{C}_1 = E$; $\mathscr{C}_2 = C_{3\delta}, C_{3\delta}^{-1}$; $\mathscr{C}_3 = C_{2b}, C_{2d}, C_{2f}$; $\mathscr{C}_4 = I$; $\mathscr{C}_5 = IC_{3\delta}, IC_{3\delta}^{-1}$; $\mathscr{C}_6 = IC_{2b}, IC_{2d}, IC_{2f}$; $\mathscr{C}_7 = \overline{\mathscr{C}}_1$; $\mathscr{C}_8 = \overline{\mathscr{C}}_2$; $\mathscr{C}_9 = \overline{\mathscr{C}}_3$; $\mathscr{C}_{10} = \overline{\mathscr{C}}_4$; $\mathscr{C}_{11} = \overline{\mathscr{C}}_5$; $\mathscr{C}_{12} = \overline{\mathscr{C}}_6$.

TABLE A.30

$\mathscr{C}_1$	$\mathscr{C}_2$	$\mathscr{C}_3$	$\mathscr{C}_4$	$\mathscr{C}_5$	$\mathscr{C}_6$
2	1	0	0	$\sqrt{3}$	0
2	1	0	0	$-\sqrt{3}$	0
2	-2	0	0	0	0

(b) The character table is given in table A.31. The notation in the first column is that of ELLIOTT [1954].

TABLE A.31

	$\mathscr{C}_1$	$\mathscr{C}_2$	$\mathscr{C}_3$	$\mathscr{C}_4$	$\mathscr{C}_5$	$\mathscr{C}_6$
L_4^+	1	−1	i	1	−1	i
L_5^+	1	−1	−i	1	−1	−i
L_6^+	2	1	0	2	1	0
L_4^-	1	−1	i	−1	1	−i
L_5^-	1	−1	−i	−1	1	i
L_6^-	2	1	0	−2	−1	0

(9) C_{6v}^D

(a) Classes: $\mathscr{C}_1=E$; $\mathscr{C}_2=C_{3z}, C_{3z}^{-1}$; $\mathscr{C}_3=IC_{2x}, IC_{2A}, IC_{2B}, \overline{IC}_{2x}, \overline{IC}_{2A}, \overline{IC}_{2B}$; $\mathscr{C}_4=C_{2z}, \bar{C}_{2z}$; $\mathscr{C}_5=C_{6z}, C_{6z}^{-1}$; $\mathscr{C}_6=IC_{2y}, IC_{2C}, IC_{2D}, \overline{IC}_{2y}, \overline{IC}_{2C}, \overline{IC}_{2D}$; $\mathscr{C}_7=\overline{\mathscr{C}}_1$; $\mathscr{C}_8=\overline{\mathscr{C}}_2$; $\mathscr{C}_9=\overline{\mathscr{C}}_5$.

(b) The character table is given in table A.30.

(10) C_{6h}^D

(a) Classes: $\mathscr{C}_1=E$; $\mathscr{C}_2=C_{6z}$; $C_3=C_{3z}$; $\mathscr{C}_4=C_{2z}$; $\mathscr{C}_5=C_{3z}^{-1}$; $\mathscr{C}_6=C_{6z}^{-1}$; $\mathscr{C}_7=I$; $\mathscr{C}_8=IC_{6z}$; $C_9=IC_{3z}$; $\mathscr{C}_{10}=IC_{2z}$; $\mathscr{C}_{11}=IC_{3z}^{-1}$; $\mathscr{C}_{12}=IC_{6z}^{-1}$; $\mathscr{C}_{(12+n)}=\overline{\mathscr{C}}_n$, $n=1, 2, \ldots, 12$.

(b) The character table is given in table A.32.

(11) D_6^D

(a) Classes: $\mathscr{C}_1=E$; $\mathscr{C}_2=C_{3z}, C_{3z}^{-1}$; $\mathscr{C}_3=C_{2x}, C_{2A}, C_{2B}, \bar{C}_{2x}, \bar{C}_{2A}, \bar{C}_{2B}$; $\mathscr{C}_4=C_{2z}, \bar{C}_{2z}$; $\mathscr{C}_5=C_{6z}, C_{6z}^{-1}$; $\mathscr{C}_6=C_{2y}, C_{2C}, C_{2D}, \bar{C}_{2y}, \bar{C}_{2C}, \bar{C}_{2D}$; $\mathscr{C}_7=\overline{\mathscr{C}}_1$; $\mathscr{C}_8=\overline{\mathscr{C}}_2$; $\mathscr{C}_9=\overline{\mathscr{C}}_5$.

(b) The character table is given in table A.30.

(12) T^D

(a) Classes: $\mathscr{C}_1=E$; $\mathscr{C}_2=C_{3\alpha}, C_{3\beta}, C_{3\gamma}, C_{3\delta}$; $\mathscr{C}_3=C_{3\alpha}^{-1}, C_{3\beta}^{-1}, C_{3\gamma}^{-1}, C_{3\delta}^{-1}$; $\mathscr{C}_4=C_{2x}, C_{2y}, C_{2z}, \bar{C}_{2x}, \bar{C}_{2y}, \bar{C}_{2z}$; $\mathscr{C}_5=\overline{\mathscr{C}}_1$; $\mathscr{C}_6=\overline{\mathscr{C}}_2$; $\mathscr{C}_7=\overline{\mathscr{C}}_3$.

TABLE A.32

$\mathscr{C}_1$	$\mathscr{C}_2$	$\mathscr{C}_3$	$\mathscr{C}_4$	$\mathscr{C}_5$	$\mathscr{C}_6$	$\mathscr{C}_7$	$\mathscr{C}_8$	$\mathscr{C}_9$	$\mathscr{C}_{10}$	$\mathscr{C}_{11}$	$\mathscr{C}_{12}$
1	θ	θ^2	θ^3	$-\theta^4$	$-\theta^5$	1	θ	θ^2	θ^3	$-\theta^4$	$-\theta^5$
1	$-\theta$	θ^2	$-\theta^3$	$-\theta^4$	θ^5	1	$-\theta$	θ^2	$-\theta^3$	$-\theta^4$	θ^5
1	θ^5	$-\theta^4$	θ^3	θ^2	$-\theta$	1	θ^5	$-\theta^4$	θ^3	θ^2	$-\theta$
1	$-\theta^5$	$-\theta^4$	$-\theta^3$	θ^2	θ	1	$-\theta^5$	$-\theta^4$	$-\theta^3$	θ^2	θ
1	i	-1	$-$i	1	i	1	i	-1	$-$i	1	i
1	$-$i	-1	i	1	$-$i	1	$-$i	-1	i	1	$-$i
1	θ	θ^2	θ^3	$-\theta^4$	$-\theta^5$	-1	$-\theta$	$-\theta^2$	$-\theta^3$	θ^4	θ^5
1	$-\theta$	θ^2	$-\theta^3$	$-\theta^4$	θ^5	-1	θ	$-\theta^2$	θ^3	θ^4	$-\theta^5$
1	θ^5	$-\theta^4$	θ^3	θ^2	$-\theta$	-1	$-\theta^5$	θ^4	$-\theta^3$	$-\theta^2$	θ
1	$-\theta^5$	$-\theta^4$	$-\theta^3$	θ^2	θ	-1	θ^5	θ^4	θ^3	$-\theta^2$	$-\theta$
1	i	-1	$-$i	1	i	-1	$-$i	1	i	-1	$-$i
1	$-$i	-1	i	1	$-$i	-1	i	1	$-$i	-1	i

$\theta = \exp(\frac{1}{6}\pi\text{i})$

(b) The character table is given in table A.33.

(13) D^D_{2h} or V^D_h

(a) Classes [for $\mathscr{G}^D_o(\boldsymbol{k})$ of N]: $\mathscr{C}_1 = E$; $\mathscr{C}_2 = C_{2x}, \bar{C}_{2x}$; $\mathscr{C}_3 = C_{2e}, \bar{C}_{2e}$; $\mathscr{C}_4 = C_{2f}, \bar{C}_{2f}$; $\mathscr{C}_5 = I$; $\mathscr{C}_6 = IC_{2x}, \overline{IC}_{2x}$; $\mathscr{C}_7 = IC_{2e}, \overline{IC}_{2e}$; $\mathscr{C}_8 = IC_{2f}, \overline{IC}_{2f}$; $\mathscr{C}_9 = \bar{\mathscr{C}}_1$; $\mathscr{C}_{10} = \bar{\mathscr{C}}_5$.

(b) The character table is given in table A.34. The notation in the first column is that used in table 8.4.

(14) C^D_{4v}

(a) Classes [for $\mathscr{G}^D_o(\boldsymbol{k})$ of Δ]: $\mathscr{C}_1 = E$; $\mathscr{C}_2 = C_{2z}, \bar{C}_{2z}$; $\mathscr{C}_3 = C_{4z}, C_{4z}^{-1}$; $\mathscr{C}_4 = IC_{2x}, IC_{2y}, \overline{IC}_{2x}, \overline{IC}_{2y}$; $\mathscr{C}_5 = IC_{2a}, IC_{2b}, \overline{IC}_{2a}, \overline{IC}_{2b}$; $\mathscr{C}_6 = \bar{\mathscr{C}}_1$; $\mathscr{C}_7 = \bar{\mathscr{C}}_3$.

(b) The character table is given in table A.35. The notations in the third column are those of ELLIOTT [1954].

TABLE A.33

$\mathscr{C}_1$	$\mathscr{C}_2$	$\mathscr{C}_3$	$\mathscr{C}_4$
2	1	1	0
2	ω	ω^2	0
2	$-\omega^2$	$-\omega$	0

$\omega = \exp(\frac{1}{3}\pi i)$

TABLE A.34

	$\mathscr{C}_1$	$\mathscr{C}_2$	$\mathscr{C}_3$	$\mathscr{C}_4$	$\mathscr{C}_5$	$\mathscr{C}_6$	$\mathscr{C}_7$	$\mathscr{C}_8$
N_5	2	0	0	0	2	0	0	0
$N_{5'}$	2	0	0	0	-2	0	0	0

TABLE A.35

			$\mathscr{C}_1$	$\mathscr{C}_2$	$\mathscr{C}_3$	$\mathscr{C}_4$	$\mathscr{C}_5$
X_6	W_6	Δ_6, T_6	2	0	$\sqrt{2}$	0	0
X_7	W_7	Δ_7, T_7	2	0	$-\sqrt{2}$	0	0

(15) D_4^D

(a) Classes: $\mathscr{C}_1 = E$; $\mathscr{C}_2 = C_{2y}, \bar{C}_{2y}$; $\mathscr{C}_3 = C_{4y}, C_{4y}^{-1}$; $\mathscr{C}_4 = C_{2x}, C_{2z}, \bar{C}_{2x}, \bar{C}_{2z}$; $\mathscr{C}_5 = C_{2c}, C_{2d}, \bar{C}_{2c}, \bar{C}_{2d}$; $\mathscr{C}_6 = \bar{\mathscr{C}}_1$; $\mathscr{C}_7 = \bar{\mathscr{C}}_3$.

(b) The character table is given in table A.35.

(16) D^D_{2d} or V^D_d

(a) Classes [for $\mathscr{G}^D_o(\boldsymbol{k})$ of W]: $\mathscr{C}_1=E$; $\mathscr{C}_2=C_{2y}, \bar{C}_{2y}$; $\mathscr{C}_3=IC_{4y}, IC_{4y}^{-1}$; $\mathscr{C}_4=IC_{2x}, IC_{2z}, \overline{IC}_{2x}, \overline{IC}_{2z}$; $\mathscr{C}_5=C_{2c}$, $C_{2d}, \bar{C}_{2c}, \bar{C}_{2d}$; $\mathscr{C}_6=\bar{\mathscr{C}}_1$; $\mathscr{C}_7=\bar{\mathscr{C}}_3$.

(b) The character table is given in table A.35. The notations in the first and second columns are those of PARMENTER [1955] and ELLIOTT [1954] respectively.

(17) C^D_{4h}

(a) Classes: $\mathscr{C}_1=E$; $\mathscr{C}_2=C_{4z}$; $\mathscr{C}_3=C_{2z}$; $\mathscr{C}_4=C_{4z}^{-1}$; $\mathscr{C}_5=I$; $\mathscr{C}_6=IC_{4z}$; $\mathscr{C}_7=IC_{2z}$; $\mathscr{C}_8=IC_{4z}^{-1}$; $\mathscr{C}_{(8+n)}=\bar{\mathscr{C}}_n$, $n=1, 2, \ldots, 8$.

(b) The character table is given in table A.36.

TABLE A.36

$\mathscr{C}_1$	$\mathscr{C}_2$	$\mathscr{C}_3$	$\mathscr{C}_4$	$\mathscr{C}_5$	$\mathscr{C}_6$	$\mathscr{C}_7$	$\mathscr{C}_8$
1	ψ	i	$-\psi^3$	1	ψ	i	$-\psi^3$
1	$-\psi$	i	ψ^3	1	$-\psi$	i	ψ^3
1	ψ^3	$-$i	$-\psi$	1	ψ^3	$-$i	$-\psi$
1	$-\psi^3$	$-$i	ψ	1	$-\psi^3$	$-$i	ψ
1	ψ	i	$-\psi^3$	-1	$-\psi$	$-$i	ψ^3
1	$-\psi$	i	ψ^3	-1	ψ	$-$i	$-\psi^3$
1	ψ^3	$-$i	$-\psi$	-1	$-\psi^3$	i	ψ
1	$-\psi^3$	$-$i	ψ	-1	ψ	i	$-\psi$

$\psi=\exp(\frac{1}{4}\pi \mathrm{i})$

(18) C^D_{3h}

(a) Classes: $\mathscr{C}_1=E$; $\mathscr{C}_2=IC_{6z}$; $\mathscr{C}_3=C_{3z}$; $\mathscr{C}_4=IC_{2z}$; $\mathscr{C}_5=C_{3z}^{-1}$; $\mathscr{C}_6=IC_{6z}^{-1}$; $\mathscr{C}_{(6+n)}=\bar{\mathscr{C}}_n$, $n=1, 2, \ldots, 6$.

(b) The character table is given in table A.37.

TABLE A.37

$\mathscr{C}_1$	$\mathscr{C}_2$	$\mathscr{C}_3$	$\mathscr{C}_4$	$\mathscr{C}_5$	$\mathscr{C}_6$
1	θ	θ^2	θ^3	$-\theta^4$	$-\theta^5$
1	$-\theta$	θ^2	$-\theta^3$	$-\theta^4$	θ^5
1	θ^5	$-\theta^4$	θ^3	θ^2	$-\theta$
1	$-\theta^5$	$-\theta^4$	$-\theta^3$	θ^2	θ
1	i	-1	$-$i	1	i
1	$-$i	-1	i	1	$-$i

$\theta = \exp(\frac{1}{6}\pi \mathrm{i})$

(19) C_{3v}^D

(a) Classes [for $\mathscr{G}_o^D(\boldsymbol{k})$ of Λ]: $\mathscr{C}_1 = E$; $\mathscr{C}_2 = C_{3\delta}, C_{3\delta}^{-1}$; $\mathscr{C}_3 = IC_{2b}, IC_{2d}, IC_{2f}$; $\mathscr{C}_4 = \overline{\mathscr{C}}_1$; $\mathscr{C}_5 = \overline{\mathscr{C}}_2$; $\mathscr{C}_6 = \overline{\mathscr{C}}_3$.

Classes [for $\mathscr{G}_3^D(\boldsymbol{k})$ of F]: $\mathscr{C}_1 = E$; $\mathscr{C}_2 = C_{3\alpha}, C_{3\alpha}^{-1}$; $\mathscr{C}_3 = IC_{2b}, IC_{2c}, IC_{2e}$; $\mathscr{C}_4 = \overline{\mathscr{C}}_1$; $\mathscr{C}_5 = \overline{\mathscr{C}}_2$; $\mathscr{C}_6 = \overline{\mathscr{C}}_3$.

(b) The character table is given in table A.38. The notations in the first column are those of PARMENTER [1955].

(20) D_3^D

(a) Classes: $\mathscr{C}_1 = E$; $\mathscr{C}_2 = C_{3z}, C_{3z}^{-1}$; $\mathscr{C}_3 = C_{2x}, C_{2A}, C_{2B}$; $\mathscr{C}_4 = \overline{\mathscr{C}}_1$; $\mathscr{C}_5 = \overline{\mathscr{C}}_2$; $\mathscr{C}_6 = \overline{\mathscr{C}}_3$.

(b) The character table is given in table A.38.

TABLE A.38

	$\mathscr{C}_1$	$\mathscr{C}_2$	$\mathscr{C}_3$
Λ_4, F_4	1	-1	i
Λ_5, F_5	1	-1	$-$i
Λ_6, F_6	2	1	0

(21) C_{3i}^D or S_6^D

(a) Classes: $\mathscr{C}_1=E$; $\mathscr{C}_2=IC_{3z}^{-1}$; $\mathscr{C}_3=C_{3z}$; $\mathscr{C}_4=I$;
$\mathscr{C}_5=C_{3z}^{-1}$; $\mathscr{C}_6=IC_{3z}$; $\mathscr{C}_{(6+n)}=\bar{\mathscr{C}}_n$, $n=1, 2, \ldots, 6$.

(b) The character table is given in table A.37.

(22) C_6^D

(a) Classes: $\mathscr{C}_1=E$; $\mathscr{C}_2=C_{6z}$; $\mathscr{C}_3=C_{3z}$; $\mathscr{C}_4=C_{2z}$;
$\mathscr{C}_5=C_{3z}^{-1}$; $\mathscr{C}_6=C_{6z}^{-1}$; $\mathscr{C}_{(6+n)}=\bar{\mathscr{C}}_n$, $n=1, 2, \ldots, 6$.

(b) The character table is given in table A.37.

(23) C_{2v}^D

(a) Classes [for $\mathscr{G}_o^D(\boldsymbol{k})$ of Σ and K]: $\mathscr{C}_1=E$; $\mathscr{C}_2=C_{2e}, \bar{C}_{2e}$;
$\mathscr{C}_3=IC_{2x}, \overline{IC}_{2x}$; $\mathscr{C}_4=IC_{2f}, \overline{IC}_{2f}$; $\mathscr{C}_5=\bar{\mathscr{C}}_1$.
Classes [for $G_o^D(\boldsymbol{k})$ of D]: $\mathscr{C}_1=E$; $\mathscr{C}_2=C_{2x}, \bar{C}_{2x}$;
$\mathscr{C}_3=IC_{2e}, \overline{IC}_{2e}$; $\mathscr{C}_4=IC_{2f}, \overline{IC}_{2f}$; $\mathscr{C}_5=\bar{\mathscr{C}}_1$.
Classes [for $\mathscr{G}_o^D(\boldsymbol{k})$ of U and S]: $\mathscr{C}_1=E$; $\mathscr{C}_2=C_{2a}, \bar{C}_{2a}$;
$\mathscr{C}_3=IC_{2z}, \overline{IC}_{2z}$; $\mathscr{C}_4=IC_{2b}, \overline{IC}_{2b}$; $\mathscr{C}_5=\bar{\mathscr{C}}_1$.
Classes [for $\mathscr{G}_o^D(\boldsymbol{k})$ of Z]: $\mathscr{C}_1=E$; $\mathscr{C}_2=C_{2y}, \bar{C}_{2y}$;
$\mathscr{C}_3=IC_{2z}, \overline{IC}_{2z}$; $\mathscr{C}_4=IC_{2x}, \overline{IC}_{2x}$; $\mathscr{C}_5=\bar{\mathscr{C}}_1$.
Classes [for $\mathscr{G}_o^D(\boldsymbol{k})$ of G]: $\mathscr{C}_1=E$; $\mathscr{C}_2=C_{2f}, \bar{C}_{2f}$;
$\mathscr{C}_3=IC_{2x}, \overline{IC}_{2x}$; $\mathscr{C}_4=IC_{2e}, \overline{IC}_{2e}$; $\mathscr{C}_5=\bar{\mathscr{C}}_1$.

(b) The character table is given in table A.39.

(24) C_{2h}^D

(a) Classes: $\mathscr{C}_1=E$; $\mathscr{C}_2=C_{2z}$; $\mathscr{C}_3=I$; $\mathscr{C}_4=IC_{2z}$;
$\mathscr{C}_{(4+n)}=\bar{\mathscr{C}}_n$, $n=1, 2, 3, 4$.

(b) The character table is given in table A.40.

TABLE A.39

$\mathscr{C}_1$	$\mathscr{C}_2$	$\mathscr{C}_3$	$\mathscr{C}_4$
2	0	0	0

TABLE A.40

$\mathscr{C}_1$	$\mathscr{C}_2$	$\mathscr{C}_3$	$\mathscr{C}_4$
1	i	1	i
1	−i	1	−i
1	i	−1	−i
1	−i	−1	i

(25) D_2^D or V^D

(a) Classes: $\mathscr{C}_1 = E$; $\mathscr{C}_2 = C_{2x}, \bar{C}_{2x}$; $\mathscr{C}_3 = C_{2y}, \bar{C}_{2y}$;
$\mathscr{C}_4 = C_{2z}, \bar{C}_{2z}$; $\mathscr{C}_5 = \bar{\mathscr{C}}_1$.

(b) The character table is given in table A.39.

(26) C_4^D

(a) Classes: $\mathscr{C}_1 = E$; $\mathscr{C}_2 = C_{4y}$; $\mathscr{C}_3 = C_{2z}$; $\mathscr{C}_4 = C_{4z}^{-1}$;
$\mathscr{C}_{(4+n)} = \bar{\mathscr{C}}_n$, $n = 1, 2, 3, 4$.

(b) The character table is given in table A.41.

(27) S_4^D

(a) Classes: $\mathscr{C}_1 = E$; $\mathscr{C}_2 = IC_{4y}$; $\mathscr{C}_3 = C_{2y}$; $\mathscr{C}_4 = IC_{4y}^{-1}$;
$\mathscr{C}_{(4+n)} = \bar{\mathscr{C}}_n$, $n = 1, 2, 3, 4$.

TABLE A.41

	$\mathscr{C}_1$	$\mathscr{C}_2$	$\mathscr{C}_3$	$\mathscr{C}_4$
W_5	1	ψ	i	$-\psi^3$
W_6	1	$-\psi$	i	ψ^3
W_8	1	ψ^3	−i	$-\psi$
W_7	1	$-\psi^3$	−i	ψ

$\psi = \exp(\frac{1}{4}\pi i)$

(b) The character table is given in table A.41. The notation in the first column is that of PARMENTER [1955].

(28) C_3^D

(a) Classes: $\mathscr{C}_1=E$; $\mathscr{C}_2=C_{3z}$; $\mathscr{C}_3=C_{3z}^{-1}$; $\mathscr{C}_{(3+n)}=\bar{\mathscr{C}}_n$, $n=1, 2, 3$.

(b) The character table is given in table A.42.

TABLE A.42

$\mathscr{C}_1$	$\mathscr{C}_2$	$\mathscr{C}_3$
1	ω	ω^2
1	$-\omega^2$	$-\omega$
1	-1	1

$\omega=\exp(\frac{1}{3}\pi i)$

(29) C_s^D or C_{1h}^D

(a) Classes: $\mathscr{C}_1=E$; $\mathscr{C}_2=IC_{2z}$; $\mathscr{C}_3=\bar{\mathscr{C}}_1$; $\mathscr{C}_4=\bar{\mathscr{C}}_2$.

(b) The character table is given in table A.43.

(30) C_2^D

(a) Classes [for $\mathscr{G}_o^D(\boldsymbol{k})$ of Q]: $\mathscr{C}_1=E$; $\mathscr{C}_2=C_{2d}$; $\mathscr{C}_3=\bar{\mathscr{C}}_1$; $\mathscr{C}_4=\bar{\mathscr{C}}_2$.

(b) The character table is given in table A.43. The notation in the first column is that used in tables 8.3 and 8.4.

TABLE A.43

	$\mathscr{C}_1$	$\mathscr{C}_2$
Q_3	1	i
Q_4	1	$-$i

(31) C_i^D

(a) Classes: $\mathscr{C}_1=E$; $\mathscr{C}_2=I$; $\mathscr{C}_3=\bar{\mathscr{C}}_1$; $\mathscr{C}_4=\bar{\mathscr{C}}_2$.

(b) The character table is given in table A.43.

(32) C_1^D

(a) Classes: $\mathscr{C}_1=E$; $\mathscr{C}_2=\bar{\mathscr{C}}_1$.

(b) $\chi(\mathscr{C}_1)=1$.

Appendix 3

TABULATIONS OF THE CHARACTERS OF THE IRREDUCIBLE REPRESENTATIONS OF SINGLE AND DOUBLE SPACE GROUPS

The relationship between the irreducible representations of a symmorphic space group $\mathscr{G}$ and the *point* groups $\mathscr{G}_o(\boldsymbol{k})$ was described in ch. 5 § 1, the corresponding relationship between double groups being described in ch. 8 § 8.2. As the character tables for the irreducible representations of the single and double point groups are given in appendix 1 and appendix 2 respectively, no further character tables need be specified for the symmorphic space groups. Table 1.4 gives a list of the space groups that are symmorphic.

The relationship between the irreducible representations of a *non-symmorphic* space group $\mathscr{G}$ and the groups $\mathscr{G}(\boldsymbol{k})$ was described in ch. 7 § 1, the corresponding relationship between double groups being described in ch. 8 § 9.1.

The character tables for all the irreducible representations of *all* the 230 space groups have been given by Kovalev [1965]. The corresponding dimensions have also been listed by Kovalev and Lyubarskii [1958], Kovalev [1960] and Kudryavtseva [1965]. However, it is worthwhile to give the following list of character tables calculated by other authors, particularly as most of the work on energy band theory uses the notations of the references of this list.

In some cases tabulations of symmetrized wave functions are noted. However, this is not a comprehensive list of such tabulations, for they are very well scattered throughout the literature on energy bands, many being given in papers dealing with specific energy band calculations, lists of which have been given by CALLAWAY [1958, 1964] and SLATER [1965a].

(1) *Triclinic symmetry system*

(a) Crystal class C_1 (i.e. space group C_1^1 only): Irreducible representations of $\mathscr{G}_o(\boldsymbol{k})$ and $\mathscr{G}_o^D(\boldsymbol{k})$ are trivial.

(b) Crystal class C_i (i.e. space group C_i^1 only): Irreducible representations of $\mathscr{G}_o(\boldsymbol{k})$ and $\mathscr{G}_o^D(\boldsymbol{k})$ are almost trivial.

(2) *Monoclinic symmetry system*

(a) Crystal class C_s (i.e. space groups C_s^1–C_s^4): Single and double: SUSHKEVYCH [1966].

(b) Crystal class C_2 (i.e. space groups C_2^1–C_2^3): Single and double: SUSHKEVYCH [1966].

(c) Crystal class C_{2h} (i.e. space groups C_{2h}^1–C_{2h}^6): Single and double: SUSHKEVYCH [1966].

(3) *Orthorhombic symmetry system*

(a) Crystal class C_{2v} (i.e. space groups C_{2v}^1–C_{2v}^{22}):
- C_{2v}^1–C_{2v}^{10}: Single and double: TOVSTYUK and BERCHA [1964].
- C_{2v}^{11}–C_{2v}^{17}: Single and double: TOVSTYUK and SUSHKEVYCH [1964].
- C_{2v}^{18}, C_{2v}^{19}: Single and double: SUSHKEVYCH [1965].

(b) Crystal class D_2 (i.e. space groups D_2^1–D_2^9).
- D_2^1–D_2^4: Single and double: TOVSTYUK and BERCHA [1964].
- D_2^5, D_2^6: Single and double: TOVSTYUK and SUSHKEVYCH [1964].
- D_2^7: Single and double: SUSHKEVYCH [1965].

(c) Crystal class D_{2h} (i.e. space groups D_{2h}^1–D_{2h}^{28}).
- D_{2h}^1–D_{2h}^{14}: Single and double: TOVSTYUK and BERCHA [1964].
- D_{2h}^{15}: Single and double: KHARTSIEV [1962], TOVSTYUK and BERCHA [1964].
- D_{2h}^{16}: Single: GASHIMZADE [1960a], KARPUS and BATARUNAS [1961], TOVSTYUK and BERCHA [1964], SLATER [1965a]; double: GASHIMZADE [1960a],

KARPUS and BATARUNAS [1961], TOVSTYUK and BERCHA [1964].

D_{2h}^{17}: Single and double: SUFFCZYNSKI [1960], TOVSTYUK and SUSHKEVYCH [1964].

D_{2h}^{18}: Single: SLATER *et al.* [1962], TOVSTYUK and SUSHKEVYCH [1964], SLATER [1965a]; double: TOVSTYUK and SUSHKEVYCH [1964]; symmetrized plane waves: SLATER *et al.* [1962].

D_{2h}^{19}–D_{2h}^{22}: Single and double: TOVSTYUK and SUSHKEVYCH [1964].

D_{2h}^{23}, D_{2h}^{24}: Single and double: SUSHKEVYCH [1965].

(4) *Tetragonal symmetry system*

(a) Crystal class D_{2d} (i.e. space groups D_{2d}^{1}–D_{2d}^{12}).

D_{2d}^{1}: Single and double: GUBANOV and GASHIMZADE [1959].

D_{2d}^{12}: Single and double: SANDROCK and TREUSCH [1964].

(b) Crystal class S_4 (i.e. space groups S_4^1 and S_4^2): none.

(c) Crystal class C_4 (i.e. space groups C_4^1–C_4^6): none.

(d) Crystal class C_{4h} (i.e. space groups C_{4h}^1–C_{4h}^6): none.

(e) Crystal class D_4 (i.e. space groups D_4^1–D_4^{10}): none.

(f) Crystal class C_{4v} (i.e. space groups C_{4v}^1–C_{4v}^{12}): none.

(g) Crystal class D_{4h} (i.e. space groups D_{4h}^1–D_{4h}^{20}).

D_{4h}^1–D_{4h}^{16}: Single and double: OLBRYCHSKI [1963b].

D_{4h}^{17}–D_{4h}^{20}: Single and double: SEK [1963] (CHEN [1967] in footnote 17 has stated that these tables contain errors); symmetrized plane waves: VAN HUONG and OLBRYCHSKI [1964].

Also:

D_{4h}^{14}: Single: SLATER [1965a].

D_{4h}^{18}: Single and double: GASHIMZADE [1960b] (for corrections see GASHIMZADE [1962]).

D_{4h}^{19} (white tin): Single: MASE [1959b], MIASEK and SUFFCZYNSKI [1961a, 1961b]; double: MASE [1959b], SUFFCZYNSKI [1961];

symmetrized L.C.A.O.s: MIASEK [1960]; symmetrized plane waves: MIASEK and SUFFCZYNSKI [1961c], MIASEK [1966].

(5) *Cubic symmetry system*

(a) Crystal class T (i.e. space groups T^1–T^5): Single and double: TOVSTYUK and TARNAVSKAYA [1964].

(b) Crystal class T_h (i.e. space groups T_h^1–T_h^7): Single and double: TOVSTYUK and TARNAVSKAYA [1964].

Also:

T_h^6 (pyrite): Single: GORZKOWSKI [1963a], SLATER [1965a]; double: GORZKOWSKI [1963a]; symmetrized plane waves: GORZKOWSKI [1963a].

T_h^7: Single: SLATER [1965a].

(c) Crystal class T_d (i.e. space groups T_d^1–T_d^6): Single and double: TOVSTYUK and TARNAVSKAYA [1964].

Also:

T_d^2 (zinc blende): Single: PARMENTER [1955]; double: PARMENTER [1955], DRESSELHAUS [1955]. (This space group is described in ch. 1 § 3.2, and its irreducible representations are referred to in ch. 5 § 7.2.)

T_d^6: Single and double: KARAVAEV *et al.* [1962].

(d) Crystal class O (i.e. space groups O^1–O^8): Single and double: TOVSTYUK and TARNAVSKAYA [1964].

(e) Crystal class O_h (i.e. space groups O_h^1–O_h^{10}): Single and double: TOVSTYUK and TARNAVSKAYA [1964].

Also:

O_h^1 (simple cubic): see O_h^9 for references.

O_h^3: Single and double: GORZKOWSKI [1963b, 1964].

O_h^4: Single: MOSKALENKO [1960], BORDURE *et al.* [1963a]; double: MOSKALENKO [1960], BORDURE *et al.* [1963b].

O_h^5 (face-centred cubic): see O_h^9 for references.

O_h^7 (diamond): Single: HERRING [1942] (for correction to table XI see ELLIOTT [1954]), DÖRING and ZEHLER [1953], SUGITA and YAMAKA [1954]; double: ELLIOTT [1954].

O_h^9 (body-centred cubic): Single: BOUCKAERT *et al.* [1936]; double: ELLIOTT [1954] (for corrections see ch. 8 § 8.5). For symmetrized wave functions see ch. 5 § 6 and ch. 8 § 8.1. The space groups O_h^5 and O_h^9 have been treated in detail as examples throughout this book.

O_h^{10}: Single and double: SLATER [1965a].

(6) *Rhombohedral and hexagonal symmetry systems*

(a) Crystal class C_3 (i.e. space groups C_3^1–C_3^4): Single: KOVALEV [1961a]; double: KOVALEV [1961b].

(b) Crystal class C_{3v} (i.e. space groups C_{3v}^1–C_{3v}^6): Single: KOVALEV [1961a]; double: KOVALEV [1961b].

(c) Crystal class D_3 (i.e. space groups D_3^1–D_3^7): Single: KOVALEV [1961a]; double: KOVALEV [1961b].

Also:

D_3^3: Single and double: ASENDORF [1957], FIRSOV [1957] (for correction see FIRSOV [1958]).

D_3^4: Single: RUDRA [1965], SLATER [1965a], FREI [1967]; double: FREI [1967].

D_3^6: Single and double: FREI [1967].

D_3^7: Single: SLATER [1965a].

(d) Crystal class S_6 (i.e. space groups S_6^1 and S_6^2): Single: KOVALEV [1961a]; double: KOVALEV [1961b].

Also:

S_6^2: Single: SLATER [1965a].

(e) Crystal class D_{3d} (i.e. space groups D_{3d}^1–D_{3d}^6): Single: KOVALEV [1961a]; double: KOVALEV [1961b].

Also:

D_{3d}^3: Single: SLATER [1965a].

D_{3d}^5 (bismuth): Single: YAMASAKI [1957], MASE [1958,

1959a], FALICOV and GOLIN [1965], SLATER [1965a]; double: MASE [1958, 1959a], FALICOV and GOLIN [1965]; symmetrized L.C.A.O.s: MASE [1958, 1959a], symmetrized plane waves: FALICOV and GOLIN [1965].

D_{3d}^6: Single: SLATER [1965a].

(f) Crystal class C_{3h} (i.e. space group C_{3h}^1 only): Single: KOVALEV [1961a]; double: KOVALEV [1961b].

(g) Crystal class D_{3h} (i.e. space groups D_{3h}^1–D_{3h}^4): Single: KOVALEV [1961a]; double: KOVALEV [1961b].

(h) Crystal class C_6 (i.e. space groups C_6^1–C_6^6): Single: KOVALEV [1961a]; double: KOVALEV [1961b].

(i) Crystal class C_{6h} (i.e. space groups C_{6h}^1 and C_{6h}^2): Single: KOVALEV [1961a]; double: KOVALEV [1961b].

Also:

C_{6h}^2: Single: MURPHY *et al.* [1964].

(j) Crystal class C_{6v} (i.e. space groups C_{6v}^1–C_{6v}^4): Single: KOVALEV [1961a]; double: KOVALEV [1961b].

Also:

C_{6v}^4 (wurtzite): Single: GLASSER [1959], RASHBA [1959], NUSIMOVICI [1965], SLATER [1965a]; double: GLASSER [1959], RASHBA and SHEKA [1959].

(k) Crystal class D_6 (i.e. space groups D_6^1–D_6^6): Single: KOVALEV [1961a]; double: KOVALEV [1961b].

Also:

D_6^4, D_6^5: Single and double: FREI [1967].

(l) Crystal class D_{6h} (i.e. space groups D_{6h}^1–D_{6h}^4): Single: KOVALEV [1961a]; double: KOVALEV [1961b].

Also:

D_{6h}^4 (close-packed hexagonal): Single: HERRING [1942], ANTONCIK and TRLIFAJ [1952]; double: ELLIOTT [1954] (for correction to table XI see CORNWELL [1966]). This space group has been treated in some detail in this book, particularly in ch. 1 § 3.3, ch. 7 and ch. 8 § 9. Symmetrized wave functions for D_{6h}^4 are mentioned in ch. 7 § 7.

APPENDIX 4

DEFINITIONS, NOTATIONS AND PROPERTIES OF MATRICES

The important definitions, notations and properties of matrices that are used in this book are gathered together here for convenience.

A $m \times n$ *matrix* $\boldsymbol{a}$ is defined to be an array of elements a_{ij}, $i=1, 2, \ldots, m$, $j=1, 2, \ldots, n$. This matrix is said to have m rows and n columns. The *product* $\boldsymbol{ab}$ of a $m \times n$ matrix $\boldsymbol{a}$ and a $n \times p$ matrix $\boldsymbol{b}$ is defined to be a $m \times p$ matrix whose elements are given by

$$(\boldsymbol{ab})_{ik} = \sum_{j=1}^{n} a_{ij} b_{jk}, \qquad i = 1, 2, \ldots, m, \quad k = 1, 2, \ldots, p. \qquad \text{(A.1)}$$

The product $\boldsymbol{ba}$ is not defined unless $p=m$, and even then it is in general not equal to $\boldsymbol{ab}$.

The *transpose* $\tilde{\boldsymbol{a}}$ of a $m \times n$ matrix $\boldsymbol{a}$ is defined to be a $n \times m$ matrix whose elements are defined by $\tilde{\boldsymbol{a}}_{ji} = a_{ij}$, $i=1, 2, \ldots, m$, $j=1, 2, \ldots, n$. A *vector*, such as $\boldsymbol{r}=(x, y, z)$, is treated as a 3×1 'column' matrix in matrix products such as (A.1) unless otherwise indicated, even though it is displayed for typographical convenience like a 1×3 'row' matrix.

The $m \times m$ *unit matrix* $\boldsymbol{E}$ is defined to be such that

$$E_{ij} = \delta_{ij} = \begin{cases} 1, & \text{if} \quad i = j, \\ 0, & \text{if} \quad i \neq j. \end{cases}$$

The *inverse* $\boldsymbol{a}^{-1}$ of a $m \times m$ matrix $\boldsymbol{a}$ is then defined by $\boldsymbol{aa}^{-1}=\boldsymbol{E}$. Then $\boldsymbol{a}^{-1}\boldsymbol{a}=\boldsymbol{E}$.

The *complex conjugate* matrix $\boldsymbol{a}^*$ of a $m \times n$ matrix $\boldsymbol{a}$ is a $m \times n$

matrix such that $(\boldsymbol{a}^*)_{jk}=a^*_{jk}$, the star indicating complex conjugation. That is, if $a_{jk}=b_{jk}+\mathrm{i}c_{jk}$, where b_{jk} and c_{jk} are real, then $a^*_{jk}=b_{jk}-\mathrm{i}c_{jk}$. If $\tilde{\boldsymbol{a}}^*=\boldsymbol{a}^{-1}$, then $\boldsymbol{a}$ is said to be *unitary*. If $\tilde{\boldsymbol{a}}^*=\boldsymbol{a}$, then $\boldsymbol{a}$ is said to be *Hermitian*. If $\boldsymbol{a}$ is a matrix with real elements and $\tilde{\boldsymbol{a}}=\boldsymbol{a}^{-1}$, then $\boldsymbol{a}$ is said to be *orthogonal*. The matrix $\boldsymbol{a}^\dagger=\tilde{\boldsymbol{a}}^*$ is called the *adjoint* of $\boldsymbol{a}$.

A $m\times n$ matrix $\boldsymbol{a}$ can be *partitioned* into 4 sub-matrices $\boldsymbol{a}^{11}$, $\boldsymbol{a}^{12}$, $\boldsymbol{a}^{21}$ and $\boldsymbol{a}^{22}$ with dimensions $m_1\times n_1$, $m_1\times n_2$, $m_2\times n_1$ and $m_2\times n_2$ respectively, where $m_1+m_2=m$ and $n_1+n_2=n$. It may then be written as

$$\boldsymbol{a}=\begin{pmatrix}\boldsymbol{a}^{11} & \boldsymbol{a}^{12}\\ \boldsymbol{a}^{21} & \boldsymbol{a}^{22}\end{pmatrix}.$$

This may be visualized more easily by indicating explicitly the dimensions, as follows

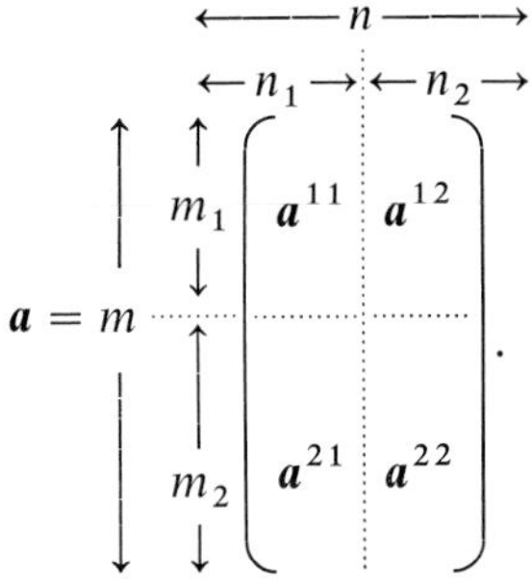

If a $n\times p$ matrix $\boldsymbol{b}$ is similarly partitioned into 4 sub-matrices $\boldsymbol{b}^{11}$, $\boldsymbol{b}^{12}$, $\boldsymbol{b}^{21}$ and $\boldsymbol{b}^{22}$ with dimensions $n_1\times p_1$, $n_1\times p_2$, $n_2\times p_1$ and $n_2\times p_2$ respectively, where $n_1+n_2=n$ and $p_1+p_2=p$, then the product $\boldsymbol{c}=\boldsymbol{ab}$ may also be partitioned into 4 sub-matrices $\boldsymbol{c}^{11}$, $\boldsymbol{c}^{12}$, $\boldsymbol{c}^{21}$ and $\boldsymbol{c}^{22}$ of dimensions $m_1\times p_1$, $m_1\times p_2$, $m_2\times p_1$ and $m_2\times p_2$ respectively, and it may be shown from (A.1) that

$$\boldsymbol{c}^{rs}=\sum_{t=1}^{2}\boldsymbol{a}^{rt}\boldsymbol{b}^{ts},\qquad r=1,2,\quad s=1,2\,. \tag{A.2}$$

That is, it is as though the sub-matrices $\boldsymbol{a}^{rt}$, $\boldsymbol{b}^{ts}$ and $\boldsymbol{c}^{rs}$ are themselves matrix elements, but the product of $\boldsymbol{a}^{rt}$ and $\boldsymbol{b}^{ts}$ is given by (A.1). A matrix may be similarly partitioned into more than 4 sub-matrices.

The property of partitional matrices used in ch. 2 § 4 is a special case of (A.2) in which the sub-matrices $\boldsymbol{a}^{12}$, $\boldsymbol{a}^{21}$, $\boldsymbol{b}^{12}$ and $\boldsymbol{b}^{21}$ consist entirely of zero elements. Then (A.2) gives that $\boldsymbol{c}^{11}=\boldsymbol{a}^{11}\boldsymbol{b}^{11}$ and $\boldsymbol{c}^{22}=\boldsymbol{a}^{22}\boldsymbol{b}^{22}$, while $\boldsymbol{c}^{12}$ and $\boldsymbol{c}^{21}$ consist entirely of zero elements.

The *direct product* of two matrices is defined and described in ch. 3 § 7.1.

BIBLIOGRAPHY

There exist a number of books and review articles on the application of group theory to quantum mechanical problems. They may be divided into two classes.

The first consists of books that deal with a number of branches of quantum physics, and they demonstrate well the very wide applicability of group theoretical concepts. Most of these contain the explicit proofs of the theorems that were stated in ch. 2. However, their wide coverage necessarily restricts their treatments of electronic energy bands in solids. The following is a list of these books: Bhagavantam and Venkatarayudu [1962], Hall [1966], Hamermesh [1962], Heine [1960], Lyubarskii [1960], Meijer and Bauer [1962], Tinkham [1964], Wigner [1959].

The other class consists of books and articles dealing with the application of group theory to various solid-state topics. Their aims, and consequently their coverage and approach differ widely from each other and from the present work. They are as follows: Johnston [1960], Jones [1960], Knox and Gold [1964], Koster [1957], Mariot [1962], Nussbaum [1962, 1966], Slater [1965a, 1965b], Sokolev and Shirokovskii [1960].

REFERENCES

Altmann, S. L., 1957, Proc. Cambridge Phil. Soc. **53** 343–367.
Altmann, S. L., 1962, Phil. Trans. A **255** 216–240.
Altmann, S. L., 1963, Rev. Mod. Phys. **35** 641–645.
Altmann, S. L., and C. J. Bradley, 1964, Phys. Rev. **135** A1253–1256.
Altmann, S. L., and C. J. Bradley, 1965a, Rev. Mod. Phys. **37** 33–45.
Altmann, S. L., and C. J. Bradley, 1965b, Proc. Phys. Soc. **86** 915–931.
Altmann, S. L., and C. J. Bradley, 1967, Proc. Phys. Soc. **92** 764–775.
Altmann, S. L., and A. P. Cracknell, 1965, Rev. Mod. Phys. **37** 19–32.
Antoncik, E., and M. Trlifaj, 1952, Czech. J. Phys. **1** 97–109.
Asendorf, R. H., 1957, J. Chem. Phys. **27** 11–16.
Balkanski, M., and M. Nusimovici, 1964, Phys. Status Solidi **5** 635–647.
Bell, D. G., 1954, Rev. Mod. Phys. **26** 311–320.
Bethe, H. A., 1929, Ann. Physik **3** 133–208.
Bhagavantam, S., and T. Venkatarayuda, 1962, Theory of Groups and its Application to Physical Problems (Andhra University, Waltair, India).
Birman, J. L., 1962, Phys. Rev. **127** 1093–1106.
Birman, J. L., 1963, Phys. Rev. **131** 1489–1496.
Birman, J. L., 1966, Phys. Rev. **150** 771–782.
Bloch, F., 1928, Z. Phys. **52** 555–600.
Bordure, G., G. Lecoy and M. Savelli, 1963a, Cahiers Phys. **17** 309–318.
Bordure, G., G. Lecoy and M. Savelli, 1963b, Cahiers Phys. **17** 382–389.
Born, M., and T. Von Kármán, 1912, Phys. Z. **13** 297–309.
Bouckaert, L. P., R. Smoluchowski and E. P. Wigner, 1936, Phys. Rev. **50** 58–67.
Bradley, C. J., 1966, J. Math. Phys. **7** 1145–1152.
Callaway, J., 1958, Electron Energy Bands in Solids. *In:* Seitz, F., and D. Turnbull, eds., Solid State Physics, Advances in Research and Applications **7** (Academic Press, New York and London) pp. 100–212.
Callaway, J., 1964, Energy Band Theory (Academic Press, New York and London).
Chen, S. H., 1967, Phys. Rev. **163** 532–547.
Cohen, M. H., and L. M. Falicov, 1960, Phys. Rev. Letters **5** 544–546.
Cornwell, J. F., 1961, Proc. Roy. Soc. London A **261** 551–564.
Cornwell, J. F., 1966, Phys. Kondens. Materie **4** 327–329.
Cornwell, J. F., D. M. Hum and K. C. Wong, 1968, Phys. Letters **26**A 365–366.
Donnay, J. D. H., and W. Nowacki, 1954, Crystal Data (Geological Society of America, Memoir 60, New York).

Döring, W., and V. Zehler, 1953, Ann. Physik **13** 214–228.
Dresselhaus, G., 1955, Phys. Rev. **100** 580–586.
Elliott, R. J., 1954, Phys. Rev. **96** 280–287.
Elliott, R. J., and R. Loudon, 1960, J. Phys. Chem. Solids **15** 146–151.
Erdmann, J., 1960, Z. Naturforsch. **15**a 524–531.
Falicov, L. M., and S. Golin, 1965, Phys, Rev. **137** A871–882.
Firsov, Yu. A., 1957, Zh. Eksperim. i Teor. Fiz. **32** 1350–1367. [Translation: Soviet Phys. JETP **5** 1101–1114 (1957)].
Firsov, Yu. A., 1958, Zh. Eksperim. i Teor. Fiz. **34** 240–242. [Translation: Soviet Phys. JETP **7** 166–168 (1958)].
Flodmark, S., 1963, Phys. Rev. **132** 1343–1348.
Frei, V., 1967, Czech. J. Phys. B **17** 147–166.
Frobenius, G., and I. Schur, 1906, Sitzgsber. Preuss. Akad. Wiss. 186–208.
Gashimzade, F. M., 1960a, Fiz. Tverd. Tela **2** 2070–2076. [Translation: Soviet Phys. Solid State **2** 1856–1862 (1961)].
Gashimzade, F. M., 1960b, Fiz. Tverd. Tela **2** 3040–3044. [Translation: Soviet Phys. Solid State **2** 2700–2704 (1961)].
Gashimzade, F. M., 1962, Fiz. Tverd. Tela **4** 2282–2283. [Translation: Soviet Phys. Solid State **4** 1681 (1963)].
Glasser, M. L., 1959, J. Phys. Chem. Solids **10** 229–241.
Glück, M., Y. Gur and J. Zak, 1967, J. Math. Phys. **8** 787–790.
Gorzkowski, W., 1963a, Phys. Status Solidi **3** 599–614.
Gorzkowski, W., 1963b, Phys. Status Solidi **3** 910–921.
Gorzkowski, W., 1964, Phys. Status Solidi **6** 521–528.
Gubanov, A. I., and F. M. Gashimzade, 1959, Fiz. Tverd. Tela **1** 1411–1416. [Translation: Soviet Phys. Solid State **1** 1294–1298 (1960)].
Hall, G. G., 1966, Applied Group Theory (Longmans, London).
Hamermesh, M., 1962, Group Theory and its Application to Physical Problems (Addison-Wesley, New York).
Harrison, W. A., 1960a, Phys. Rev. **118** 1182–1189.
Harrison, W. A., 1960b, Phys. Rev. **118** 1190–1208.
Harrison, W. A., 1966, Pseudopotentials in the Theory of Metals (W. A. Benjamin, New York).
Heine, V., 1960, Group Theory in Quantum Mechanics (Pergamon Press, Oxford).
Herring, C., 1937a, Phys. Rev. **52** 361–365.
Herring, C., 1937b, Phys. Rev. **52** 365–373.
Herring, C., 1940, Phys Rev. **57** 1169–1177.
Herring, C., 1942, J. Franklin Inst. **233** 525–543.
Hodges, L., H. Ehrenreich and N. D. Lang, 1966, Phys. Rev. **152** 505–526.
Howarth, D. J., and H. Jones, 1952, Proc. Phys. Soc. A **65** 355–368.
Hund, F., 1936, Z. Phys. **99** 119–136.
International Tables for X-ray Crystallography, Vol. I, 1965. Ed. by N. F. M. Henry and K. Lonsdale (International Union of Crystallography, Kynoch Press, Birmingham, England).

Johnston, D. F., 1958, Proc. Roy. Soc. London A **243** 546–554.
Johnston, D. F., 1960, Repts. Progr. Phys. **23** 66–153.
Jones, H., 1960, The Theory of Brillouin Zones and Electronic States in Crystals (North-Holland, Amsterdam).
Karavaev, G. F., 1965, Fiz. Tverd. Tela **6** 3676–3683. [Translation: Soviet Phys. Solid State **6** 2943–2948 (1965)].
Karavaev, G. F., N. V. Kudryavtseva, and V. A. Chaldyshev, 1962, Fiz. Tverd. Tela **4** 3471–3481. [Translation: Soviet Phys. Solid State **4** 2665–2671 (1963)].
Karpus, A. S., and I. B. Batarunas, 1961, Litov. Fiz. Sbornik **1** 315–328.
Khartsiev, V. E., 1962, Fiz. Tverd. Tela **4** 983–991. [Translation: Soviet Phys. Solid State **4** 721–728 (1962)].
Kittel, C., 1963, Quantum Theory of Solids (John Wiley and Sons, New York and London) pp. 182–184.
Kitz, A., 1965, Phys. Status Solidi **8** 813–829.
Knox, R. S., and A. Gold, 1964, Symmetry in the Solid State (W. A. Benjamin, New York).
Kohn, W., and N. Rostoker, 1954, Phys. Rev. **94** 1111–1120.
Koster, G. F., 1957, Space Groups and their Representations. *In:* Seitz, F., and D. Turnbull, eds., Solid State Physics, Advances in Research and Applications **5** (Academic Press, New York and London) pp. 173–256.
Koster, G. F., J. O. Dimmock, R. E. Wheeler and H. Statz, 1964, Properties of the Thirty-two Point Groups (M.I.T. Press, Cambridge, Mass., U.S.A.).
Kovalev, O. V., 1960, Fiz. Tverd. Tela **2** 2557–2566. [Translation: Soviet Phys. Solid State **2** 2279–2288 (1961)].
Kovalev, O. V., 1961a, Ukr. Fiz. Zh. **6** 353–365.
Kovalev, O. V., 1961b, Ukr. Fiz. Zh. **6** 366–375.
Kovalev, O. V., 1965, Irreducible Representations of the Space Groups (Gordon and Breach, New York). [Translation of book originally published by the Academy of Sciences USSR Press, Kiev, 1961].
Kovalev, O. V., and G. Ya. Lyubarskii, 1958, Zh. Tekhn. Fiz. **28** 1151–1158. [Translation: Soviet Phys. Tech. Phys. **3** 1071–1077 (1958)].
Kudryavtseva, N. A., 1965, Fiz. Tverd. Tela **7** 998–1007. [Translation: Soviet Phys. Solid State **7** 803–810 (1965)].
Kudryavtseva, N. V., 1967, Fiz. Tverd. Tela **9** 2364–2368. [Translation: Soviet Phys. Solid State **9** 1850–1853 (1968)].
Lax, M., 1962, Influence of Time-reversal on Selection Rules connecting different Points in the Brillouin Zone. *In:* Proceedings of the International Conference on Physics of Semiconductors, Exeter (The Institute of Physics and the Physical Society, London) pp. 333–340.
Lax, M., 1965, Phys. Rev. **138** A793–802.
Lax, M., and J. J. Hopfield, 1961, Phys. Rev. **124** 115–123.
Loucks, T. L., 1967, Augmented Plane Wave Method (W. A. Benjamin, New York).
Löwdin, P. O., 1950, J. Chem. Phys. **18** 365–375.
Luehrmann, A. W., 1968, Advan. Phys. **17** 1–77.

Lyubarskii, G. Ya., 1960, The Application of Group Theory in Physics (Pergamon Press, Oxford) pp. 18–40 and pp. 348–372.
Maradudin, A. A., and S. H. Vosko, 1968, Rev. Mod. Phys. **40** 1–37.
Mariot, L., 1962, Group Theory and Solid State Physics (Prentice-Hall, New Jersey).
Mase, S., 1958, J. Phys. Soc. Japan **13** 434–445.
Mase, S., 1959a, J. Phys. Soc. Japan **14** 584–589.
Mase, S., 1959b, J. Phys. Soc. Japan **14** 1538–1550.
McIntosh, H. V., 1963, J. Mol. Spectry. **10** 51–74.
Meijer, P. H. E., and E. Bauer, 1962, Group Theory. The Application to Quantum Mechanics (North-Holland, Amsterdam).
Miasek, M., 1957a, Acta Phys. Polon. **16** 343–368.
Miasek, M., 1957b, Acta Phys. Polon. **16** 447–469.
Miasek, M., 1958, Acta Phys. Polon. **17** 371–387.
Miasek, M., 1960, Bull. Acad. Polon. Sci. Ser. Sci. Math. Astron. Phys. **8** 89–93.
Miasek, M., 1966, Acta Phys. Polon. **29** 141–161.
Miasek, M., and M. Suffczynski, 1961a, Bull. Acad. Polon. Sci. Ser. Sci. Math. Astron. Phys. **9** 477–482.
Miasek, M., and M. Suffczynski, 1961b, Bull. Acad. Polon. Sci. Ser. Sci. Math. Astron. Phys. **9** 483–487.
Miasek, M., and M. Suffczynski, 1961c, Bull. Acad. Polon. Sci. Ser. Sci. Math. Astron. Phys. **9** 609–615.
Mitra, S. S., 1964, Phys. Letters **11** 119–120.
Moskalenko, S. A., 1960, Fiz. Tverd. Tela **2** 1755–1765. [Translation: Soviet Phys. Solid State **2** 1587–1596 (1961)].
Mueller, F. M., 1967, Phys. Rev. **153** 659–669.
Murphy, J., H. H. Caspers and R. A. Buchanan, 1964, J. Chem. Phys. **40** 743–753.
Nusimovici, M., 1965, J. Phys. **26** 689–696.
Nussbaum, A., 1962, Proc. Inst. Radio Engrs. **50** 1762–1781.
Nussbaum, A., 1966, Crystal Symmetry, Group Theory, and Band Structure Calculations. *In:* Seitz, F., and D. Turnbull, eds., Solid State Physics, Advances in Research and Applications **18** (Academic Press, New York and London) pp. 165–272.
Olbrychski, K., 1963a, Phys. Status Solidi **3** 1868–1875.
Olbrychski, K., 1963b, Phys. Status Solidi **3** 2143–2154.
Onodera, Y., and M. Okazaki, 1966, J. Phys. Soc. Japan, **21** 2400–2408.
Opechowski, W., 1940, Physica **7** 552–562.
Parmenter, R. H., 1955, Phys. Rev. **100** 573–579.
Phillips, J. C., 1956, Phys. Rev. **104** 1263–1277.
Pincherle, L., 1960, Rep. Progr. Phys. **23** 355–394.
Raghavacharyulu, I. V. V., 1961, Can. J. Phys. **39** 830–840.
Raghavacharyulu, I. V. V., and I. Bhavanacharyulu, 1962, Can. J. Phys. **40** 1490–1495.
Rashba, E. I., 1959, Fiz. Tverd. Tela **1** 407–421. [Translation: Soviet Phys. Solid State **1** 368–380 (1959)].

Rashba, E.I., and V.I. Sheka, 1959, Fiz. Tverd. Tela, supplement II, 162–176.
Reitz, J.R., 1955, Methods of the One-electron Theory of Solids. *In:* Seitz, F., and D. Turnbull, eds., Solid State Physics, Advances in Research and Applications **1** (Academic Press, New York and London) pp. 1–95.
Rudra, P., 1965, J. Math. Phys. **6** 1278–1282.
Sandrock, R., and J. Treusch, 1964, Z. Naturforsch. **19a** 844–850.
Schiff, L.I., 1955, Quantum Mechanics, 2nd ed. (McGraw-Hill, New York) p. 333.
Schlosser, H., 1962, J. Phys. Chem. Solids **23** 963–969.
Schönfliess, A., 1923, Theorie der Kristallstruktur (Borntraeger, Berlin).
Seitz, F., 1936, Ann. Math. **37** 17–28.
Sek, Z., 1963, Phys. Status Solidi **3** 2155–2165.
Sheka, V.I., 1960, Fiz. Tverd. Tela **2** 1211–1219. [Translation: Soviet Phys. Solid State **2** 1096–1104 (1960)].
Slater, J.C., 1965a, Quantum Theory of Molecules and Solids **2** (McGraw-Hill, New York).
Slater, J.C., 1965b, Rev. Mod. Phys. **37** 68–83.
Slater, J.C., and G.F. Koster, 1954, Phys. Rev. **94** 1498–1524.
Slater, J.C., G.F. Koster and J.H. Wood, 1962, Phys. Rev. **126** 1307–1317.
Slechta, J., 1966, Cesk. Casopis Fys. **A16** 1–5.
Sokolev, A.V., and V.P. Shirokovskii, 1960, Usp. Fiz. Nauk. **71** 485–513. [Translation: Soviet Phys. Usp. **3** 551–566 (1961)].
Stern, F., 1963, Elementary Theory of the Optical Properties of Solids. *In:* Seitz, F., and D. Turnbull, eds., Solid State Physics, Advances in Research and Applications **15** (Academic Press, New York and London) pp. 365–366.
Suffczynski, M., 1960, J. Phys. Chem. Solids **16** 174–176.
Suffczynski, M., 1961, Bull. Acad. Polon. Sci. Ser. Sci. Math. Astron. Phys. **9** 489–495.
Sugita, T., and Yamaka, E., 1954, Denki Tsushin Kenkyusho Kenkyu Jitsuyoka Hokoku **2** No. 8.
Sushkevych, T.M., 1965, Ukr. Fiz. Zh. **10** 861–866.
Sushkevych, T.M., 1966, Ukr. Fiz. Zh. **11** 739–744.
Teleman, E., and A. Glodeanu, 1967, Rev. Roumaine Sci. Tech. Electrotech. Energet. **12** 551–560.
Tinkham, M., 1964, Group Theory and Quantum Mechanics (McGraw-Hill, New York).
Tovstyuk, K.D., and D.M. Bercha, 1964, Fiz. Tverd. Tela **6** 662–679. [Translation: Soviet Phys. Solid State **6** 517–532 (1964)].
Tovstyuk, K.D., and T.M. Sushkevych, 1964, Ukr. Fiz. Zh. **9** 933–942.
Tovstyuk, K.D., and M.V. Tarnavskaya, 1963, Fiz. Tverd. Tela **5** 819–838. [Translation: Soviet Phys. Solid State **5** 597–613 (1963)].
Tovstyuk, K.D., and M.V. Tarnavskaya, 1964, Ukr. Fiz. Zh. **9** 629–641.
Treusch, J., and R. Sandrock, 1966, Phys. Status Solidi **16** 487–497.
Van Hove, L., 1953, Phys. Rev. **89** 1189–1193.
Van Huong, N., and K. Olbrychski, 1964, Phys. Status Solidi **7** 57–65.

Von der Lage, F. C., and H. A. Bethe, 1947, Phys. Rev. **71** 612–622.
Warren, J. L., 1968, Rev. Mod. Phys. **40** 38–76.
Wigner, E. P., 1930, Nachr. Kgl. Ges. Wiss. Göttinger Math.-Physik. Kl. pp. 133–146. [Translation: Knox, R. S., and A. Gold, 1965, Symmetry in the Solid State (W. A. Benjamin, New York) pp. 173–181)].
Wigner, E. P., 1932, Nachr. Kgl. Ges. Wiss. Göttingen Math.-Physik. Kl. pp. 546–559.
Wigner, E. P., 1959, Group Theory and its Application to the Quantum Mechanics of Atomic Spectra (Academic Press, New York and London).
Wohlfarth, E. P., and J. F. Cornwell, 1961, Phys. Rev. Letters **7** 342–343.
Wondratschek, H., and J. Neubüser, 1967, Acta Cryst. **23** 349–352.
Wood, J. H., 1962, Phys. Rev. **126** 517–527.
Wyckoff, R. W. G., 1963, Crystal Structures **1** (Interscience, New York).
Wyckoff, R. W. G., 1964, Crystal Structures **2** (Interscience, New York).
Wyckoff, R. W. G., 1965, Crystal Structures **3** (Interscience, New York).
Yamasaki, M., 1957, J. Chem. Phys. **27** 746–751.
Zak, J., 1960, J. Math. Phys. **1** 165–171.
Zak, J., 1962, J. Math. Phys. **3** 1278–1279.

AUTHOR INDEX

SUBJECT INDEX